Women
in their Element

Selected Women's Contributions to the Periodic System

IUPAC Periodic Table of the Elements

Key:

| atomic number |
| Symbol |
| name |
| conventional atomic weight; standard atomic weight |

1	2		3	4	5	6	7	8	9	10	11	12	13	14	15	16	17	18
1 H hydrogen [1.0078, 1.0082]																		2 He helium 4.0026
3 Li lithium 6.94 [6.938, 6.997]	4 Be beryllium 9.0122												5 B boron 10.81 [10.806, 10.821]	6 C carbon 12.011 [12.009, 12.012]	7 N nitrogen 14.007 [14.006, 14.008]	8 O oxygen 15.999 [15.999, 16.000]	9 F fluorine 18.998	10 Ne neon 20.180
11 Na sodium 22.990	12 Mg magnesium 24.305 [24.304, 24.307]												13 Al aluminium 26.982	14 Si silicon 28.085 [28.084, 28.086]	15 P phosphorus 30.974	16 S sulfur 32.06 [32.059, 32.076]	17 Cl chlorine 35.45 [35.446, 35.457]	18 Ar argon 39.95 [39.792, 39.963]
19 K potassium 39.098	20 Ca calcium 40.078(4)	21 Sc scandium 44.956	22 Ti titanium 47.867	23 V vanadium 50.942	24 Cr chromium 51.996	25 Mn manganese 54.938	26 Fe iron 55.845(2)	27 Co cobalt 58.933	28 Ni nickel 58.693	29 Cu copper 63.546(3)	30 Zn zinc 65.38(2)	31 Ga gallium 69.723	32 Ge germanium 72.630(8)	33 As arsenic 74.922	34 Se selenium 78.971(8)	35 Br bromine 79.904 [79.901, 79.907]	36 Kr krypton 83.798(2)
37 Rb rubidium 85.468	38 Sr strontium 87.62	39 Y yttrium 88.906	40 Zr zirconium 91.224(2)	41 Nb niobium 92.906	42 Mo molybdenum 95.95	43 Tc technetium	44 Ru ruthenium 101.07(2)	45 Rh rhodium 102.91	46 Pd palladium 106.42	47 Ag silver 107.87	48 Cd cadmium 112.41	49 In indium 114.82	50 Sn tin 118.71	51 Sb antimony 121.76	52 Te tellurium 127.60(3)	53 I iodine 126.90	54 Xe xenon 131.29
55 Cs caesium 132.91	56 Ba barium 137.33	57-71 lanthanoids	72 Hf hafnium 178.49(2)	73 Ta tantalum 180.95	74 W tungsten 183.84	75 Re rhenium 186.21	76 Os osmium 190.23(3)	77 Ir iridium 192.22	78 Pt platinum 195.08	79 Au gold 196.97	80 Hg mercury 200.59	81 Tl thallium 204.38 [204.38, 204.39]	82 Pb lead 207.2	83 Bi bismuth 208.98	84 Po polonium	85 At astatine	86 Rn radon
87 Fr francium	88 Ra radium	89-103 actinoids	104 Rf rutherfordium	105 Db dubnium	106 Sg seaborgium	107 Bh bohrium	108 Hs hassium	109 Mt meitnerium	110 Ds darmstadtium	111 Rg roentgenium	112 Cn copernicium	113 Nh nihonium	114 Fl flerovium	115 Mc moscovium	116 Lv livermorium	117 Ts tennessine	118 Og oganesson

57 La lanthanum 138.91	58 Ce cerium 140.12	59 Pr praseodymium 140.91	60 Nd neodymium 144.24	61 Pm promethium	62 Sm samarium 150.36(2)	63 Eu europium 151.96	64 Gd gadolinium 157.25(3)	65 Tb terbium 158.93	66 Dy dysprosium 162.50	67 Ho holmium 164.93	68 Er erbium 167.26	69 Tm thulium 168.93	70 Yb ytterbium 173.05	71 Lu lutetium 174.97
89 Ac actinium	90 Th thorium 232.04	91 Pa protactinium 231.04	92 U uranium 238.03	93 Np neptunium	94 Pu plutonium	95 Am americium	96 Cm curium	97 Bk berkelium	98 Cf californium	99 Es einsteinium	100 Fm fermium	101 Md mendelevium	102 No nobelium	103 Lr lawrencium

INTERNATIONAL UNION OF PURE AND APPLIED CHEMISTRY

For notes and updates to this table, see www.iupac.org. This version is dated 1 December 2018.
Copyright © 2018 IUPAC, the International Union of Pure and Applied Chemistry.

IUPAC

2019 IYPT — International Year of the Periodic Table of Chemical Elements

UNESCO — United Nations Educational, Scientific and Cultural Organization

Women
in their Element

Selected Women's Contributions to the Periodic System

Edited by

Annette Lykknes
Norwegian University of Science and Technology, Norway

Brigitte Van Tiggelen
Science History Institute, USA

World Scientific

NEW JERSEY · LONDON · SINGAPORE · BEIJING · SHANGHAI · HONG KONG · TAIPEI · CHENNAI · TOKYO

Published by

World Scientific Publishing Co. Pte. Ltd.

5 Toh Tuck Link, Singapore 596224

USA office: 27 Warren Street, Suite 401-402, Hackensack, NJ 07601

UK office: 57 Shelton Street, Covent Garden, London WC2H 9HE

Library of Congress Control Number: 2019030149

British Library Cataloguing-in-Publication Data
A catalogue record for this book is available from the British Library.

Cover photo: The Wilson College Chemistry Club, Pennsylvania, circa 1937. Kindly provided by the Science History Institute, Philadelphia.

WOMEN IN THEIR ELEMENT
Selected Women's Contributions To The Periodic System

ISBN 978-981-120-628-3
ISBN 978-981-120-768-6 (pbk)

For any available supplementary material, please visit
https://www.worldscientific.com/worldscibooks/10.1142/11442#t=suppl

Typeset by Stallion Press
Email: enquiries@stallionpress.com

Contents

Foreword from the President of the European Chemical Society

It is a wonderful privilege for me to be able to write about this book in my role as President of EuChemS, the European Chemical Society. EuChemS is a supranational association that represents over forty chemical societies and organisations from across Europe, and by extension, some 160 000 active researchers, chemists, historians, educators — all involved in the many fields of chemistry.

As an organisation, we have long been attentive and sensitive to the role of women in chemistry. In 2011, a year decreed the International Year of Chemistry, we co-edited the book *European Women in Chemistry* with editors Jan Apotheker and Livia Sarkadi, a member of the EuChemS Executive Board. This is a book that explores the outstanding female chemists who today continue to be role models for chemists of all ages.

The approach of the present book is different however, not least because of the context in which it is being published. The year 2019, as announced by the United Nation's General Assembly and UNESCO, was proclaimed the International Year of the Periodic Table, in recognition of the 150-year anniversary of scientist Dmitri Mendeleev's discovery of the periodic system. It has been an honour for us to support and be involved in the creation of this book from the very beginning, and to be able to count it amongst our core initiatives for this year that we are celebrating.

The book is all the more timely as we witness a shift away from histories of grand narratives and great men, to exploring processes and cultural circumstances, and approaching subjects from new perspectives. The publication of this book also coincides with a shifting environment, notably following outcries over enduring gender inequality and discrimination — in academia, industry and in all parts of society more generally.

To look at the history of science is to look at a story largely dominated by renowned men and their discoveries, with names like Isaac Newton, Dmitri Mendeleev, Thomas Edison, Albert Einstein or Stephen Hawking that roll off the tongue. The story that most of us know does also involve some notable women here and there, of which Marie Skłodowska-Curie is undeniably the most well recognised. At first glance, the story of the Periodic Table yields a similar picture. For centuries, women's access to science was complicated, limited, discouraged or simply forbidden, with the result that the number of women taking part in scientific investigations could never match that of men. But this book shows a different side to the story.

Women in Their Element is a compendium of the many women who contributed to the development of the Periodic Table through deeds large and small. Some of the names one may indeed recognise, but many are less well known, having failed to be acknowledged as they should have been. This book plays a unique and significant role in identifying historical silences and finally spotlighting these extraordinary women.

The editors, Brigitte van Tiggelen (who is also Chair of the EuChemS Working Party on the History of Chemistry) and Annette Lykknes (Vice-Chair of the Working Party), in addition to having written remarkable chapters themselves, have done an excellent job in bringing together a wide variety of authors from different countries around the world, and from different fields and disciplines. Whether chemists, physicists or historians, the authors have brought with them their own particular expertise and perspectives, giving us a rich and diverse book and taking us on a kaleidoscopic journey and discovery of the women behind the Periodic Table.

The book is organised in chronological order, taking us from the eighteenth century and earlier, through the nineteenth century and eventually to the many discoveries and important events of the twentieth century. It exemplifies the extent to which women played their part through the centuries, whether in the shadows or for all to see.

The fact that the chapters and their contents are of very different character and cover a wide variety of topics is a noteworthy feature of the whole book. As you will come to discover, some chapters profile women who either discovered or played significant roles in the identification and isolation of particular elements, such as polonium, radium or cobalt, to mention but a few, whilst other sections look at women who had particular skills in different techniques such as X-ray crystallography or chromatography. Some chapters are dedicated to the women who worked with the superheavy elements, whilst others explore the women who, in different historical periods and contexts, played a vital role in the popularisation of science. Each and every one of the scientists referred to in the thirty-eight chapters of the book contributed, in their own special way, to the development of the Periodic Table as we know it today, and they deserve our recognition and appreciation.

It is my hope that this book sparks curiosity and the realisation that there is always more behind the stories we think we know. It is a book for those with an interest in history, in science, in the Periodic Table, and in the stories of women who through their actions, often in challenging circumstances, shaped the state of science. It successfully fills a void in our knowledge of one of the most iconic and enduring scientific emblems and teaches us to seek out what lies beneath the surface — the result could indeed surprise us.

Pilar Goya

EuChemS President

Research Professor at the CSIC, Spanish Research Council

February 2019

Foreword from the Co-chair of the IYPT Management Committee

It is my great pleasure and honour to introduce readers to the book *Women in Their Element.* This book originated from the European Chemical Society (EuChemS) project under the leadership of the EuChemS Working Party on the History of Science and will be published during the International Year of the Periodic Table of Chemical Elements (IYPT 2019).

On 20 December 2017, during its 74th Plenary Meeting, the United Nations (UN) General Assembly 72nd Session proclaimed 2019 as the International Year of the Periodic Table of Chemical Elements. In proclaiming an International Year focusing on the Periodic Table of Chemical Elements and its applications, the United Nations has recognized the importance of raising global awareness of how chemistry promotes sustainable development and provides solutions to global challenges in energy, education, agriculture and health. Indeed, the resolution was adopted as part of a more general agenda item on Science and technology for development. This International Year will bring together many different stakeholders including UNESCO, scientific societies and unions, educational and research institutions, technology platforms, non-profit organizations and private sector partners to promote and celebrate the significance of the Periodic Table of Elements and its applications for society during 2019.

The development of the Periodic Table of Elements is one of the most significant achievements in science and a uniting scientific concept, with broad implications in astronomy, chemistry, physics, biology and other natural sciences. The International Year of the Periodic Table of Chemical Elements in 2019 coincides with the 150th anniversary of the discovery of the Periodic System by Dmitri Mendeleev in 1869, as well as with the centenaries of the International Union of Pure and Applied Chemistry (IUPAC) and the International Astronomical Union (IAU). The IYPT is supported by the IUPAC, IAU, EuChemS, the International Union of Pure and Applied Physics (IUPAP), the International Union of History and Philosophy of Science and Technology (IUHPST), and the International Science Council (ISC).

Chemistry, as well as other natural and mathematical sciences, has long and honourable traditions of participation by highly creative women contributors. However, the percentages of women scientists remain shockingly low and there is a significant gender gap at all levels between women and men. Barriers to achievement by women persist, especially in developing countries. While some forms of discrimination against women and girls are diminishing, gender inequality continues to hold women back and deprives them of basic rights and opportunities. Empowering women requires addressing structural issues such as unfair social norms and attitudes, as well as developing progressive legal frameworks that promote equality between women and men. Gender inequalities are still deep-rooted in every society. Women suffer from lack of access to decent work and face occupational segregation and gender wage gaps. In many situations, they are denied access to basic education and health care and are victims of violence and discrimination. They are under-represented in political and economic decision-making processes. In this regard, the 2030 Agenda for Sustainable Development, as the agreed framework for international development, has a stand-alone Goal 5 on gender equality and the empowerment of women and girls. In addition, other Goals include gender equality targets and a more consistent call for sex disaggregation of data across many indicators. The celebration of the 150th anniversary of the Periodic Table of Chemical Elements in 2019 will provide an unparalleled opportunity to highlight the continuous nature of scientific discovery in different contexts, with particular emphasis on promoting science education at all levels among women and girls, especially in developing countries, including Africa.

The book *Women in Their Element* offers a wide scope of stories highlighting the outstanding role of women in the development of the Periodic Table of Chemical Elements. The examples of Marie Curie, who was awarded Nobel Prizes in 1903 and 1911 for the discovery of radium (Ra) and polonium (Po), Berta Karlik for the discovery of stable isotopes of astatine (At), Lise Meitner, who discovered protactinium (Pa), Ida Noddack, who discovered rhenium (Re) and Marguerite Perey, who discovered francium (Fr) highlight the women role models who contributed significantly to the discovery of elements of the Periodic Table. These examples, as well as many others who are less widely known, are in line with the gender equality priority of UNESCO in view of the advancement of the 2030 Agenda for Sustainable Development.

The interested reader will find information about the development of new fields of research and advanced instrumental methods in the twentieth century, the search for new elements in the eighteenth and nineteenth centuries, and the contribution of women to education and astronomy.

Dmitri Mendeleev wrote: "The most all penetrating spirit before which will open the possibility of tilting not tables, but planets, is the spirit of free human inquiry. Believe only in that". The book *Women in Their Element* confirms this thought of the great scientist. I can highly recommend reading all the articles in this book.

IUPAC Past President
Natalia P. Tarasova
Co-chair of the IYPT2019 Management Committee
Corresponding member of the Russian Academy of Sciences
Director of the Institute of Chemistry and Problems of
Sustainable Development
D.I. Mendeleev University of Chemical Technology of Russia

February 2019

Preface and Acknowledgements

The idea of a collective volume dedicated to women's contributions to knowledge about elements and the periodic system arose during one of our many inspirational discussions back in spring 2017. As historians of chemistry interested in women's history, we wanted to take the opportunity of the 150th anniversary of Dmitri Mendeleev's presentation of the periodic system to offer a somewhat different perspective, one that would spotlight some of the many women who have contributed to the history of the periodic system and to the history and knowledge of the elements.

While some volumes about women in the fields of chemistry and radioactivity have appeared in the last twenty years, to date no collection exists of women's work on the elements and the periodic system. Likewise, whereas short, popular accounts of the elements directed at a general public are available, none of these pay particular attention to women's contributions in these stories. Furthermore, historical accounts often focus on discoveries, including the controversies and priority disputes associated with these. Since only a handful women have discovered elements, and a few more isotopes thereof, we wanted to go beyond this short list. Indeed, our main aim is not to celebrate individual heroines or to restrict the knowledge of the elements to mere discoveries or isolations of elements; rather we wanted to shed light on the multifaceted character of scientific work centred on elements, on collaboration, and on the articulation of individual and sometimes almost imperceptible, but never-theless crucial, steps within the greater scientific endeavour.

While this volume is thicker than anticipated, we realized that many more women could have been showcased as work on the edition unfolded. It is our hope that this book will both add to the attention to and recognition of the role of women in the understanding of the periodic system and its elements, and trigger a deeper exploration of the part played by both men and women often left in the shadows. By spotlighting women's work on elements and the periodic system, therefore, we aim to reveal a fuller picture of the nature of science and all the people involved in the scientific enterprise, from unpaid assistants and technicians to full professors and leaders of laboratories.

We foresee two primary readerships for this book: professional historians of science and historically interested, science-curious adult audiences. In style and form we address the latter, which is why we decided to publish short stories of women and elements, where the scholarship is informed yet accessible in style. Our authors have been asked to address readers who have some basic knowledge of chemistry, equivalent to upper secondary school level, but for whom it may no longer be fresh in their minds. To the extent possible, therefore, our chapters explain scientific detail so that the general, science-interested reader should be able to follow. Yet, in contrast with many books in the popular genre, we include reference lists and select bibliographies, and it was our intention all along that the articles in this volume should rely on recent scholarship. The contributors come from a wealth of diverse backgrounds and with a variety of expertise, from history of science to the disciplines of chemistry and physics, from the field of science education to the practice of Wikipedians and activists, and represent the community at large: independent scholars, established academics, practising scientists, educators, and researchers at different stages in their careers.

While a few of the articles deal with women who have contributed knowledge on phenomena of importance to the periodic system as such, most of the stories are about women and their research on *elements*. The periodic system is a system of elements, and acquiring knowledge on the elements is pivotal for positioning them in the right place in the system, and in turn to improving the system and its representations. Also, a system

	Gruppe I. R'O	Gruppe II. RO	Gruppe III. R'O'	Gruppe IV. RH⁴ RO'	Gruppe V. RH³ R²O⁵	Gruppe VI. RH² RO³	Gruppe VII. RH R²O⁷	Gruppe VIII. RO⁴
1	H = 1							
2	Li = 7	Be = 9.4	B = 11	C = 12	N = 14	O = 16	F = 19	
3	N = 23	Mg = 24	Al = 27.3	Si = 28	P = 31	S = 32	Cl = 35.5	
4	K = 39	Ca = 40	— = 44	Ti = 48	V = 51	Cr = 52	Mn = 55	Fe = 56 Co = 59 Ni = 60, Cu = 63.
5	(Cu = 63)	Zn = 65	— = 68	— = 72	As = 75	Se = 78	Br = 80	
6	Rb = 85	Sr = 87	?Yt = 88	Zr = 90	Nb = 94	Mo = 56	— = 100	Ru = 104, Rh = 104, Pd = 106, Ag = 104.
7	(Ag = 104)	Cd = 112	In = 113	Sn = 118	Sb = 122	Te = 125	J = 127	
8	Cs = 133	Ba = 137	?Di = 138	?Ce = 140	—	—	—	— — —
9	(—)	—	...	—	—	—	—	
10	—	—	?Er = 178	?La = 180	Ta = 182	W - 184	—	Os = 195, Ir = 197, Pt = 198, Au = 199.
11	(Au = 199)	Hg = 200	Tl = 204	Pb = 207	Bi = 208	—	—	
12	—	—	—	Th = 231	—	U = 240	—	— — —

Mendeleev's Periodic Table of 1871, redrawn by Jeff Moran, 2013

Figure 1. Mendeleev's periodic system from 1871, redrawn in 2013 by Jeff O. Moran. The lines represent elements that were predicted to exist, but had not yet been discovered. Among them are francium (two places below caesium in the left-hand part of Group 1), radium (two places below barium in the left-hand part of Group II), protactinium (two places between thorium and uranium, bottom of Group V), polonium (two places below tellurium in the right-hand part of Group VI), rhenium (two places below manganese in the left-hand part of Group VII), and astatine (two places below iodine in the right-hand part of Group VII) — all elements or their isotopes discovered by women featured in this volume. (Reproduced by kind permission of Jeff O. Moran.)

based on 63 elements (as were Mendeleev's from 1869 and 1871, Figure 1) is not necessarily transferable to a jungle of 92 or even 118 elements which it has grown into, especially since the understanding of matter and its basic constituents has evolved continuously over the last 150 years. Let us remember that at the time Mendeleev proposed his system, many chemists did not assume the existence of atoms in their chemical practice and writings. Continued efforts to understanding the elements and providing data about them are therefore intertwined with the development of the periodic system itself. The 38 articles in this volume aim to provide insight into the complexity of that uninterrupted work on the periodic system which has brilliantly survived many new developments, but also into the different human aspects of that work.

To address a general audience that is interested in science and its history, we opted to use the introduction to place the women in our volume in the larger context of the history of science and chemistry. Overall narratives of the developments in chemistry may be found in other books, but rarely with sustained emphasis on women's contributions. Based on the research on each of the women in the volume, we believe we have presented a novel, inclusive history of the developments in chemistry, radioactivity and nuclear science from the perspective of women's contributions (and also mentioning forgotten men). Thus our introductory chapter is not a traditional introduction to the essays in the volume, but rather a historiography of chemistry with women embedded. Naturally, as is the case with all histories, not every aspect can be included. But as the reader will see, the women presented in this volume have taken part in surprisingly varied scientific discussions and contexts, from the nature of elements in the eighteenth century to the manufacture of superheavy elements in our times.

When preparing the cover of this book, we first discussed which women to highlight and how to present them. But instead of displaying the women featured between the covers, we decided to show a different group of "ordinary" women, that of the Wilson College Chemistry Club, Pennsylvania, circa 1937 (Figure 2). The picture nicely shows nine women gathered around a table that is filled with glassware and a scale, watching one of them perform a chemical gesture — most probably pipetting a liquid. Shelves of bottled and labelled substances on the right and the periodic system (a Van Nostrand Company periodic table) on the left both have their natural place in the laboratory setting. Instead of finding the women behind the periodic system, our volume aims to put them and their work front stage. In many ways, this image captures the main idea of this volume: women doing analytical-chemical work, using physical or chemical methods, performing experiments and analysing results, collaborating and contributing knowledge of importance for the periodic system, which is basic chemistry condensed.

When the book was still in gestation, the European Chemical Society (EuChemS) wholeheartedly embraced it as one of their projects for the IYPT. This first response from representatives of the European chemical community was a huge encouragement, and we are grateful for that early endorsement. We, furthermore, wish to express our gratitude to Yugarani Thanabalasingam at World Scientific for taking on this project with faith and enthusiasm, to Ernst Homburg for supporting the idea and Hattie

Figure 2.   The Wilson College Chemistry Club, Pennsylvania, circa 1937. Left to right: Elizabeth Buckley (Mrs. Harry J. Washington), Dr. Joan Humphrey-Long (Mrs. Edwin S. Gault, Jr.), Peggy Walker (Mrs. S.H. Carl Bear), Zeuter [additional information unknown], Barbara L. Bittner (Mrs. Kenneth M. Houck), Raquelita Wistar, Sarah Ebersol (Mrs. Daniel Van Meter, Jr.), Betty R. Baumgartner (Mrs. Lynn Parker), Betty J. Fast (Mrs. Robert A. Whitmore) (Courtesy of Science History Institute).

Edmondson for sharing constructive ideas when the book project was in its initial phase, to all the authors who have contributed articles and institutions and individuals who have given permission to reproduce their images, to Mentz Indergaard and Jonas Persson for assistance with some of the images, Ingrid Nuse who helped us in the last phase of the work, and last but not least to our competent and flexible colleague and copy editor, Anke Timmermann, who has made sure that all contributions are in a good shape in every possible way. Annette Lykknes wishes to acknowledge Geert Vanpaemel, Kaat Wils, and the rest of the research unit for Cultural History since 1750 at Katholieke Universiteit Leuven for accepting her for some months during her sabbatical, NTNU-Norwegian

University of Science and Technology for a research grant, and the NTNU University Library for help tracking down sources. Brigitte Van Tiggelen wishes to thank Kristian H. Nielsen, head of the Centre for Science Studies at Aarhus University, for a short stay of intensive writing in Aarhus. She also wishes to acknowledge the Donald F. and Mildred Topp Othmer Library of Chemical History of the Science History Institute for providing professional and timely help with documentation and access to photographic documents. Both of us would also like to thank our families for being patient with us, women in the history of chemistry, when we needed to work on the volume at odd hours and during weekends and vacations.

March, 2019

Annette Lykknes
Brigitte Van Tiggelen

Introduction

Annette Lykknes* and Brigitte Van Tiggelen[†]

*Department of Teacher Education,
NTNU-Norwegian University of Science and Technology,
7491 Trondheim, Norway
[†]Mémosciences asbl, Voie du Vieux Quartier 18,
1348 Louvain-la-neuve, Belgium
Science History Institute, 315 Chestnut Street,
Philadelphia, PA 19106, USA

The periodic system of chemical elements is one of the best-known and most used icons of modern science. It appears everywhere: on the internet, on sweaters and mugs, in textbooks — and in lecture halls all over the world. Although not everybody is familiar with the periodic system and understands what it represents, most will have an idea of it as an important symbol for and tool in the science of chemistry. Many may have heard of its most celebrated creator, the Russian chemist Dmitri Ivanovich Mendeleev (1834–1907), who came up with the idea of a classification when working on a chemistry textbook for his students at St. Petersburg University. The first published form of his periodic table, which Mendeleev presented as a system from the start, was dated 17 February 1869 (Gregorian calender), and appeared in 150 Russian and 50 French copies [Gordin, 2004, p. 29]. To celebrate the 150th anniversary of the discovery of this periodic system, the United Nations' General Assembly has declared 2019 the International Year of the Periodic Table of Chemical Elements (IYPT). On the website for this International Year, the periodic system is celebrated as "one of the most significant achievements in science, capturing the essence not only of chemistry, but also of physics and biology" [IYPT, 2019]. The historian Michael D. Gordin, Mendeleev's biographer, has even described it as "the single most important discovery of inorganic chemistry in the nineteenth century — and quite possibly of chemistry in general, in any century" [Gordin, 2015, p. 53].

As is the case with most developments in science, there is much more to the discovery of the periodic system than Mendeleev's first presentation of a periodic table in February 1869. The development and refining of a periodic law describing the regularities among the chemical elements was not one of those "eureka" moments often presented in textbooks in chemistry and in popular accounts of the history of the periodic table; in fact, it took Mendeleev roughly nine years to work it out [Bensaude-Vincent, 1986]. And as Gordin [2004, p. xvii] reminds us in the preface to his biography of Mendeleev, "the periodic system actually presents one of the more complicated cases, emerging independently during the 1860s in England, France, the United States, Germany, and Russia," respectively developed by the British chemists John Alexander Newlands (1837–1898) and William Odling (1829–1921), the French geologist and mineralogist Alexandre Béguyer de Chancourtois (1820–1886), the Danish-born[1] American professor of physics and chemistry Gustavus Detlef Hinrichs (1836–1923), the German chemist Julius Lothar Meyer (1830–1895), and Dmitri Mendeleev. There are, indeed, good reasons for granting Mendeleev a prominent position in the historiography of the periodic system — his periodic law being one of them[2] — but this should not

---

[1] Hinrichs was born in Lunden, Holstein, which is now part of Germany but belonged to Denmark in 1836.

[2] Popular accounts of the history of science (and sometimes "old-fashioned" scholarly work) often mention the gaps in Mendeleev's periodic system (which he left blank for undiscovered elements) and his accurate descriptions of their properties as reasons for honouring him with the title of the discoverer of the periodic system. For a critical review see [Scerri and Worrall, 2001]. The fact that he was strategic thinker who continuously developed and improved his system while insisting on publishing along the way is another reason worth mentioning here. Bensaude Vincent [1986, 2019] has interpreted Mendeleev's

obscure the complexity of its discovery history. In J. W. Van Spronsen's seminal book, published for the centenary of the periodic system [Van Spronsen, 1969], we further learn about early classifications and precursors to the system, as well as a number of discoverers and contributions from 1871 onwards.

The history of the development of the periodic system encompasses, for instance, early nineteenth-century numerical try-outs based on the average atomic weights within groups of three elements (triads); circular and three-dimensional versions of a periodic law from the late nineteenth century onwards; turn-of-the-century challenges in implementing the group of noble gases; and also a range of newly discovered "radioelements" (many of which turned out to be isotopes of known elements) in the early twentieth century. Also central to the understanding of the modern periodic system are twentieth-century developments such as the discoveries of subatomic particles, the introduction of the atomic number which determines an element's place in the periodic system, and the interpretation of the periodic system using quantum theoretical principles. Intertwined in the story of the periodic system are the many histories of the discoveries of the elements, the discussions around the concept of the element itself, and the development of analytical methods for the detection of new elements.

Common to the rich corpus of literature on the elements and the periodic system, however, is the general absence of female historical figures. This is, however, not due to women's absence from this kind of scientific endeavour. On the contrary, as the 38 articles in this book will demonstrate, women have contributed to important work on the elements and the periodic system throughout the past centuries. Although this book will certainly give attention to these women's contributions, its aim is not mainly to rehabilitate those women, but rather to shed light on the complexity of scientific work, its multifaceted character, and the roles of different forms of collaboration in scientific work — and ultimately, to contribute to the understanding of the nature of science.

---

success as a result, first, of his creation of a law of nature — the periodic law — that made the system into a tool not just for describing, but also predicting properties of the elements; and second, of Mendeleev's distinction between an element as an abstract entity, and simple substances, which are the simplest forms of the elements found in nature.

This introduction is organised into three main parts. In the first part, "Women in Science," we introduce the general history of the participation of women in science (or natural philosophy), and the different roles women have taken as learners, practitioners and disseminators in science broadly construed throughout the centuries. This part is not mainly about the women featured in this volume, although many of them feature as examples in the historiographical discussion. The second part of the introduction, "Women Discovering and Mastering Elements," tells the story of chemistry and radioactivity, with the stories of some women featured in this volume embedded within. While this part emphasises work on elements and the scientific endeavour characteristic for early modern times and into the early twentieth century, the third part, "Women and the Modern Scientific Enterprise," takes modern developments in the twentieth century, such as the instrumental revolution in chemistry and Big Science, as a point of departure, and presents women's contributions to knowledge about atoms, nuclear reactions, elements and isotopes.

Women in Science

Women of all eras have been interested in natural philosophy, mathematics and medicine. As early as around 400 CE, Hypatia of Alexandria (c. 350–415 AD), who had been taught by her father, became a leading philosopher, mathematician and teacher in Alexandria. Another famous pre-modern woman who engaged in scientific learning is the abbess Hildegard von Bingen (1098–1179), who wrote about causes and treatments of illnesses in the twelfth century [Watts, 2007, pp. 18–26]. However, women interested in acquiring education were not admitted to the schools and then universities when these were first established in Europe in the eleventh and twelfth centuries [Schiebinger, 1987]. Instead, women were trained at home, by their fathers and brothers, or in monasteries and courts — in fact, they were educated like many men of the time, since only few were able to attend university. The early-eighteenth-century chymist (chemist/alchymist) Dorothea Juliana Wallich (1657–1725) of Saxony, who is featured in this book, is a counterexample: a woman who was part of the court circles of Saxe-Weimar, and who made the best use of her time, her husband's wealth and both their social connections to gain knowledge and the reputation of an erudite person.[3]

The Age of Amateurs

Alongside the universities other networks manifested in the epistolary interchanges connecting local communities, since those personal letters were often shared with or read out loud to friends and acquaintances. While women were not excluded from correspondence — noble and powerful women like Queen Christine of Sweden (1626–1689) come to mind — most of these epistolary exchanges were initiated and led by men [Ogilvie, 2016, pp. 361–362]. The establishment of learned societies and academies during the seventeenth century created yet another forum for learned networks, and apart from a few exceptions, women were excluded from membership. However, nothing prevented those who wanted to learn about the scientific developments accomplished or shared in those contexts from attending public meetings, or reading the scientific journals or essays published by these academies. Margaret Cavendish (1623–1673), the Duchess of Newcastle, became fascinated with studying and writing, especially after marrying into a family which was at the centre of intellectual life, and she encouraged women in her writing to follow in her footsteps. Never afraid to challenge authorities, neither in her works nor as a public persona, Cavendish used her position to force the Royal Society of London to admit her to their meetings [Watts, 2007, pp. 49–50].

Scientific conversations and exchanges also occurred in the so-called salons — private gatherings facilitating conversation and discussion among the intellectuals of the upper classes that appeared during the seventeenth century mainly in France. Since women were an integral social part of the salon, they enjoyed and at times influenced the exchange of knowledge in this context [Ogilvie, 2016]. The salon became especially important during the French Enlightenment. Emilie Du Châtelet (1706–1749) and Marie-Anne Lavoisier (1758–1836), whom we will get to know better in this book, both frequented the French salons of the eighteenth century, and Madame Lavoisier even used hers as a vehicle for

---

[3] See Kraft in this volume.

disseminating the new chemistry devised by her husband and his collaborators.[4] Very few women were given formal opportunities to educate, and they include Laura Bassi (1711–1778), who was appointed professor of anatomy and experimental philosophy at the University of Bologna as early as in 1732 [Watts, 2007, pp. 71–72]. Others, like the anatomist Marie Thiroux d'Arconville (1720–1805), were self-taught [Bret and Van Tiggelen, 2011], or, like Emilie Du Châtelet, received a private education from the most highly reputed and current savants.[5] Formal intellectual training, however, remained elusive for women.

But in the course of the nineteenth century women increasingly took roles in education (formal and informal), both as teachers and as learners [Woyshner and Tai, 1997]. Some women wrote books on natural philosophy and on progressive educational ideas. Others made a living as translators, continuing a previous trend. Drawing on their social roles, women could be mediators of knowledge, and this was especially encouraged since it was in line with their duties as wives and mothers [Shteir, 1997].[6] But some went beyond such roles and gained a scientific reputation in the public scene. In 1806, the Swiss-born British gentlewoman Jane Marcet (1769–1858), who will be profiled in this book, published her *Conversations in Chemistry*: chemistry lessons in the form of fictitious dialogues between a female teacher and her female students.[7] The book appeared in sixteen British and at least 23 American editions, was translated into French, German and Italian, and Marcet became known as an influential science writer as she turned to the popularisation of other subjects such as economy and the natural sciences [Dreifuss and Sigrist, 2012]. Interestingly Marcet was also a *salonnière*, and her social life connected her to the community of savants. Marcet intended her *Conversations* for a general public, particularly girls and the women responsible for their upbringing, and encouraged women to attend public lectures about new developments in science. Mid-seventeenth-century Paris had already cultivated a passion

---

[4] See Bret and Kawashima in this volume.
[5] See Van Tiggelen on Du Châtelet in this volume.
[6] For a more recent example, see [Baldwin, 2009] on Kathleen Lonsdale, who is also featured in this volume.
[7] See Childs in this volume.

for chemistry, and women would attend the public lectures delivered by the French apothecary Guillaume-François Rouelle (1703–1770).[8] Marcet, however, aimed for more: by educating mothers, she was educating the public in a most useful science that had undergone profound changes [Watts, 2007, pp. 90–98].

The Quest for a Professional Standing

The nineteenth century witnessed the emergence of science as a profession and with it, a shift from "amateur" to "professional" science [Opitz *et al.*, 2012]. Academic institutions and professional societies became the new arenas for scientific work, and access to the science profession was usually dependent on a secondary school diploma. Women had been able to take part in the scientific endeavour as amateurs, and continued to be active in salons or through teaching at lower levels, but with the advent of professionalisation they were no longer able to do scientific work, or be included in the professional community of science, mainly because they were excluded from the necessary qualification of a higher education. This started to change in the 1860s, 1870s and 1880s, around the time when the periodic system was developed, when universities and colleges were opened up to women in Europe and the United States [Rogers, 2006; Rossiter, 1982, pp. 1–28; Schiebinger, 1987]. The progress was slow, and at first mostly "foreign" female students, from Russia or Eastern Europe, were enrolled. Indeed, the first two women receiving doctoral degrees in chemistry came from the Russian Empire, both graduating in 1874: Lydia Sesemann (1845–1925)[9] and Julia Lermontova (1847–1919) earned their titles in Zurich (Switzerland) and in Göttingen (Prussia), respectively. Lermontova worked on the separation of the elements of the platinum group, of which we will learn more in this book;[10] but even having a PhD

---

[8] Madame d'Arconville herself mentions how remarkable a lecturer Rouelle was [Bret and Van Tiggelen, 2011]. On the English scene, see [Higgitt and Whithers, 2008]. See also [Golinski, 1992, pp. 102 and 241].

[9] Sesemann, who was from an area belonging to Finland today, was awarded a doctoral degree six months prior to Julia Lermontova [Creese, 1998, p. 22].

[10] See Boeck on Lermontova in this volume.

was not enough to earn credit for scientific contributions. Indeed, she never published her results and her work, therefore, exemplifies the "grey" zone of privately circulated research reports which are rarely credited and difficult to spot for lesser known figures of science.

Not long afterwards, in 1890, Ida Freund (1863–1914), another of the women featured in this book, became a lecturer in chemistry at Newnham College, one of the two pioneering women's college at Cambridge, England, which had been founded in 1871 [Newnham College Cambridge, 2019].[11] She was the first woman to become a lecturer at a university in the United Kingdom. In her teaching, she paid particular attention to the periodic table which, in her view, needed to be studied thoroughly by students. Ironically, while Freund was keen for chemistry for women not to be confined to domestic science, she is especially remembered for her cupcake version of the periodic table.

But the path is not always straightforward, as the example of the Swedish scientist Astrid Cleve (1875–1968), to whom one article in this book is dedicated, demonstrates.[12] At first, it seemed that life had dealt her good cards: a supportive family, including a father who was a chemist and university professor, and who let her take part in his investigations from an early age; a PhD in botany (1898), marking her as the first woman to be awarded a doctoral degree in the natural sciences in Sweden; and, from 1902 onwards, a chemist husband with an excellent academic career, who allowed her to conduct chemical research in his laboratory, on subjects specially selected by him. Their divorce after ten years of marriage, however, also led to her separation from scientific work. Without a position or funding, she was unable to follow a line of research and build on her former work; she had to earn a living for herself and her children, and arrange her research projects to fit the circumstances. It is noteworthy that she here built on some of the traditional roles available to women: translation, teaching and natural history allowed her to create her masterwork *Die Diatomeen von Schweden und Finnland I–V* (1951–1955) through a

---

[11] See Opitz in this volume. Cecilia Payne-Gaposchkin, another woman featured in this book, studied at Newnham from 1919 onwards. See Shindell on Payne-Gaposchkin in this volume.

[12] See Espmark and Nordlund in this volume.

progressive accumulation of field work that did not require a costly laboratory setting, and could be pursued in spite of interruptions [Espmark and Nordlund, 2012].

Although she is celebrated as the first female student at Massachusetts Institute of Technology (MIT), Ellen Swallow Richards' (1842–1911) career also faced difficulties: upon graduating from Vassar College she had applied for work in companies opposed to hiring a woman, and this eventually led her to pursuing an education at MIT.[13] Her marriage with an MIT professor granted her a continued association with the laboratory, and thus continued access to publishing, but as the "wife of a scientist," not as a scientist in her own right. But even marriage was not always a solution, as witnessed by Harriet Brooks (1876–1933): when she announced that she was engaged to be married, Barnard College, where she had a (paid) teaching position at the time, informed her that she would have to resign once she was married.[14] Although Brooks subsequently did not get married, she nevertheless resigned, and crossed the Atlantic to work as an independent researcher in Marie Skłodowska Curie's (1867–1934) laboratory. When she did get married some time later and started a family, Brooks abandoned the subject of radioactivity, and indeed research, altogether. We will get to know both Richards and Brooks, and their work on the elements, in this volume.

Marriage was an even more problematic event for another woman featured in this book, the German-Jewish chemist Clara Immerwahr (1870–1915). Immerwahr's marriage with Fritz Haber (1868–1934) brought an end to her research career, which had been off to an excellent start and was supported by the highly regarded chemist Richard Abegg (1869–1910). Immerwahr's suicide following the 1915 deployment of chlorine on the Ypres front has been considered by historians a sign of protest against her husband's role in the development of chemical warfare. While the Ypres incident might well have triggered Immerwahr's suicide, a closer look at historical documents reveals that the causes may have been more complex, including the slow but constant decline of her personal

---

[13] See Charbonneau and Rice in this volume.

[14] See Rayner-Canham and Rayner-Canham on Brooks in this volume.

intellectual life due to her family duties, and, in parallel, the demands of her husband's burgeoning career.[15]

This series of glorious yet sometimes difficult careers should not, however, obscure the fact that in many cases, traditional roles and expectations still influenced women's roles in the scientific endeavour. The lives of Louise Hammarström (1849–1917) and Naima Sahlbom (1871–1957), two Swedish women who we will meet in this book, attest to the fact that the metaphor of the kitchen legitimised laboratory work for female practitioners.[16] Since analytical tasks are meticulous, repetitive, and in a way subcontracted, the names of those who executed these analyses are often not known and hardly ever mentioned in scientific publications. Historical research has not yet sufficiently investigated and documented these second- and third-tier figures, among them also men, without whom scientific progress could not have been achieved. In fact, many women featured in this volume started their careers thanks to their analytical skills or experimental abilities: Julia Lermontova,[17] Ellen Swallow Richards,[18] Ellen Gleditsch (1879–1968),[19] Stefanie Horovitz (1877–1942),[20] Elizabeth Róna (1890–1985),[21] Berta Karlik (1904–1990), Traude Bernert (1915–1998)[22] and, last but not least, Marguerite Perey (1909–1975),[23] who discovered francium (element 87) while employed as a technician, only acquiring a PhD and a career after this feat.

More recently, the Japanese-American chemist Toshiko Mayeda[24] (1923–2004) and the English scientists Mary Almond (b. 1928)[25] and Barbara Bowen (b. 1932),[26] all featured in this volume, held roles in scien-

---

[15] See Friedrich and Hoffmann in this volume.
[16] See Lundgren in this volume.
[17] See Boeck on Lermontova in this volume.
[18] See Charbonneau and Rice in this volume.
[19] See Lykknes on Gleditsch in this volume.
[20] See Rayner-Canham and Rayner-Canham on Horovitz in this volume.
[21] See Rentetzi in this volume.
[22] See Forstner (on both Karlik and Bernert) in this volume.
[23] See Rayner-Canham and Rayner-Canham on Perey in this volume.
[24] See Shindell on Mayeda in this volume.
[25] See Horrocks, Lean and Merchant on Almond in this volume.
[26] See Horrocks, Lean and Merchant on Bowen in this volume.

tific teams that allowed them to develop specialised skills. They are visible examples of the many invisible technicians and lab assistants that were frequently women. Almond's and Bowen's supporting roles in teams furthermore made it possible for them to combine a professional with a family life, enabling them also to go beyond their initial training and education.

Women Discovering and Mastering Elements

Women have been involved in work on the elements in many different ways throughout history, from investigations on matter in early modern times through to the present-day production of the short-lived superheavy elements. Women's contributions to science include the discovery of elements and their properties, the weighing, separation and manufacture of elements and their isotopes, and the production of knowledge on the effects of elements on humans and their environment.

Old and New Understandings of the Elements

In 1735 Georg Brandt (1694–1768), the director of the Swedish Board of Mines discovered that cobalt was a metal distinct from bismuth. Retrospectively, this is regarded as one of the earliest element discoveries after the Middle Ages. Thirty years previously, Dorothea Juliana Wallich published three books about chymical (early chemical/alchemical) experiments using a cobalt-containing ore.[27] While intending to produce the philosophers' stone, she provided a detailed description of what we today would call the thermochromic reactions of cobalt compounds. This is the earliest example of a woman's work relating to the elements presented in this volume, and shows that women were active in alchemical quests as well as mineralogical-chemical work.[28] In her understanding of matter, though, Wallich did not consider cobalt an element, nor for that matter bismuth. This is not the case for the two other eighteenth-century women featured in this book. Their work relied on

---

[27] See Kraft in this volume.

[28] For a recent example of historical research on women alchemists, see [Nummedal, 2019].

their contemporary understanding of elements, and this concept requires a short detour through the history of chemical elements.

Ever since the beginning of civilisation humans have reflected on the order of nature and the question of what constitutes matter. The ancient Greek philosopher Thales of Milet (c. 620–555 BC) thought water was the essence of matter. His successors, by contrast, suggested that the ultimate constituents of matter were fire, air and what they called *apeiron*, an indefinite first substance [Ball, 2004, p. 6]. The most famous theory of primary matter is, however, that of the four elements earth, air, fire and water, which originated with Empedocles (c. 490–430 BC), and which Aristotle (384–322 BC) borrowed, adding the four qualities (hot, cold, wet and dry). One element could turn into another when its qualities were altered. This four-element theory of matter would be influential for centuries, and well into the early modern period.

Other theories of matter were developed as well. The writings attributed to the Arabic alchemist Jābir ibn Hayyān (c. 721–c. 815) presented the idea that all metals were mixtures of the principles of sulphur (hot and dry) and mercury (cool and moist). (These principles were not to be confused with elementary sulphur and mercury, i.e. sulphur lumps or mercury drops.) Later a third principle, salt, was added to the mixture, creating a three-principle theory for the constitution of matter, and this is usually associated with the Swiss alchemist Paracelsus (1493–1541). Paracelsus called these principles the *tria prima* which formed a trinity of qualities, representing the body (salt), the soul (sulphur) and the spirit (mercury).

The influential alchemist Johann Becher (1635–1682) of Speyer in the Rhineland rejected Paracelus' *tria prima* but acknowledged the existence of three forms of earth: *terra fluida*, i.e. the mercurious earth which represented volatility and metallicity; *terra pinguis*, which was the principle of combustion; and *terra lapidea*, the principle of fusibility. Becher's pupil Georg Ernst Stahl (1659–1734), developed the idea of the *terra pinguis* further into that of "phlogiston," a principle which all flammable substances would give off during combustion. Across the channel, in England, the chymist and co-founder of the Royal Society in London, Robert Boyle (1627–1691) had only recently criticised the four-element and three-principle theories and introduced small bodies named "corpuscles" as the constituents of matter instead.

Chymists were not alone in their interest in the nature of elements. As the foundations of mechanical philosophy laid by the Frenchman René Descartes (1596–1650) and the Englishman Isaac Newton (1642/43–1727) were gaining momentum, the nature of the corpuscles of fire and the laws governing their interactions with matter in general were also debated by physicists (or rather, "natural philosophers," in the terminology of the time). The French noblewoman Emilie Du Châtelet took advantage of the request for anonymous submissions for an essay for competition at the *Académie Royale des Sciences* (French Academy of Sciences), and submitted an essay on the nature and propagation of fire, alongside her lover and intellectual partner Voltaire's (1694–1778) submission.[29] Neither of them were awarded the prize, but both contributions were published. This was the first time that a female savant had an essay published by the Academy. Beyond this accomplishment, Du Châtelet's essay is also striking because of its content, its method, and its broad knowledge not only of natural philosophy but also of chymistry. We now know that fire is not an element. The nature of fire, however, was a riddle that would not be solved until the following century.

The phlogiston theory, based on Stahl's concept of the principle of fire, developed into a very successful chemical theory that provided a framework for chemical investigations for most of the eighteenth century. When in the 1780s and 1790s the process of combustion was explained as a reaction involving the element oxygen, which had been previously known as "fire air" (thus named by Carl W. Scheele, 1742–1786) or "dephlogisticated air" (Joseph Priestley, 1733–1804), it became the cornerstone of a reform of chemistry — or a chemical revolution — orchestrated by the French aristocrat and member of the French Academy of Sciences Antoine Lavoisier (1743–1794). The reform also included the launch of a new nomenclature, with an emphasis on terms that would distinguish the new chemistry from the old phlogiston paradigm. Lavoisier worked closely with his wife, Marie-Anne, née Paulze, who translated English and Italian scholarly works and scientific correspondence into French, illustrated Lavoisier's books, assisted him in the laboratory, and regularly invited savants to her salon.[30] In this volume we will read about

[29] See Van Tiggelen on Du Châtelet in this volume.
[30] See Bret and Kawashima in this volume.

the role Mme Lavoisier had in translating and commenting on a classical work in defence of phlogiston by the Irish chemist Richard Kirwan (1733–1812), which was part of the propaganda for the new chemistry.

Lavoisier and his colleagues also insisted on a new understanding of the elements as simple substances which could not be decomposed further during chemical analyses — a so-called "negative-empirical" definition of element. By demonstrating that water could be produced from oxygen and hydrogen, and that it could be decomposed back into its elements, Lavoisier and colleagues insisted that water could not be an element in the four-element theory, and neither could air, since different kinds of "airs" (gases) existed, e.g. oxygen and hydrogen. As the historian of early modern chemistry Hjalmar Fors [2014, 2015] has shown, however, a similar concept of elements had already been developed in the first half of the eighteenth century by "mineralogical chemists" at the Swedish Board of Mines. While at the time metals were regarded as composite materials, these mineralogical chemists developed the idea that metals were simple substances, basic species of nature and end products of chemical analyses. The discovery of cobalt by Georg Brandt in 1735 played an important role in the discussions on metals as simple substances, and cobalt was established as distinct from bismuth, together with which it was often found in ores. By then, Dorothea Juliana Wallich's comprehension of matter was completely outdated.

The Glory of Analytical Chemistry: The Elements Multiply

In 1789 Lavoisier presented the then-known elements in the form of a table (Figure 1). His list included substances that, according to Lavoisier, belonged to all the kingdoms of nature (oxygen, hydrogen, nitrogen, even caloric — the substance of heat — and light), followed by general chemical families (non-metals, metals, and earths that can form salts). These families included elements that had been known prior to the Middle Ages (copper, lead, gold, silver, iron, carbon, tin, sulphur, mercury, arsenic and antimony), as well as elements discovered or isolated in the seventeenth and eighteenth centuries, such as zinc, cobalt, bismuth and nickel. Even with its hierarchy, Lavoisier's table was not a systematic overview of the elements, and surely not a full classification of all elements.

TABLE OF SIMPLE SUBSTANCES.

Simple fubftances belonging to all the kingdoms of nature, which may be confidered as the elements of bodies.

New Names.				Correfpondent old Names.
Light	-	-	-	Light.
Caloric	-	-	-	{ Heat. Principle or element of heat. Fire. Igneous fluid. Matter of fire and of heat.
Oxygen	-	-	-	{ Dephlogifticated air. Empyreal air. Vital air, or Bafe of vital air.
Azote	-	-	-	{ Phlogifticated air or gas. Mephitis, or its bafe.
Hydrogen	-	-	-	{ Inflammable air or gas, or the bafe of inflammable air.

Oxydable and Acidifiable fimple Subftances not Metallic.

New Names.				Correfpondent old names.
Sulphur	-	-	-	}
Phofphorus	-	-	-	} The fame names.
Charcoal	-	-	-	}
Muriatic radical	-			}
Fluoric radical	-	-		} Still unknown.
Boracic radical	-	-		}

Oxydable and Acidifiable fimple Metallic Bodies.

New Names.					Correfpondent Old Names.
Antimony	-				{ Antimony.
Arfenic	-	-			Arfenic.
Bifmuth	-	-			Bifmuth.
Cobalt	-	-			Cobalt.
Copper	-	-			Copper.
Gold	-	-			Gold.
Iron	-	-	-	of	Iron.
Lead	-	-	-	Regulus	Lead.
Manganefe	-	-	-		Manganefe.
Mercury	-	-			Mercury.
Molybdena	-	-			Molybdena.
Nickel	-	-	-		Nickel.
Platina	-	-			Platina.
Silver	-	-			Silver.
Tin	-	-			Tin.
Tungftein	-	-			Tungftein.
Zinc	-	-			Zinc.

Salifiable fimple Earthy Subftances.

New Names.	Correfpondent old Names.
Lime	{ Chalk, calcareous earth. Quicklime.
Magnefia	{ Magnefia, bafe of Epfom falt. Calcined or cauftic magnefia.
Barytes	Barytes, or heavy earth.
Argill	Clay, earth of alum.
Silex	Siliceous or vitrifiable earth.

Figure 1. Lavoisier's table of elements, 1789, from the English translation (by Robert Kerr) *Elements of Chemistry in a New Systematic Order* (Edinburgh: William Creech, 1790), pp. 175–6. (Compiled by Michael McBride for the The Lavoisier on Elements and Oxidation Website, reproduced by kind permission of Michael McBride.)

From the early eighteenth century onwards, and long before Lavoisier's time, chemists had also drawn up tables in attempts to systematise their knowledge about chemical reactions, in particular displacement reactions, and these were known as "affinity tables" (Figure 2). Tabulating all known displacement reactions naturally involved a choice of substances for the top of the columns, and the choice was mainly guided by the laboratory use of chemical reagents rather than the substances' assumed elementary nature. Affinity tables were developed until the end of the eighteenth

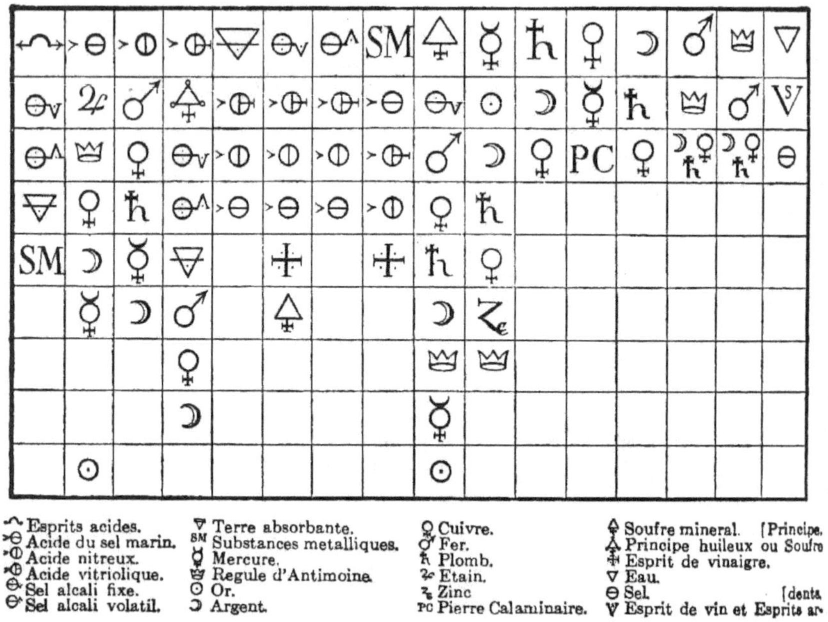

Esprits acides. ▽ Terre absorbante. ♀ Cuivre. ♀ Soufre mineral. [Principe.
Acide du sel marin. SM Substances metalliques. ♂ Fer. ♠ Principe huileux ou Soufre
Acide nitreux. ☿ Mercure. ♄ Plomb. ♯ Esprit de vinaigre.
Acide vitriolique. Regule d'Antimoine. ♃ Etain. ▽ Eau.
Sel alcali fixe. ☉ Or. Zinc ⊖ Sel. [denta
Sel alcali volatil. ☽ Argent. PC Pierre Calaminaire. Esprit de vin et Esprits ar-

Figure 2. The first known affinity table, devised by the French physician and chymist Étienne François Geoffroy (1672–1731) in 1718. From M. Geoffroy, Table des différents rapports observés en Chimie entre différentes substances, printed in *Histoire de l'Académie royale des sciences, avec les mémoires de mathématique et de physique* (Paris: Académie royale des sciences, 1741) (public domain). The table is presented between page 212 and 213. The affinity table summarises different substances' reactions with one another (or their "affinity" for, i.e. being "attracted" to — one another). For example, the second column shows that the (in modern terms) hydrochloric acid at the top will have more affinity with tin than with antimony, copper, silver, mercury or gold (the order in which they appear in the column); and thus, copper would displace silver and mercury in their combination with hydrochloric acid.

century, and the Swedish mineralogical chemist Torbern Bergman's (1735–1784) table from 1775 (enlarged edition 1783), with 59 columns and 50 rows, was the largest of these tables [Duncan, 1962].

Apart from the elementary gases discovered in the 1760s and 1770s (hydrogen, oxygen, nitrogen), when studies of air had become popular, most new elements discovered between 1700 and 1800 were discovered via chemical analyses of minerals. Women too, participated in chemical analyses; these were, in fact, as we have described above, the roles they could take on easily. This volume presents many examples for this, starting with Julia Lermontova, whom Mendeleev knew personally and whom he is likely to have encouraged to devise improved methods of element separation in the platinum group, so that the elements' atomic weights could be determined.[31] Accurate atomic weights were clearly essential for Mendeleev's task of ordering the elements; and in the case of the platinum group they were even more essential, since chemists struggled to distinguish the elements in the group. They appeared in the same ores, had similar chemical properties and were also of close atomic weight (Figure 3). While other families of elements exhibited very characteristic behaviours that would allow for speculation and the reordering of atomic weights — this was the case, for instance, for iodine and tellurium — the elements of the platinum group required experimental precision on the atomic weight that could not be deduced from the chemical behaviour as easily. This demanded experimental expertise, which Lermontova could offer.

From the eighteenth into the early twentieth century, new elements were found in newly discovered mineral ores, and in ores that had previously not been sufficiently described. Its mineral richness made Scandinavia a prime site for mineralogical investigations centuries ago. Sweden is more frequently represented in the discoveries of new elements in mineral sources than most other countries; in fact, 40 per cent of all element discoveries between the Middle Ages and 1886 were made in Sweden [Fors, 2015, p. 100]. Swedish discoverers include the chemical authority Jöns Jacob Berzelius (1779–1848), who discovered cerium, selenium and thorium, isolated zirconium, and prepared silicon in its pure form; the professor of chemistry and pharmacy Carl Gustaf Mosander

---

[31] See Boeck on Lermontova in this volume.

	Gruppe I. $R'O$	Gruppe II. RO	Gruppe III. $R'O^3$	Gruppe IV. RH^4 RO^7	Gruppe V. RH^3 R^2O^5	Gruppe VI. RH^5 RO^3	Gruppe VII. RH R^2O^7	Gruppe VIII. — RO^4
1	H = 1							
2	Li = 7	Be = 9.4	B = 11	C = 12	N = 14	O = 16	F = 19	
3	N = 23	Mg = 24	Al = 27.3	Si = 28	P = 31	S = 32	Cl = 35.5	
4	K = 39	Ca = 40	— = 44	Ti = 48	V = 51	Cr = 52	Mn = 55	Fe = 56 Co = 59 Ni = 60, Cu = 63.
5	(Cu = 63)	Zn = 65	— = 68	— = 72	As = 75	Se = 78	Br = 80	
6	Rb = 85	Sr = 87	?Yt = 88	Zr = 90	Nb = 94	Mo = 56	— = 100	Ru = 104, Rh = 104, Pd = 106, Ag = 104.
7	(Ag = 104)	Cd = 112	In = 113	Sn = 118	Sb = 122	Te = 125	J = 127	
8	Cs = 133	Ba = 137	?Di = 138	?Ce = 140	—	—	—	— — —
9	(—)		—	—	—	—	—	—
10	—	—	?Er = 178	?La = 180	Ta = 182	W - 184	—	Os = 195, Ir = 197, Pt = 198, Au = 199.
11	(Au = 199)	Hg = 200	Tl = 204	Pb = 207	Bi = 208	—	—	
12	—	—	—	Th = 231	—	U = 240	—	— — — —

Mendeleev's Periodic Table of 1871, redrawn by Jeff Moran, 2013

Figure 3.   Mendeleev's periodic system from 1871, redrawn in 2013 by Jeff O. Moran (The Internet Database of Periodic Tables, reproduced by kind permission of Jeff O. Moran). The lines represent elements whose existence had been predicted, but which had not yet been discovered. In Group VIII, to the right, we see the platinum metals. It is clear from the atomic weights used when this table was devised that the values of their atomic weights were very close to each other. As a first step Julia Lermontova separated the platinum metals (ruthenium, rhodium, palladium, osmium, iridium and platinum) in two samples, looking for their correct order in the periodic system.

(1797–1858), who discovered lanthanum, erbium and terbium; the professor of general and agricultural chemistry Per Teodor Cleve (1840–1905), who discovered holmium and thulium; and Lars Fredrik Nilson (1840–1899) of the Royal Swedish Academy of Agriculture and Forestry, who in 1879 discovered scandium, Mendeleev's undiscovered eka-boron. As the vitae of Louise Hammarström and Naima Sahlbom demonstrate, women had the chemical skills needed for investigating the elements in nineteenth-century Sweden: analytical-chemical skills that were comparable with kitchen work.[32] Indeed, throughout the nineteenth century chemistry was often carried out in homes, close to the kitchen or even in the kitchen, which could serve as a laboratory. Women, who were housekeepers, wives, daughters or sisters, often assisted in this work. Astrid Cleve — the daughter of Per T. Cleve, for example, had access to her father's collections

---

[32] See Lundgren in this volume.

of materials and his laboratory, and became part of his extended group of students and collaborators.[33] She worked mainly on the properties of the rare earth metal ytterbium, and thereby contributed to the understanding of the rare earth metals and the relationship between them. With their chemical analyses women helped confirm and map out the different elements in the periodic system; indeed, women constituted an important workforce in the various analytical work on geological and industrial samples. As is the case for housekeepers, the work of the female analysts was expected to be invisible.

At MIT, Ellen Swallow Richards had access to the laboratory of her husband, who was also her former mineralogy professor.[34] With his support she continued conducting chemical analyses on minerals, and even published articles jointly with one of her students at the Women's Laboratory, the laboratory she had helped establish in 1876. Her detailed analysis of the mineral samarskite yielded an insoluble residue, which was later revealed to contain the then-unknown elements samarium and gadolinium. Interestingly, Swallow did not instigate the investigation, but was given the task of analysing the mineral by Thomas Bouvé (1815–1896), an amateur mineralogist who needed an expert in analytical chemistry to perform this task for him. As was the case for the work of many little-known chemists, however, Richards' solid results paved the way for future, more highly acknowledged discoveries.

The tradition of analytical chemistry continued well into the twentieth century. Mineral analysis was the main analytical tool for Marie and Pierre Curie's (1859–1906) attempts to separate salts from their mineral sources in the newly discovered elements polonium and radium.[35] It was also instrumental for Harriet Brooks,[36] May Sybil Leslie (1887–1937),[37] Ellen Gleditsch,[38] Marguerite Perey,[39] Berta Karlik and Traude Bernert,[40] who

---

[33] See Espmark and Nordlund in this volume.

[34] See Charbonneau and Rice in this volume.

[35] See Roqué on Curie in this volume.

[36] See Rayner-Canham and Rayner-Canham on Brooks in this volume.

[37] See Rae on Leslie in this volume.

[38] See Lykknes on Gleditsch in this volume.

[39] See Rayner-Canham and Rayner-Canham on Perey in this volume.

[40] See Forstner (on both Karlik and Bernert) in this volume.

were trying to solve problems in the early phase of radioactivity research (see also below). At the beginning of the twentieth century, when this research was based on mineral analysis and wet chemical analytical methods, in Spain José Casares Gil (1866–1961) initiated a research line on fluorine analyses in mineral waters. In the 1920s and 1930s, many students, including a number of women, came to do fluorine analyses.[41] Their research involved a careful comparison of the use of different reagents in the determination of fluorine, and was still embedded in the practice of wet chemistry. Here women opened up new avenues of research on applied analyses, of organic products as well as mineral waters.

As regards radioactivity research, although the work was entirely dependent on mineral analyses, new substances had to be detected in the radioactive minerals by other means, i.e. electricity and spectroscopy, the two primary detection methods which were both new to the nineteenth century. Indeed, the nineteenth century was the century of electricity, starting with Alessandro Volta's (1745–1827) invention of the battery in 1800, and Humphry Davy's (1778–1829) use of a battery to split compounded substances by means of electrolysis. Between 1807 and 1811 Davy separated a range of elements by electrolysis, and thus discovered or confirmed their existence; these elements included sodium, potassium, calcium, magnesium, strontium, barium and boron. Davy demonstrated the use of electricity in his many public and spectacular lectures in the Royal Institution in London, to audiences which included women [Golinski, 2016, pp. 74–96]. Jane Marcet, who was among them, soon incorporated the newly discovered metals in her textbook *Conversations in Chemistry*.[42]

Women were also included in other means of dissemination of knowledge not to the general public, but to the scientific community. The increasing knowledge about the chemical reactions of new and old elements and their compounds prompted the production of handbooks that grew in size and often appeared in successive updated editions with further supplements to keep up with the ever accelerating developments in chemistry. For instance John Newton Friend's (1881–1966) *Textbook of*

---

[41] See Suay-Matallana in this volume.

[42] See Childs in this volume.

Inorganic Chemistry, which he compiled in England between 1912 and 1937, ran to as many as twenty-one volumes which were organised according to the groups of elements in the periodic system. Four women authors were recruited to write on some of these elements — a sign that they were considered members of the community.[43] While we know little about the work and career of most of these women, the profile on May Sybil Leslie in this book provides details of her work on the disintegration of thorium.[44]

New Fields and Instrumental Methods

While element discoveries well into the nineteenth century (and beyond) were characterised by mineral analyses and the use of electricity, from the second half of the century onwards, and especially in the early 1900s, a range of new methods using instruments to detect and identify elements would emerge. One of these methods, spectroscopy, had been in use since the mid-nineteenth century when the German chemist Robert Bunsen (1811–1899) and the physicist Gustav Robert Kirchhoff (1824–1887) invented the spectroscope. In the following year, in 1860, they used it to detect the elements caesium and rubidium in mineral waters by studying their emission lines. Spectroscopic analyses were useful for the detection of elements that were present in such small amounts that they could not be identified by chemical means. In the 1880s, for example, spectroscopy revealed that the mineral samarskite, which Ellen Swallow Richards had investigated, contained two new elements, samarium and gadolinium.[45] Marie and Pierre Curie also turned to spectroscopy to confirm the existence of the new elements polonium and radium that were identified in pitchblende mineral in 1898, but a unique line was only found for radium, not for polonium. It was the radioactivity in itself, together with a careful examination of chemical reactions, that had identified the elements in the first place.[46]

---

[43] See Rae on Four Women Chemists in this volume.

[44] See Rae on Leslie in this volume.

[45] See Charbonneau and Rice in this volume.

[46] See Roqué on Curie in this volume.

Shortly after Bunsen's and Kirchhoff's discoveries, as early as in 1868, spectroscopy was used to identify a yellow line in the solar spectrum that did not belong to a known element. The French astronomer Jules Janssen (1824–1907) and the English astronomer Norman Lockyer (1836–1920) are usually credited separately as the discoverers of "helium," named after its presence in the sun (from Greek *helios* — sun). It was, however, the British-born American astronomer and astrophysicist Cecilia Payne-Gaposchkin (1900–1979) who, in the course of her studies of stellar spectra and ionisation theory, discovered that helium and hydrogen were the most abundant elements in the stars, a result she could hardly believe herself.[47] Her results were confirmed a few years later, and, thanks to her supportive male supervisors Payne-Gaposchkin's research earned her memberships in the American Philosophical Society and the National Academy of Sciences.

X-ray spectroscopy provided new opportunities. In 1913 the British physicist Henry Moseley (1887–1915) found a link between the X-ray emission spectrum of an element and the number of protons in its atomic nucleus, and this new technique was developed into "finger print" detection for unknown elements. Based on his findings, Moseley justified the concept of the "atomic number," which would later replace atomic weights as the crucial factor for the elements' position in the periodic system.[48] The French chemist Georges Urbain (1872–1938) used X-ray spectroscopy to confirm the existence of the rare earth element lutetium (no. 71) that he had isolated [Kragh, 1996]. For Ida Noddack (1896–1978) and Walter Noddack (1893–1960), who together searched for undiscovered manganese homologues, X-ray spectroscopy constituted one of two lines of research by which they might identify the new elements they had named rhenium (no. 75) and masurium (no. 43); the other method was the enrichment and separation of the elements based on their thorough knowledge of the periodic system and the chemistry of the elements.[49] As the article on the Noddacks' work in this book shows, precise physical

---

[47] See Shindell on Payne-Gaposchkin in this volume.

[48] From the time of Moseley's introduction of atomic numbers, in 1913, it would take another eight years before the German Atomic Weight Commission accepted to base the new periodic system on atomic numbers [Kragh, 2012, p. 34]. The final approval from the International Union of Pure and Applied Chemistry was given two years later, in 1923.

[49] See Van Tiggelen on Noddack in this volume.

methods could be a double-edged sword: if the results were not supported by physical evidence like the sufficient quantities of an element or one of its compounds, it would not convince the scientific community. For Marie and Pierre Curie, too, the production of sufficient quantities of their elements was imperative, although it would take them years to obtain even a minute amount of radium salts, and for the Noddacks to obtain sufficient amounts of pure rhenium. Masurium was even expected to be rarer in nature than rhenium, and the Noddacks never succeeded in obtaining it from the ores. When they were further not able to reproduce the spectra of masurium, and none of their peers succeeded either, it was hard to argue that a new element had been discovered.

A few years later, in 1932, while still a PhD student, the French chemical physicist Yvette Cauchois' (1908–1999) improvement of the X-ray technique made it possible to measure the emission lines of low-intensity elements, such as the rare earth elements and the heavy elements.[50] Cauchois, who later became professor at the Sorbonne in Paris, was credited as the inventor of the "Cauchois spectrograph." The spectrograph was adopted with enthusiasm by scientists. She, too, experienced difficulties in using the spectral lines as the sole evidence for an element discovery, and together with her colleague Horia Hulubei (1896–1972), she suggested the existence of the elements 85, 87 and 93. More importantly, by use of spectra taken with her spectrograph, Cauchois was able to confirm the atomic number of polonium, and even to gain deeper insights into the electronic configuration of atoms and ions.

X-rays were also utilised in new research fields such as crystallography, where the structure of a crystalline substance could be determined by the way in which it scattered a beam of X-rays. In 1929 the Irish crystallographer Kathleen Lonsdale (1903–1971) provided crystallographic evidence for the structure of the benzene molecule, which brought prestige to the field of crystallography as well as to Lonsdale herself.[51] She also found similarities between the carbon in benzene and the carbon in graphite. Her X-ray studies of artificial and natural diamonds contributed variously to the growing knowledge of carbon.

---

[50] See Van Tiggelen on Cauchois in this volume.
[51] See Wilson in this volume.

Women also excelled in emerging fields and in solving theoretical and applied problems. One of the new fields in chemistry in the nineteenth century was physical chemistry, which was announced in 1887 by the Swedish chemist Svante Arrhenius (1859–1927), the Baltic German chemist Wilhelm Oswald (1853–1932), and the Dutch chemist Jacobus Henricus van't Hoff (1852–1911) as a new discipline at the intersection between chemistry and physics. New fields presented new problems that needed to be solved. The realm of physical chemistry included investigations of the electrical properties of salts in solution. Clara Immerwahr, the first woman to be awarded (in 1900) a PhD at the University of Breslau, gained the support of one of the pioneers in the new field of physical chemistry, Richard Abegg, who had proposed the concept of electroaffinity.[52] Her doctoral research immersed Immerwahr in this tradition, and she conducted experiments on the solubility of salts of cadmium, copper, lead, mercury and zinc, among other things.

Erika Cremer (1900–1996), who is best known for her contribution to the development of the gas chromatograph from 1944 onwards, worked a decade earlier on the quantum chemistry of molecular hydrogen, in particular on the catalytic interconversion of the so-called ortho- and para-forms of hydrogen.[53] The ortho-parahydrogen conversion was regarded as a test case for the new theories of quantum mechanics that were developed towards the end of the 1920s. Knowledge about the conversion from ortho to para also became important because of hydrogen's use in military applications (bombs and rocketry), where hydrogen needed to be converted to the para form. Cremer worked on other quantum mechanical problems as well, but as an unpaid researcher she did not receive much recognition for her early work.

Clusters of Women in Radioactivity

Even though around the turn of the twentieth century chemists had become accustomed to using physical methods like spectroscopy to detect elements, with radioactivity it was not just the measurements that

---

[52] See Friedrich and Hoffmann in this volume.
[53] See Johnson in this volume.

were "invisible," but also the main property, the emitted radiation, was undetectable by the human eye. When Marie Curie received her second Nobel Prize, in chemistry in 1911, she talked about an "entirely separate kind of chemistry for which the current tool we use is the electrometer, not the balance, and which we might well call the chemistry of the imponderable" [Curie, 1911]. Indeed, in radioactivity measurements revolved around the detection of the invisible by using invisible methods. Ironically, however, this was an early-twentieth-century field that would have the opposite effect for women: women clustered together to do research in this field, and thus became more visible [Rentetzi, 2007, chapter 4]. (Another field which attracted women in the early twentieth century was crystallography, and it was here that Kathleen Lonsdale[54] and Isabella Karle (1921–2017)[55] worked [Rayner-Canham and Rayner-Canham, 1997, p. 13].)

In Marie Curie's Paris laboratory women constituted between 8.7 and 40 percent of the researchers between 1906 and 1934, i.e. in the years in which Curie headed the laboratory [Pigeard Micault, 2013]. Likewise, at the Institute for Radium Research (The Radium Institute) in Vienna, which was founded in 1910, between 1919 and 1934 around 38% of the staff members were women. In Lise Meitner (1878–1968) and Otto Hahn's (1879–1968) department at the Kaiser Wilhelm Institute for Chemistry in Berlin women made up 14 percent of the workers between 1912 and 1938 [Rentetzi, 2007, chapter 4]. Ernest Rutherford (1871–1937) also had female co-workers in Canada and in Manchester. The large share of women in radioactivity research has been attributed to support of their mentors or the laboratories' directors as well as to the meticulous and routine work often associated with the field of radioactivity [Rayner-Canham and Rayner-Canham, 1997, pp. 17–24]. However, the fact that some women were granted access to the field because they were willing to take on tasks that were not very exciting does not mean that they did routine work forever. On the contrary, the women working at The Radium Institute in Vienna were highly productive researchers [Rentetzi, 2003], and this was also the case for many of the women at the Curie laboratory, as we will see in this volume.

---

[54] See Wilson in this volume.
[55] See Robinson on Karle in this volume.

Another factor contributing to the large number of women in radioactivity in the early decades of the twentieth century was the fact that the field was new and not yet established, thus lacking strong male hierarchies and not specifically designed to support men. Their entrance into a completely new research field must have been exciting and full of promise for ambitious young people wishing to build their careers, and who, at the same time, were ready to take on the risk of embarking on an unestablished line of research. Further, the interdisciplinary nature of the field — its situation between physics and chemistry in particular, but also between biology and medicine — enabled women to take up positions in hospitals, medical laboratories and technical companies due to its close connection with industry [Rentetzi, 2004].[56] This volume testifies to the fact that many women were attracted to radioactivity: fourteen articles deal with women engaged in the science of radioactivity or nuclear science, from Harriet Brooks,[57] the first woman entering Marie Curie's laboratory in 1906 via Rutherford's laboratory in Montreal to one of today's role models, Dawn Shaughnessy, who is Group Leader, Experimental Nuclear and Radiochemistry at Lawrence Livermore National Laboratory, California.[58] The group of women in radioactivity featured in this volume include students, assistants, technicians, full-fledged researchers, professors and leaders of research laboratories.

The first generation of researchers in radioactivity, in the early twentieth century, were all working on fundamental problems in the new science, trying to understand what radioactivity was, what the decay products of the radioactive substances were, how the different substances were related to each other, and how long they would continue giving off rays, to mention just a few of the questions. Starting with Marie Curie herself, the early women "radioactivists" were all part of this joint venture to understand the nature of radioactivity, the radioactive substances and the radioactive processes. Discoveries of elements and isotopes were entwined in this tradition, which was embedded in analytical chemistry,

---

[56] However, as Maria Rentetzi [2007, chapter 4] has pointed out, local contexts, including institutional settings and political movements, must also be considered in an analysis of women in radioactivity in different laboratories.

[57] See Rayner-Canham and Rayner-Canham on Brooks in this volume.

[58] See Murray and Wade in this volume.

the so-called "wet" bench-work chemistry that had also been so characteristic for the nineteenth century.

Since Marie Curie had access to the highly sensitive piezo-electric quartz electrometer devised by her husband Pierre and his brother Jacques (1855–1941), she was able to choose radioactivity as her dissertation topic in 1897, and to discover polonium and radium with her husband at the end of the subsequent year.[59] In order to obtain substantial quantities of the new elements, a mutually beneficial cooperation with industry was established. With a steady supply of material at hand, the Curies and the many students and researchers who came to their laboratory could pursue their investigations into the nature of radioactivity.

Once a widow, from 1906 onwards, Marie Curie became the leader of the laboratory to which young people from all over the world would flock. As noted, the first woman who came to Marie Curie's laboratory in the first year of Curie's leadership was Harriet Brooks.[60] She had been Ernest Rutherford's first graduate student and had worked with him on the nature of the so-called radium "emanation," the substance which radium (and also thorium) emitted when giving off rays (Figure 4). In 1901 Rutherford and Brooks suggested that it was a gas. This was the first mention of what was later identified as radon gas, and furthermore, provided evidence that radioactive processes involve the transmutation of one element into another. In 1902, together with his colleague Frederick Soddy (1877–1956), Rutherford proposed the theory of radioactive disintegration (the transformation theory), which explained that radioactivity was an atomic property.

As the disintegration of one radioactive substance was being studied, another was uncovered, since most radioactive substances release a new radioactive substance. A series of radioactive disintegrations would follow the first, always ending in lead, which was stable. Figure 4 shows the sequences of the radioactive series which the radioactivists were able to establish by 1904. There were the substances named "uranium X," "thorium X," "thorium A," "thorium B," etc., which were simply named after their ultimate mother substance. In the early 1910s more than 30 such substances had been described. As Curie put it in her Nobel Lecture in

---

[59] See Roqué on Curie in this volume.
[60] See Rayner-Canham and Rayner-Canham on Brooks in this volume.

Product.	T.	λ (sec)$^{-1}$.	Some physical and chemical properties.
URANIUM	10^9 years	$2 \cdot 2 \times 10^{-17}$	Soluble in excess of ammonium carbonate.
$\downarrow$ Uranium X	22 days	$3 \cdot 6 \times 10^{-7}$	Insoluble in excess of ammonium carbonate.
$\downarrow$ Final product.	—	—	
THORIUM	3×10^9	7×10^{-18}	Insoluble in ammonia.
$\downarrow$ Thorium X	4 days	$2 \cdot 00 \times 10^{-6}$	Soluble in ammonia.
$\downarrow$ Thorium emanation	1 minute	$1 \cdot 15 \times 10^{-2}$	Chemically inert gas; condenses about $-120°$ C.
$\downarrow$ Thorium A	11 hours	$1 \cdot 75 \times 10^{-5}$	Behaves as solid; insoluble in ammonia; volati-
$\downarrow$ Thorium B	55 minutes	$2 \cdot 1 \times 10^{-4}$	lized at a white heat; soluble in strong acids; Thorium A can be separated from B by electrolysis.
$\downarrow$ Thorium C (final product)	—	—	
ACTINIUM	—	—	
$\downarrow$ Actinium X ?	—	—	
$\downarrow$ Actinium emanation	$3 \cdot 7$ seconds	$1 \cdot 87 \times 10^{-1}$	Behaves as a gas.
$\downarrow$ Actinium A	41 minutes	$2 \cdot 80 \times 10^{-4}$	Behaves as solid; soluble in strong acids;
$\downarrow$ Actinium B	$1 \cdot 5$ minutes	$7 \cdot 7 \times 10^{-3}$	A can be partially separated from B by electrolysis.
$\downarrow$ Actinium C (final product)	—	—	
RADIUM	800 years	$2 \cdot 8 \times 10^{-11}$	
$\downarrow$ Radium emanation	4 days	$2 \cdot 00 \times 10^{-6}$	Chemically inert gas; condenses about $-150°$ C. Definite spectrum; volume diminishes with time.
$\downarrow$ Radium A	3 minutes	$3 \cdot 8 \times 10^{-3}$	Behaves as solid; soluble in strong acids;
$\downarrow$ Radium B	21 minutes	$5 \cdot 38 \times 10^{-4}$	volatilized at a white heat; B is more volatile
$\downarrow$ Radium C	28 minutes	$4 \cdot 13 \times 10^{-4}$	than A or C and can thus be temporarily separated from them.
$\downarrow$ Radium D	About 40 years	—	Gives out only β rays. Soluble in strong acids.
$\downarrow$ Radium E $\downarrow$	About 1 year	—	Probably active constituent of radio-tellurium; soluble in strong acids; volatilized at a red heat: deposited on bismuth in solution.

Figure 4. Scheme for the radioactive decay series from 1904. From Ernest Rutherford (1904). Bakerian Lecture: The Succession of Changes in Radioactive Bodies, *Phil. Trans.*, 204, p. 206 (public domain).

1911, "[t]his hypothesis [that radioactivity is an atomic property of matter] has led to present-day theories of radioactivity, according to which we can predict with certainty the existence of about 30 new elements which we cannot generally either isolate or characterize by chemical methods" [Curie, 1911]. Indeed, the new radioactive substances were thought to be new elements, but which positions would they occupy in the periodic system? There were only a few known blank spaces.

This challenge would be solved with the concept of "isotopes," which was introduced in 1913 by Frederick Soddy to describe atoms of chemical elements with identical chemical properties (but not radioactive properties) and different atomic weights. This meant that all the new radioactive substances would not necessarily be new elements, but rather isotopes of known elements. Although Soddy had introduced the concept, it was a young woman — the medically trained Margaret Todd (1859–1918) — who suggested the name "isotope" while attending a dinner party at which Soddy was also present. Her story is told in this volume.[61] "Isotope" derives from the Greek for "same place", a language which Todd mastered. The term implies that the isotopes of one element occupy the same box in the periodic system, i.e. that they refer to one and the same element.

In 1913, at about the same time as he introduced the concept of isotopes, Soddy further proposed the so-called "Group Displacement Law," which was also independently discovered by Kasimir Fajans (1887–1975) [Kauffmann, 1982a and 1982b]. The law clarified the genetic relationship between elements (Figure 5): an alpha (α-) emitter would decay into an element two places to its left in the periodic table and have an atomic weight of four units lower, while a beta (β-) emitter would transmute into an element one place to its right in the periodic system (compare Figures 4 and 5) and preserve its atomic weight. The Group Displacement Law was applied to all known radioactive elements, and their atomic weights were predicted [Kauffmann, 1982a].

Based on the Group Displacement Law, Soddy (and also Fajans) predicted that lead stemming from the disintegration of uranium would have a lower, and lead stemming from thorium decay a higher atomic weight than common lead. Physical evidence for significantly different atomic

---

[61] See Hudson in this volume.

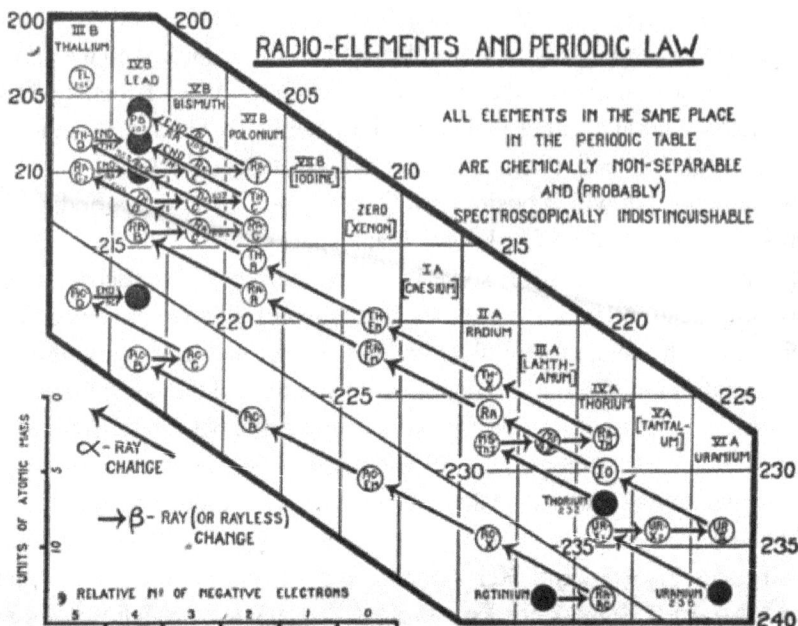

Figure 5. The Group Displacement Law, explaining the genetic relationship between the elements. From Frederick Soddy (1913). Radioactivity, *Annual Reports on the Progress of Chemistry*, 10, p. 264 (public domain).

weights in leads of different origins would support not only the Group Displacement Law, but also the concept of the "isotope" itself. Soddy carried out the initial atomic weight determinations himself, but his results were met with some scepticism. In order to continue the investigations, sufficient samples and skilled expertise were needed.

In a very short period of time experimental results were presented by Maurice Curie (1888–1975) in Paris, Otto Hönigschmid (1878–1945) and Stefanie Horovitz in Vienna, and Theodore William Richards (1868–1928), who was working with Fajans' assistant Max E. Lembert, at Harvard [Kauffmann, 1982b; Conant, 1970]. Richards, who had been awarded the 1914 Nobel Prize in chemistry for his work, was the unrivalled world authority on atomic weight determinations at the time; and Hönigschmid, who had trained under him, was considered an expert in the field across Europe. All of the scientists who presented their determinations found differences between the atomic weight of common lead and that of lead of radioactive origin. Richards and his students continued their investigations beyond the initial results, and so did Hönigschmid and Horovitz at the Institute for Radium Research.[62] In their last joint work on the atomic weight of lead, which was published in 1915, Hönigschmid and Horovitz arrived at differences of more than one weight unit, which was indeed beyond experimental error. In the chapter on Horovitz in this volume, their experimental results are identified as the first authoritative evidence for the concept of isotopes. The 1918 discovery of a long-lived isotope for actinium's mother substance, protactinium, by Lise Meitner and Otto Hahn further confirmed the Group Displacement Law, as Soddy had determined two possible places in the periodic system for this mother substance with the help of his law, one of them being an α-emitter positioned in group V — "eka-tantalum" (see Figure 6, where the element is called "brevium") [Kauffmann, 1982b; Sime, 1996, chapter 3]. Meitner and Hahn's protactinium fitted this description.[63]

The question of variations in atomic weights branched out beyond the field of radioactivity. Once the concept of isotopes had been established in the scientific community, it was relatively easy to accept that the atomic weight of lead would vary from source to source, since the average atomic weight of lead in a sample would depend on the amount of radioactive uranium or thorium in it. However, in 1919, when Francis Aston (1877–1945), who was using his newly invented mass spectrograph, reported that he had also found two different isotopes of chlorine, this was met with some concern, as chlorine was not known to be the product of a radioactive process, and this could lead to a reconsideration of the concept of atomic weight. The Norwegian chemist Ellen Gleditsch, who had trained with Curie in Paris, was well-connected in the radioactive community and, at this point, associate professor at her home university in Oslo, demonstrated through joint work with her assistants that chlorine did, in fact, have a constant atomic weight, i.e. a weighted average of the masses of each isotope based on their relative abundance — and that this weight was independent of the sample's origin.[64]

---

[62] See Rayner-Canhamn and Rayner-Canham on Horovitz in this volume.

[63] See Roqué on Meitner in this volume.

[64] See Lykknes on Gleditsch in this volume.

Tabelle I.   Periodisches System der Elemente.

Wertigkeit bez. Gruppennummer
Symbol
Atom-Gewicht
Ordnungszahl (Kernladung)

Die Atomvolumina nehmen von der Mitte nach beiden Seiten zu.

	I	II	III	IV	V	VI	VII	VIII			I	II	III	IV	V	VI	VII	0
1	H 1.008 1																	He 4.00 2
	Li 6.94 3	Be 9.1 4	B 11.0 5	C 12.0 6	N 14.0 7	O 16.0 8	F 19.0 9											Ne 20.2 10
	Na 23.0 11	Mg 24.3 12	Al 27.1 13	Si 28.3 14	P 31.0 15	S 32.1 16	Cl 35.5 17											Ar 39.9 18
4	K 39.1 19	Ca 40.1 20	Sc 45.1 21	Ti 48.1 22	V 51.0 23	Cr 52.0 24	Mn 54.9 25	Fe 55.8 26	Co 59.0 27	Ni 58.7 28	Cu 63.6 29	Zn 65.4 30	Ga 69.9 31	Ge 72.5 32	As 75.0 33	Se 79.2 34	Br 79.9 35	Kr 82.9 36
5	Rb 85.5 37	Sr 87.6 38	Y 88.7 39	Zr 90.6 40	Nb 93.5 41	Mo 96.0 42	? 43	Ru 101.7 44	Rh 102.9 45	Pd 106.7 46	Ag 107.9 47	Cd 112.4 48	Jn 114.8 49	Sn 118.7 50	Sb 120.2 51	Te 127.5 52	J 126.9 53	X 130.2 54
6	Cs 132.8 55	Ba 137.4 56	seltene Erden 139.0 bis 178 57 s. unten bis 72		Ta 181.5 73	W 184.0 74	? 75	Os 190.9 76	Jr 193.1 77	Pt 195.2 78	Au 197.2 79	Hg 200.6 80	Tl 204.0 81	Pb 207.2 82	Bi 208.0 83	Po 210 84	? 85	Em 222 86
7	? 87	Ra 226.0 88	Ac 89	Th 232.1 90	Bv 234 91	U 238.2 92												

La 139.0 57	Ce 140.3 58	Pr 140.9 59	Nd 144.3 60	? 61	Sm 150.4 62	Eu 152.0 63	Gd 157.3 64	Tb 159.2 65	Dy 162.5 66	Ho 163.5 67	Er 167.7 68	TuI 168.5 69	TuII 173.5 70	Ad 71	Cp 175.0 72

The placement of all of the radioactive substances into their respective decay series was a task that occupied many of the first-generation radio-activists. This included pain-staking "wet chemical" analyses including atomic weight determinations, occasional spectroscopic analyses, and almost always the measurement of the radiation given off from a radioactive substance — and women (and men) working in laboratories all over Europe and in North America took part in this venture. The laboratory procedures involved, first, meticulous separations of the substances with analytical-chemical methods (i.e. by means of their chemical reactions), and then, in some cases, also the concentration of radioactively pure substances on a small surface. Here the Hungarian Jewish chemist Elizabeth Róna's experience and skills in preparing polonium sources at the Radium Institute in Vienna, where she worked in the 1920s, earned her the reputation of being the "polonium woman."[65] More than a decade earlier, in 1908–09, when uranium had not yet been established as radium's ultimate mother substance (see Figure 4), Ellen Gleditsch was in Paris and took on the task of determining the ratio of radium to uranium in minerals. She thereby provided evidence of the relationship between uranium and radium in the natural decay series. In 1913–14, while working in Bertram Boltwood's (1870–1927) laboratory at Yale, she determined the half-life

←————————————————————————————

Figure 6. Periodic system from 1918 by Stefan Meyer (1872–1949), leader of the Institute for Radium Research, reproduced from Stefan Meyer (1918). Periodische Systeme der Elemente *Physikalische Zeitschrift* 19, p. 179 (public domain). It is worth noting the use (text to the left of the diagram) of the term "Ordnungszahl (Kernladung)" (atomic number). Although the concept of atomic numbers had been substantiated by Henry Moseley's X-ray spectroscopic experiments five years earlier, atomic numbers had not yet been officially introduced to new periodic systems. The space between thorium (Th) and uranium (U) (element 91) is here labelled "Bv," short for "brevium," the short-lived isotope of the element discovered by Fredrick Soddy, and named by Kasimir Fajans and Ostwald Göhring (1889–c. 1915). In 1918, i.e. in the same year in which this periodic system was devised, Lise Meitner and Otto Hahn discovered a much longer-lived isotope, and called the element "protactinium." Positions 43, 61, 72, 75, 85 and 87 were still vacant in 1918 (in this system, however, no. 72 is denoted as Tu (thulium) II, with reference to the supposed discovery of this element by Carl Auer von Welsbach, 1858–1939).

————————————————————————————

[65] See Rentetzi in this volume.

of radium (radium-226), and thereby solved a discrepancy problem that had been known since the first determinations of radium's half-life had been made.[66]

At about the same time, the British chemist May Sybil Leslie, who worked in Marie Curie's laboratory for two years from 1909 onwards, and then for one year with Ernest Rutherford in Manchester, separated different fractions (quantities with specific properties) from the mineral thorite, found thorium in several of them, and investigated its disintegration products (the thorium series).[67] She also determined the half-life of "radiothorium" (thorium-228), thorium emanation (radon-220) and actinium emanation (radon-219), and with this she contributed to resolving discrepancies in values established by different laboratories.

At Vienna, Stefanie Horovitz put her own evidence for the concept of isotopes to good use and disproved the existence of the element "ionium" (thorium-230). By means of atomic weight determinations conducted in 1914 and 1916, and chemical and spectroscopic analyses, she noted that ionium chemically behaved like thorium and differed only in its atomic weight.[68] Ionium was part of the uranium-radium series. The thorium series was also well studied. But the actinium series (Figure 4) was more mysterious. Actinium had been discovered in 1899 by André Debierne (1874–1949), who would become one of Marie Curie's closest collaborators, and it was known to decay into an "emanation" (a radon gas, like thorium and radium in their respective series), and further into other radioactive substances. In 1918, as discussed above, Lise Meitner and Otto Hahn discovered its mother substance, protactinium. The substance called "uranium Y" (thorium-231) was known to decay into protactinium. During World War I, while working with Georg von Hevesy (1885–1966) Elizabeth Róna separated uranium Y from the interfering elements, proved that it was a beta emitter and confirmed its half-life value. Laborious repetitions and improving of experimental procedures were absolutely necessary in order to arrive at consensus values which, again, could be used to determine new values. The fact that actinium's ultimate

---

[66] See Lykknes on Gleditsch in this volume.

[67] See Rae on Leslie in this volume.

[68] See Rayner-Canham and Rayner-Canham on Horovitz in this volume.

ancestor was a rare isotope of uranium (uranium-235) was not confirmed until the 1930s, and Ellen Gleditsch was one of the scientists investigating this question by looking at the ratio of actinium to uranium in minerals of various ages [Lykknes *et al.*, 2005].

Another woman working on the actinium decay series was the French laboratory technician Marguerite Perey.[69] In 1929, as the best graduate from her class, she was hired by Marie Curie as her personal assistant. When Curie died, André Debierne, who was then in charge of the laboratory, and Irène Joliot-Curie (1897–1956) independently asked Perey to determine the half-life of actinium (actinium-227). In the course of this demanding work Perey discovered radiation that was unaccounted for, which in 1939 led to her discovery of the heaviest of the alkali metals, element number 87; it was named "francium," after France. Since Debierne and Joliot-Curie could not agree who Perey was working for, Perey, the technician, was given the sole credit for this discovery, and subsequently enjoyed unexpected career opportunities: she received her doctorate in 1946, and was appointed to Chair of Nuclear Chemistry at the University of Strasbourg a few years later. Francium was the last natural element discovered that could be extracted from minerals.

The last of the naturally occurring elements to be discovered was element number 85, astatine, which only exists in minute amounts. In 1939 Yvette Cauchois and Horia Hulubei claimed to have detected it in a radon sample, but this was controversial.[70] Four to five years later, in 1943 and 1944, Berta Karlik and Traude Bernert, who were working at the Radium Institute in Vienna, found three naturally occurring if short-lived isotopes of astatine (215, 216 and 218).[71] But since other isotopes of element 87 had been produced via induced nuclear reactions a few years earlier (in 1940), Karlik and Bernert were not credited as discoverers of the element. At this time, the wet chemical approach to radioactivity research was no longer frontline research. In the 1920s radiochemistry had met its saturation point, and few novel insights were expected from the field. The concept of isotopes had been established, the disintegration theory and the Group Displacement

---

[69] See Rayner-Canham and Rayner-Canham on Perey in this volume.

[70] See Van Tiggelen on Cauchois in this volume.

[71] See Forstner in this volume.

Law had been accepted, most of the radioactive nuclides had found their place in the radioactive decay series, half-lives had been determined, and the periodic system had stabilised following the critically challenge posed by the phenomenon of radioactivity. Lawrence Badash has called this the "suicidal success of radiochemistry" [Badash, 1979]. The initial success of radioactivity research had left it without a future in top research. But starting with the discovery of the neutron particle by James Chadwick (1891–1974) in 1932, the 1930s saw a number of discoveries which presented radioactivity research with new methods and problems and rung in the new area of nuclear science.

Women and the Modern Scientific Enterprise

While Marie Curie pioneered radioactivity research within the wet chemical tradition, her daughter Irène was among those who prepared a new line of research into nuclear reactions. In the early 1930s, even before Chadwick discovered the neutron, the married couple of scientists Irène and Frédéric Joliot-Curie (1900–1958) used highly penetrating alpha particles emitted by polonium to bombard light nuclei.[72] When aluminium nuclei were used as a target, the couple detected not only neutrons but also positrons (positive electrons), and in 1934 they discovered that the positrons resulted from a separate, induced nuclear reaction. This discovery of induced, or artificial, radioactivity earned them the Nobel Prize in chemistry in 1935.

Manufacturing Elements:
From Artificial Radioactivity to Big Science

Following the discovery of artificial radioactivity Enrico Fermi (1901–1954) in Rome immediately started experimenting with neutrons as projectiles; their lack of an electrical charge was advantageous compared to alpha or proton particles, since they did not have to overcome repulsive electrical charges [Fermi, 1938]. In 1934, when bombarding uranium nuclei with neutrons, Fermi and his team concluded that they were the first to obtain elements heavier than uranium: the transuranium elements 93 and 94, which they named "ausonium" and "hesperium."

---

[72] See Jacquemond in this volume.

The German chemist Ida Noddack, however criticised Fermi's conclusions.[73] Together with her husband Walter Noddack (and with the help of Otto Berg) she had discovered the element rhenium nine years earlier, and had ample experience in evaluating the new discoveries of elements. In her opinion, Fermi ought to have ruled out any known elements that were lighter than lead before claiming to have discovered transuranium elements. Indeed, she boldly proposed that heavy nuclei might split — the phenomenon later known as nuclear fission. Noddack's insights were dismissed and ridiculed by the scientific community, and when Otto Hahn, Lise Meitner and their team finally discovered nuclear fission more than four years later, Noddack was not acknowledged as the first to have proposed it. However, Ida Noddack was not the only woman who was not acknowledged in the history of nuclear fission: Lise Meitner,[74] who interpreted Hahn's experimental results while in exile in Sweden, never received her due share in the Nobel Prize awarded in chemistry for 1944, and neither did the two male collaborators Fritz Strassmann (1902–1980) and Otto Frisch (1904–1979).

Ida Noddack's involvement with the elements would cross the path of nuclear reactions a second time: element 43, "masurium," which the Noddacks claimed to have discovered along with rhenium in 1925, was first produced in 1937 in an induced nuclear reaction carried out by Carlo Perrier (1886–1948) and Emilio Segrè (1905–1989), and by 1940 was also identified by Otto Hahn and Fritz Strassmann as one of the fission products of uranium. Because it was the first artificially made element, it was named "technetium" after the Greek for "craft." In the same year element 85, astatine, was produced by Emilio Segrè, Dale Corson (1914–2012) and Kenneth MacKenzie (1912–2002) via the bombardment of bismuth nuclei with alpha particles, and at the Radium Institute in Vienna, Berta Karlik and Traude Bernert discovered naturally occurring isotopes of the element three and four years earlier.

The change of focus in radioactivity research from chemical analyses of radioelements to the study of nuclear processes prompted the invention and use of large devises such as the cyclotron, which made it possible to accelerate nuclear particles to very high velocities without the use of high voltages [N. N., 1939]. Ernest O. Lawrence (1901–1958) was awarded the Nobel Prize in physics in 1939 for his invention and the

---

[73] See Van Tiggelen on Noddack in this volume.

[74] See Roqué on Meitner in this volume.

PERIODIC TABLE SHOWING HEAVY ELEMENTS AS MEMBERS OF AN ACTINIDE SERIES

Arrangement by Glenn T. Seaborg, 1945

I	II	III	IV	V	VI	VII	VIII	VIII	VIII	I	II	III	IV	V	VI	VII	0
1 H 1.008																	2 He 4.003
3 Li 6.940	4 Be 9.02											5 B 10.82	6 C 12.010	7 N 14.008	8 O 16.000	9 F 19.00	10 Ne 20.183
11 Na 22.997	12 Mg 24.32											13 Al 26.97	14 Si 28.06	15 P 30.98	16 S 32.06	17 Cl 35.457	18 A 39.944
19 K 39.096	20 Ca 40.08	21 Sc 45.10	22 Ti 47.90	23 V 50.95	24 Cr 52.01	25 Mn 54.93	26 Fe 55.85	27 Co 58.94	28 Ni 58.69	29 Cu 63.57	30 Zn 65.38	31 Ga 69.72	32 Ge 72.60	33 As 74.91	34 Se 78.96	35 Br 79.916	36 Kr 83.7
37 Rb 85.48	38 Sr 87.63	39 Y 88.92	40 Zr 91.22	41 Cb 92.91	42 Mo 95.95	43	44 Ru 101.7	45 Rh 102.91	46 Pd 106.7	47 Ag 107.880	48 Cd 112.41	49 In 114.76	50 Sn 118.70	51 Sb 121.76	52 Te 127.61	53 I 126.92	54 Xe 131.3
55 Cs 132.91	56 Ba 137.36	57 La 58-71 SEE LA SERIES	72 Hf 178.6	73 Ta 180.88	74 W 183.92	75 Re 186.31	76 Os 190.2	77 Ir 193.1	78 Pt 195.23	79 Au 197.2	80 Hg 200.61	81 Tl 204.39	82 Pb 207.21	83 Bi 209.00	84 Po	85	86 Rn 222
87	88 Ra	89 Ac SEE AC SERIES	90 Th 232.12	91 Pa 231	92 U 238.07	93 Np 237	94 Pu	95	96								

LANTHANIDE SERIES

57 La 138.92	58 Ce 140.13	59 Pr 140.92	60 Nd 144.27	61	62 Sm 150.43	63 Eu 152.0	64 Gd 156.9	65 Tb 159.2	66 Dy 162.46	67 Ho 163.5	68 Er 167.2	69 Tm 169.4	70 Yb 173.04	71 Lu 174.99

ACTINIDE SERIES

89 Ac	90 Th 232.12	91 Pa 231	92 U 238.07	93 Np 237	94 Pu	95	96

development of the cyclotron, and its application in studies of artificial elements. In the second half of the 1930s he helped laboratories in the United States and Europe build their own cyclotrons. One physicist inspired by Lawrence's work was Manne Siegbahn (1886–1978), whose Swedish laboratory provided the exiled Lise Meitner with a home for almost a decade. Siegbahn was so absorbed with his passion for developing and using his own cyclotron that he neglected Meitner to the extent that she did not get access to the necessary resources for her investigations, and did not get to use the cyclotron once it had been set up [Friedman, 2001, pp. 225–239; Sime, 1994].

In 1940, using Lawrence's cyclotron, Edwin McMillan (1907–1991) and Philip H. Abelson (1913–2004) discovered the first two transuranium elements, neptunium (93) and plutonium (94), by bombarding uranium nuclei with neutrons [Holl, 1997, p. 177]. They thereby, at Berkeley, started a tradition of manufacturing elements. Between 1941 and 1955 Glenn T. Seaborg (1912–1999) and his co-workers discovered nine more new elements: americium (95), curium (96), berkelium (97), californium (98), einsteinium (99), fermium (100), mendelevium (101), nobelium (102) and seaborgium (106) [Kauffman, 2019] (Figure 7). In 1951 Seaborg and McMillan shared the Nobel Prize in chemistry "for their discoveries in the chemistry of the transuranium elements" [N. N., 1951]. Ten years earlier the

---

Figure 7. Glenn T. Seaborg's periodic system from 1945, taken from Seaborg's Priestley Medal Address, "The Periodic Table: Tortuous Path to Man-Made Elements," printed in C&EN on 16 April 1979, p. 48. Seaborg and his collaborators found that plutonium and neptunium had chemical properties similar to uranium, rather than being "transition metals" (metals in the middle of the periodic system), as had been previously suggested. Seaborg furthermore proposed the actinides series as a parallel group to the lanthanide series. It is worth noting that the spaces for elements 43, 85 and 87 are still blank, four to six years after they had been discovered, as these elements were not acknowledged by the International Union of Pure and Applied Chemistry and given their names (technetium, astatine, and francium) until 1947. The places for elements 61, 95 and 96 are also vacant. These latter two elements were announced to have been discovered by Seaborg and his team in November 1945, and 61 in 1949; all of them had been observed earlier, but the announcement was postponed because of the strategic nature of the research projects in which they were isolated. (The Lawrence Berkeley National Labroatory, © 2010 The Regents of the University of California, through the Lawrence Berkeley National Laboratory.)

majority of the work had moved to the Manhattan Project's Metallurgical Laboratory (Met Lab) at the University of Chicago. The Manhattan Project was the joint nuclear weapon programme, directed by the USA and supported by Canada and the United Kingdom. A major task for Seaborg and his group was the production of pure plutonium, which was difficult because the samples produced in the nuclear reactions were small and impure. Isabella L. Karle was recruited to Seaborg's group immediately after the completion of her PhD, during which she had gained experience in glass blowing.[75] She used her skills to create a vacuum apparatus, a vessel with higher melting point than glass, and furthermore a copper furnace which she used for high temperature experiments; the subsequent cooling allowed for crystals of plutonium chloride to grow. That these were, indeed, plutonium chloride crystals was later confirmed by means of X-ray diffraction.

Another woman participating in the Manhattan Project, Chien-Shiung Wu (1912–1997), had completed her doctoral research under Lawrence at Berkeley in 1940, in which she investigated so-called "Bremsstrahlung" and the fission products of uranium.[76] By bombarding uranium nuclei with neutrons using the cyclotron, Wu and Emilio Segrè (who was also advising her) first extracted iodine and then, some hours later, xenon from the iodine. They found two isotopes of xenon, one of which had never been detected before, and by 1940 Wu had identified the two isotopes. Secrecy surrounded all research associated with the making of bombs, so Wu and Segrè's publications did not reveal many details. In 1944 Wu was recruited to the Substitute Alloy Materials (SAM) Laboratories at Columbia University, which was also part of the Manhattan Project. Her knowledge of the isotope xenon-135 solved a problem in the production of plutonium for use in atomic bombs, as it turned out that xenon-135 was shutting down the chain reaction which produced plutonium from uranium. Wu's research helped the team devise new techniques for controlling the reaction.

Elizabeth Róna, too, became engaged in the war effort in the USA, after fleeing from Austria to Sweden and Norway.[77] It was her expertise on polonium, which she had acquired after her first training with Irène

---

[75] See Robinson on Karle in this volume.

[76] See Robinson on Wu in this volume.

[77] See Rentetzi in this volume.

Curie, and then working for 13 years at the Radium Institute in Vienna, that led her to the US Atomic Energy Commission's projects on isotope separation and projects at other big US laboratories. Her preparation method for polonium was even used for the mass production of polonium sources by the Canadian Radium and Uranium Company, which was contracted by the US Office of Scientific Research and Development.

A new tendency in particle physics and nuclear science to install large particle accelerators with which experiments could be conducted on particle level, which escalated in the 1950s, 1960s and 1970s, was part of a general trend in Europe and North America now referred to as "Big Science." Big Science involved large-scale research plants, large research groups, enormous budgets and increasing frequencies in publishing, as well as cross-national collaboration [Nye, 1996, p. xiii; Kragh, 1999, pp. 312–311]. The manufacture of transuranium elements, and in particular the short-lived superheavy elements, is part of this tradition.

After Edwin McMillan and Philip Abelson's discovery of neptunium in 1940, as noted above, Glenn Seaborg and his collaborators produced a series of transuranium elements, identified them, and became renowned for this work worldwide. In the 1960s the Soviet Union started programmes for the production of transuranium elements, and by the 1980s the Soviet Union and the USA had become equal competitors.[78] The researchers at the GSI Helmholtz Centre for Heavy Ion Research in Darmstadt, Germany, also joined the race in the 1980s; and from the 1990s onwards, the Riken Nishina Centre for Accelerator-Based Science in Japan gained "full momentum" in this line of research [Riken Nishina, 2019]. All four countries have contributed to the discovery of superheavy elements, as is evident from the elements' names, which were approved by the International Union of Pure and Applied Chemistry (IUPAC), and include darmstadtium (no. 110, after Darmstadt), nihonium (no. 113, Japanese for "Japan"), flerovium (no. 114, after the Flerov Laboratory of Nuclear Reactions in Dubna, Russia), moscovium (no. 115, after Moscow), livermorium (no. 116, after the Lawrence Livermore National Laboratory), tennessine (no. 117, after Tennessee) and oganesson (no. 118, named after the Russian scientist Yuri Oganessian, b. 1933).

---

[78] See Murray and Wade in this volume.

Darleane Hoffman became the first female team leader in superheavy elements research.[79] After completing her PhD in nuclear and radiochemistry in 1951, she worked at the Oak Ridge National Laboratory for one year and then moved to Los Alamos, where she remained for the next three decades. One of her (and her team's) most celebrated achievements from this period is the finding that the nuclei of fermium-257 could split via symmetric fission, i.e. into two fragments of equal mass. In 1984 she was offered a full professorship at the University of California, Berkeley, where she directed the Heavy Element Nuclear and Radiochemistry Group which had previously been led by Glenn Seaborg. At Berkeley Hoffman carried out difficult experiments on short-lived radioactive elements. Her PhD student, Dawn Shaugnessy, is now the group leader for the Nuclear and Radiochemistry Group at the Lawrence Livermore National Laboratory at Berkeley.[80] Together with the Russian Flerov Laboratory of Nuclear Reactions at Dubna her team has contributed to the identification of elements 113 to 118.

Instrumental Revolution
and the Interface between Chemistry and Industry

The move away from laboratory bench work in radioactivity research was symptomatic of the changes affecting the entire field of analytical chemistry from the 1930s onwards and until the 1950s and 1960s, when advanced instrumental methods gradually took over parts of the meticulous and demanding wet chemical laboratory bench work [Baird, 2002]. While in the nineteenth and early twentieth centuries, chemical constituents in a sample would be determined from its chemical reactions (and partly via spectroscopy), in the new era physical methods took over as instruments for the direct detection of elements and isotopes. The new physical instruments also changed analytical practice in that the new methods were more sensitive and accurate, had lower limits of detection, and required less material to be viable for analysis [Baird, 2002]. Typical representatives for this so-called "instrumental revolution" in chemistry

---

[79] See Murray and Wade in this volume.
[80] See Murray and Wade in this volume.

[Morris, 2002] are mass spectrometry (MS) and nuclear magnetic resonance (NMR), which originated in physics and high technology, and became common tools for structure determinations in chemistry [Reinhardt, 2006, pp. 1–9]. Although, as mentioned above, the mass spectrograph was first invented by Francis Aston in 1919, mass spectrometry did not find widespread use in chemistry until the 1960s, when it became a common method for determining structure in organic chemistry [Morris, 2015, p. 309], and commercial mass spectrometers became common in research laboratories.

Prior to this development, in the 1950s, the Japanese-American chemist Toshiko Mayeda, who held a bachelor degree in chemistry, became one of the most talented and experienced mass spectrometer operators of her time.[81] Since the instruments were home-made, skilled technicians were considered as important for their successful use as the instruments themselves. As a technician for the Nobel laureate Harold C. Urey (1893–1981) Mayeda started as a washer of glassware, and would eventually take on more responsibilities. Eventually her aptitude for mass spectrometry work marked her as a crucial member of Urey's team, where she was acknowledged and credited as a co-author in all published work. Urey and his team developed a so-called "oxygen thermometer" to determine relative amounts of stable isotopes of oxygen trapped in marine fossils. Sensitive mass spectrometers were needed, and the opportunity to specialise in these instruments gave Mayeda the opportunity to develop as a scientist beyond her formal education.

While Urey and his team aimed to determine temperatures for the prehistoric oceans, a small team working in England in the early 1950s and led by the physicist Patrick Blackett (1897–1974) tested the theory of continental drift — a theory far from accepted in the scientific community at the time. Mary Almond, PhD, was part of the team studying the magnetism of iron in rocks using a magnetometer.[82] Like Mayeda for the mass spectrometer, Almond was regarded as the most efficient operator of the complex magnetometer. Within the team she contributed to revolutionary results in the earth sciences, although retrospectively — not uncommonly

---

[81] See Shindell on Mayeda in this volume.
[82] See Horrocks, Lean and Merchant on Almond in this volume.

for the women of her time — she did not consider her own role in the scientific endeavour very important.

Another example of a woman playing a crucial role in teams investigating environmental problems is Barbara Bowen, who was active in England in the late 1970s.[83] Working on atmospheric ozone, she digitised geophysical and atmospheric data including data collected using the so-called Dobson spectrometer. Awareness of the depletion of the ozone layer by pollutants increased at the time, and Bowen's work contributed to the finding of the ozone hole over Antarctica. Bowen characterised the digitisation work as "boring" work yet "very important," and moreover, this work made it possible for her to combine a career with having a family. Like Almond, Bowen did not seek credit for her scientific achievements. Nevertheless, the examples of Mayeda, Almond and Bowen all help to shed light on the invaluable work of team workers operating in the shadow of the main (and often, acknowledged) actors, which was essential to the whole team's results.

Research and Development (R&D) laboratories have co-existed with university research laboratories since the periodic system was first developed, and shed light on yet other roles taken on by women in research on the elements. The first purpose-built research laboratory was the Badische Anilin- und Soda-Fabrik (BASF) in Ludwigshafen, Germany, which was built in the mid-1870s [Morris, 2015, p. 234]; the first R&D laboratory in the USA was set up in 1901 at the General Electric Schenectady plant in New York [Carlson, 1997]. In Norway, at the beginning of the twentieth century, power-intensive electrochemical and electrometallurgical industries were established thanks to the country's easy access to hydro-electric power. The Norwegian Electrochemical Industry Corporation, Elkem, which was founded in 1904, was one of the companies which early on established research stations for the purpose of experimenting and researching processes of interest for the company, such as the production of metal alloys. Sonja Smith-Meyer Hoel (1920–2004), a chemical engineer graduating from the Norwegian Institute of Technology in 1947, was later in the same year employed at one of Elkem's research stations, at Fiskaa in the south of Norway.[84] At the time Elkem specialised in the

---

[83] See Horrocks, Lean and Merchant on Bowen in this volume.

[84] See Lykknes on Smith-Meyer Hoel in this volume.

production of ferro-silicon alloys, and shortly afterwards started to produce mineral wool and aluminium. After a few years of working at the research station, Smith-Meyer Hoel took up a position at the company's patent office in Oslo, and became its director in 1965. Her engagement with nation-wide and international patent work in combination with numerous other offices she held turned her into an early influential woman in the Norwegian industry. Smith-Meyer Hoel is probably unique in that she had practical-industrial competence in working with metals and metal products, both from production sites and research stations, and gathered extensive experience in helping colleagues to file patents on metal elements and alloys.

The chemist Reatha Clark King, PhD (b. 1938) was the first African-American woman to work at the National Bureau of Standards in Washington, DC (today the National Institute of Standards and Technology, which "promotes U.S. innovation and industrial competitiveness" [NIST, 2019]).[85] In 1963 she joined the calorimetry laboratory in the heat division of the bureau, working on materials science and high-temperature chemistry. As part of a contract project for the National Aeronautics and Space Administration (NASA), King was assigned to the study of fluorine, which was proposed for use in rocket propellants, and took up work that others had failed to complete with success. Her investigations required her to design calorimetry equipment. She also devised procedures for controlling and measuring the combustion of dangerous gases in the lab. Although fluorine was assessed to be too dangerous to be used as fuel, the equipment and methods King developed continue to be used to study potential gaseous propellants.

Social Activism, Sisters in Arms

King was not the first scientist to work with dangerous materials. In radioactivity research, although Marie Curie herself had warned of the dangers of handling radium and suggested that researchers have regular blood tests [Rentetzi, 2017], many of those working in radioactivity research laboratories became ill from exposure to radiation, including Marie Curie and her daughter Irène Joliot-Curie, and Marguerite Perey.

---

[85] See Ockerbloom on King in this volume.

As early as in 1908, the professor of pathology Alice Hamilton (1869–1970) wrote an article on industrial poisons, long before a general awareness about the industrial exposure to poisons had developed.[86] As a doctor for immigrants in her neighbourhood, she learned about the daily lives of people working in dangerous trades. She gained knowledge of many instances of occupational lead poisoning, and took responsibility for investigations into lead, reporting (and confirming with the use of a diagnostic indicator) more than 500 cases of lead poisoning. Hamilton gave her reports to companies together with recommendations for improvements. She even used publicity as a weapon. Indeed, her work had a great impact on the working conditions of many workers. When the new field of industrial medicine was established at Harvard in 1919, she became the first female professor there, in recognition of the work she had done to reduce exposure to industrial hazards.

At around the same time, in Bern, Switzerland, the head of the physico-chemical biology laboratory at the University of Bern, Gertrud Woker (1878–1978), advocated against chemical warfare, and wrote about the body's reaction to chemical and nuclear poisons including tetraethyl lead, which she suggested should be replaced in petrol with non-toxic compounds.[87] Simultaneously active in the Women's International League for Peace and Freedom, Woker exemplifies the woman scientist as activist. Others, like Irène Joliot-Curie, were politically engaged against fascism, and were passionate advocates for peace.

An organisation which was established after World War I in reaction to the devastating experiences of war was the International Federation of University Women (IFUW). It was thought that fostering women's friendships across nations would contribute to peace [Gildersleeve, 1954; Oertzen, 2014]. Ellen Gleditsch was active both in the Norwegian branch of the organisation and in the IFUW itself, and was the latter's president from 1926 to 1929. During her term she established scholarships for university-educated women to go abroad. Berta Karlik and Isabella Karle were among those who benefited from these arrangements: A fellowship from the IFUW enabled Karlik to study in London for one year, while Karle received a scholarship from the American Association of University Women to complete her graduate studies.

---

[86] See Ockerbloom on Hamilton in this volume.

[87] See Boeck on Woker in this volume.

In addition to the official networks of women in organisations, informal networks were formed among women who worked together. As we have seen, radioactivity and crystallography became scientific niches for women desiring to understand matter. Proximity on the bench in several of the laboratories that welcomed young female students created a long-term camaraderie, a sort of "sisters in arms" spirit, and a support network that turned out to be crucial to the fight against misogyny once back home. Early on in the period discussed in this book, pioneers like Ida Freund and Ellen Swallow Richards devoted time and energy to providing young women with the best education in science possible; and Marie Curie, in her professional beginnings, was keen to offer an outstanding training to her female students at l'École de Sèvres. These remarkable figures knew only too well from experience that only education could equip the next generation of women to conquer an even greater part of the pursuit of science. Ida Freund, for instance, insisted that the quality and level of chemistry lectures for women should not be any different from those for young men. In this context her lessons on the periodic table were most informative.[88] The pioneers were not shy to defend the women's cause in science, or in wider society. Knowledge has always been a key factor to emancipation.

Many male leaders opened up opportunities for women. Dmitri Mendeleev, Stefan Meyer, Ernest Rutherford, Otto Hönigschmid, Harold C. Urey and others were eager to hire and promote their female students in an egalitarian manner. John Newton Friend, in recruiting female authors for his multivolume *Textbook of Inorganic Chemistry*, also contributed to making women in chemistry visible to the wider chemical community. And while cases like those of Julia Lermontova and Toshiko Mayeda underline how crucial the recognition of their male colleagues proved to be in their career, men's support in the domestic realm was equally important. Marie Curie had the opportunity to pursue her passion for science because of arrangements made at home, firstly with her parents, and secondly when her father-in-law helped out with child care so that Marie could spend time in the laboratory. An awareness of the support provided and needed from family members as well as mentors is as important in the

---

[88] See Opitz in this volume.

history of scientific endeavour as the concepts of genius, determination and passion.

Elements of Diversity

The 38 vignettes in this book showcase the diversity of women's involvement with the elements and the periodic system. While, over time, the roles accessible to women in the sciences evolved, many of the traditional tasks were often allotted to women in laboratory or in scientific teams: women were experts in analytical chemistry, technicians, assistants, supporting wives and sometimes heads of the laboratory or professors, and more than often pioneers in a male dominated community. Only recently more women have had the chance to lead groups that hunt elements, as exemplified by the cases of Darleane Hoffman and Dawn Shaughnessy, and this trajectory needs to expand and continue to show that science is for all. Women have also occupied other roles pertaining to science in society, in political and social activism, equal opportunity and solidarity engagement, as educators and teachers, data analysts and patent specialists. The story has just begun; indeed access to interviews and oral history, which some essays in this book are based on, help keep the stories alive while paper archives as historians' primary material is starting to vanish.

Interestingly, the different roles attributed to women throughout history also reflect the diversity of the elements. Across the contributions in this volume, elements from all over the periodic system are represented, from hydrogen and the alkali metals through the transition metals, light and heavy metals, gases, halogens, rare earth elements and superheavy, manufactured elements — from their extraction and their discovery to their applications and the evaluation of their impact on humans and the environment.

References

Badash, L. (1979). The Suicidal Success of Radioactivity, *British Journal for the History of Science*, 12, pp. 245–256.

Baird, D. (2002). Analytical Chemistry and the "Big" Scientific Instrumentation Revolution, in *From Classical to Modern Chemistry: The Instrumental*

Revolution (2002), ed. Morris, P. J. T. (Royal Society of Chemistry, London; Science Museum, London; and Chemical Heritage Foundation, Philadelphia, PA) [originally published as Baird, D. (1993), Analytical Chemistry and the "Big" Scientific Instrumentation Revolution, *Annals of Science*, 50, pp. 267–290].

Baldwin, M. (2009). "Where Are Your Intelligent Mothers to Come From?" Marriage and Family in the Scientific Career of Dame Kathleen Lonsdale FRS (1903–71), *Notes and Records of the Royal Society*, 63, pp. 81–94.

Ball, P. (2004). *The Elements: A Very Short Introduction* (Oxford University Press, Oxford).

Bensaude-Vincent, B. (1986). Mendeleev's Periodic System of Chemical Elements, *British Journal for the History of Science*, 19, pp. 3–17.

Bensaude-Vincent, B. (2019). Reconceptualizing Chemical Elements Through the Construction of the Periodic System, *Centaurus*, forthcoming.

Bret, P. and Van Tiggelen, B., eds. (2011). *Madame d'Arconville (1720–1805): Une femme de lettres et de sciences au siècle des Lumières* (Hermann, Paris).

Carlson, W. B. (1997). Innovation and the Modern Corporation: From Heroic Invention to Industrial Science, in *Science in the Twentieth Century*, eds. Krige, J. and Pestre, D. (Harwood, Amsterdam), pp. 203–226.

Conant, J. B. (1970). Theodore William Richards and the Periodic Table, *Science*, 168, pp. 425–428.

Creese, M. R. S. (1998). Early Women Chemists in Russia: Anna Volkova, Iulia Lermontova, and Nadezhda Ziber-Shumova, *Bull. Hist. Chem.*, 21, pp. 19–24.

Curie, M. (1911). Radium and the New Concepts in Chemistry (Nobel Lecture, https://www.nobelprize.org/prizes/chemistry/1911/marie-curie/lecture/, last accessed 26 March 2019).

Dreifuss, J.-J. and Sigrist, N. T. (2012). The Making of a Bestseller: Alexandre and Jane Marcet's *Conversations on Chemistry*, in *For Better or For Worse: Collaborative Couples in the Sciences*, eds. Lykknes, A., Opitz, D. L. and Van Tiggelen, B. (Springer Birkhäuser, Heidelberg), pp. 91–32.

Duncan, A. M. (1962). Some Theoretical Aspects of Eighteenth-Century Tables of Affinity, *Annals of Science*, 18, pp. 177–194.

Espmark, K. and Nordlund, C. (2012). Married for Science, Divorced for Love: Success and Failure in the Collaboration Between Astrid Cleve and Hans von Euler-Chelpin, in *For Better or For Worse: Collaborative Couples in the Sciences*, eds. Lykknes, A., Opitz, D. L. and Van Tiggelen, B. (Springer Birkhäuser, Heidelberg), pp. 81–102.

Fermi, E. (1938). Artificial Radioactivity Produced by Neutron Bombardment (Nobel Lecture, https://www.nobelprize.org/uploads/2018/06/fermi-lecture. pdf, last accessed 26 March 2019).

Fors, H. (2014). Elements in the Melting Pot: Merging Chemistry, Assaying, and Natural History, Ca. 1730–1760, *Osiris*, 29, pp. 230–244.

Fors, H. (2015). *The Limits of Matter* (University of Chicago Press, Chicago, IL).

Friedman, R. M. (2001). *The Politics of Excellence: Behind the Nobel Prize in Science* (Times Books, New York, NY).

Gildersleeve, V. (1954). *Many a Good Crusade* (MacMillan, New York, NY).

Golinski, J. (1992). *Science as Public Culture: Chemistry and Enlightenment in Britain, 1760–1820* (Cambridge University Press, Cambridge).

Golinski, J. (2016). *The Experimental Self: Humphry Davy and the Making of a Man of Science* (University of Chicago Press, Chicago, IL).

Gordin, M. D. (2004). *A Well-Ordered Thing: Dmitrii Mendeleev and the Shadow of the Periodic Table* (Basic Books, New York, NY).

Gordin, M. D. (2015). *Scientific Babel: The Language of Science from the Fall of Latin to the Rise of English* (University of Chicago Press, Chicago, IL).

Higgitt, R. and Withers, C. W. J. (2008). Science and Sociability: Women as Audience at the British Association for the Advancement of Science, 1831–1901, *Isis*, 99, pp. 1–27.

Holl, J. M. (1997). *Argonne National Laboratory 1946–96* (University of Illinois Press, Urbana and Chicago, IL).

IYPT (2019). International Year of the Periodic Table of Chemical Elements. About IYPT2019 (https://www.iypt2019.org/about, last accessed, 29 March 2019).

Kauffman, G. B. (1982a). The Atomic Weight of Lead of Radioactive Origin: A Confirmation of the Concept of Isotopy and the Group Displacement Laws. Part I, *J. Chem. Ed.*, 59, pp. 3–8.

Kauffman, G. B. (1982b). The Atomic Weight of Lead of Radioactive Origin: A Confirmation of the Concept of Isotopy and the Group Displacement Laws. Part II, *J. Chem. Ed.*, 59, pp. 119–123.

Kauffman, G. B. (2019). Glenn T. Seaborg: American Chemist, *Encyclopaedia Britannica* (https://www.britannica.com/biography/Glenn-T-Seaborg, last accessed 26 March 2019).

Kragh, H. (1996). Elements No. 70, 71 and 72: Discoveries and Controversies, in *Episodes from the History of the Rare Earth Elements: Chemists and Chemistry,* vol. 15, ed. Evans C. H. (Springer, Dordrecht), pp. 67–89.

Kragh, H. (1999). *Quantum Generations: A History of Physics in the Twentieth Century* (Princeton University Press, Princeton, NJ).

Kragh, H. (2012). *Niels Bohr and the Quantum Atom: The Bohr Model of Atomic Structure 1913–1925* (Oxford University Press, Oxford).

Lykknes, A., Kvittingen, L. and Børresen, A. K. (2005). Ellen Gleditsch: Duty and Responsibility in a Research and Teaching Career, 1916–1946, *HSPS*, 36, pp. 131–188.

Morris, P. J. T., ed. (2002). *From Classical to Modern Chemistry: The Instrumental Revolution* (Royal Society of Chemistry, London; Science Museum, London; and Chemical Heritage Foundation, Philadelphia, PA).

Morris, P. J. T. (2015). *The Matter Factory: A History of the Chemistry Laboratory* (Reaktion Books, London).

Newnham College Cambridge (2019). Women's Education: Newnham College, University of Cambridge (https://www.newn.cam.ac.uk/about/history/womens-education/, last accessed, 15 February 2019).

NIST (2019). National Institute of Standards and Technology (https://www.usa.gov/federal-agencies/national-institute-of-standards-and-technology, last accessed 26 March 2019).

N. N. (1939). Nobel Prize: Ernest Lawrence. Biographical (https://www.nobelprize.org/prizes/physics/1939/lawrence/biographical/, last accessed 26 March 2019).

N. N. (1951). The Nobel Prize in Chemistry 1951 (https://www.nobelprize.org/prizes/chemistry/1951/summary/, last accessed 26 March 2019).

Nummedal, T. (2019). *Anna Zieglerin and the Lion's Blood: Alchemy and End Times in Reformation Germany* (University of Pennsylvania Press, Philadelphia, PA).

Nye, M. J. (1996). *Before Big Science: The Pursuit of Modern Chemistry and Physics, 1800–1940* (Harvard University Press, Cambridge, MA).

Oertzen, C. von (2014). *Science, Gender, and Internationalism: Women's Academic Networks, 1917–1955* (Palgrave, Macmillan, New York, NY).

Ogilvie, B. (2016). Correspondence Networks, in *A Companion to the History of Science*, ed. Lightman, B. (Wiley, London), pp. 358–371.

Opitz, D. L., Lykknes, A. and Van Tiggelen, B. (2012). Introduction, in *For Better or For Worse: Collaborative Couples in the Sciences*, eds. Lykknes, A., Opitz, D. L. and Van Tiggelen, B. (Springer Birkhäuser, Heidelberg), pp. 1–15.

Pigeard Micault, N. (2013). Le laboratoire Curie et ses femmes (1906–1934), *Annals of Science*, 70, pp. 71–100.

Rayner-Canham, M. F. and Rayner-Canham, G. W. (1997). Pioneer Women of Radioactivity, in *A Devotion to Their Science: Pioneer Women of Radioactivity*, eds. Rayner-Canham, M. F. and Rayner-Canham, G. W. (McGill-Queen's

University Press, Montreal; and Chemical Heritage Foundation, Philadelphia, PA), pp. 12–28.

Reinhardt, C. (2006). *Shifting and Rearranging: Physical Methods and the Transformation of Modern Chemistry* (Science History Publications, Sagamore Beach, MA).

Rentetzi, M. (2003). Women Physicists in the Institute for Radium Research in Vienna, 1920–1938: A Statistical Report, *PCNews*, February 2003, pp. 14–16.

Rentetzi, M. (2004). Gender, Politics, and Radioactivity Research in Interwar Vienna, *Isis*, 95, pp. 359–393.

Rentetzi, M. (2007). *Trafficking Materials and Gendered Experimental Practices: Radium Research in Early Twentieth Century Vienna* (Columbia University Press, New York, NY) (http://www.gutenberg-e.org/rentetzi/, last accessed 26 March 2019).

Rentetzi, M. (2017). Marie Curie and the Perils in Radium, *Physics Today*, 7 November 2017 (https://physicstoday.scitation.org/do/10.1063/PT.6.4.20171107a/full/, last accessed 26 March 2019).

Riken Nishina (2019). Riken Nishina Center for Accelerator-Based Science (history at https://www.nishina.riken.jp/about/history_e.html, last accessed 26 March 2019).

Rogers, R. (2006). Learning to Be Good Girls and Women: Education, Training and Schools, in *The Routledge History of Women in Europe since 1700*, ed. Simonton, D. (Routledge, London and New York, NY), pp. 93–133.

Rossiter, M. W. (1982). *Women Scientists in America, vol. 1: Struggles and Strategies to 1940* (Johns Hopkins University Press, Baltimore, MD and London).

Scerri, E. and Worrall, J. (2001). Prediction and the Periodic Table, *Studies in History and Philosophy of Science Part A*, 32, pp. 407–452.

Schiebinger, L. (1987). The History and Philosophy of Women in Science: A Review Essay, *Signs: Journal of Women in Culture and Society*, 12, pp. 305–332.

Shteir, A. B. (1997). Elegant Recreations? Configuring Science Writing for Women, *Victorian Science in Context*, ed. Lightman, B. (University of Chicago Press, Chicago, IL), pp. 236–255.

Sime, R. L. (1994). Lise Meitner in Sweden 1938–1960: Exile From Physics, *Am. J. Phys.*, 62, pp. 695–701.

Sime, R. L. (1996). *Lise Meitner: A Life in Physics* (University of California Press, Berkeley, CA, Los Angeles, CA and London).

The Internet Database of Periodic Tables (conservator: Mark R. Leach) (https://www.meta-synthesis.com/webbook/35_pt/pt_database.php, last accessed 10 April 2019).

The Lavoisier on Elements and Oxidation Website (Michael McBride) (http://chem125-oyc.webspace.yale.edu/125/history99/2Pre1800/Lavoisier/Nomenclature/Lavoisier_on_Elements.html, last accessed 23 April 2019).

Van Spronsen, J. W. (1969). *The Periodic System of Chemical Elements: A History of the First Hundred Years* (Elsevier, Amsterdam, London and New York, NY).

Watts, R. (2007). *Women in Science: A Social and Cultural History* (Routledge, London and New York, NY).

Woyshner, C. and Tai, B. H. K. (1997). Symposium: The History of Women in Education, *Harvard Educational Review*, 67, pp. v–xiv.

Select Bibliography

Aldersey-Williams, H. (2012). *Periodic Tales: The Curious Lives of the Elements* (Penguin Books, London).

Apotheker, J. and Sarkadi, L. S., eds. (2011). *European Women in Chemistry* (Wiley-VCH, Weinheim).

Badash, L. (1979). *Radioactivity in America: Growth and Decay of a Science* (Johns Hopkins University Press, Baltimore, MD and London).

Brock, W. H. (2016). *The History of Chemistry: A Very Short Introduction* (Oxford University Press, Oxford).

Hargittai, M. (2015). *Women Scientists: Reflections, Challenges, and Breaking Boundaries* (Oxford University Press, New York, NY).

Kean, S. (2011). *The Disappearing Spoon: And Other True Tales of Madness, Love, and the History of the World from the Periodic Table of the Elements* (Doubleday, London, Toronto, Sydney, Auckland and Johannesburg).

Lykknes, A., Opitz, D. L. and Van Tiggelen, B., eds. (2012). *For Better or For Worse: Collaborative Couples in the Sciences* (Springer Birkhäuser, Heidelberg).

Morris, P. T. J., ed. (2002). *From Classical to Modern Chemistry: The Instrumental Revolution* (Royal Society of Chemistry, London; Science Museum, London; and Chemical Heritage Foundation, Philadelphia, PA).

Quinn, S. (1995). *Marie Curie: A Life* (Perseus Books, Cambridge, MA).

Rayner-Canham, M. F. and Rayner-Canham, G. W., eds. (1997). *A Devotion to Their Science: Pioneer Women of Radioactivity* (McGill-Queen's University Press, Montreal; and Chemical Heritage Foundation, Philadelphia, PA).

Rayner-Canham, M. F. and Rayner-Canham, G. W. (1998). *Women in Chemistry: Their Changing Roles from Alchemical Times to the Mid-Twentieth Century* (American Chemical Society, Washington, DC).

Rentetzi, M. (2007). *Trafficking Materials and Gendered Experimental Practices: Radium Research in Early Twentieth Century Vienna* (Columbia University Press, New York, NY) (http://www.gutenberg-e.org/rentetzi/, last accessed 26 March 2019).

Rossiter, M. W. (1982). *Women Scientists in America, vol. 1: Struggles and Strategies to 1940* (Johns Hopkins University Press, Baltimore, MD and London).

Rossiter, M. W. (1995). *Women Scientists in America, vol. 2: Before Affirmative Action 1940–1972* (Johns Hopkins University Press, Baltimore, MD and London).

Rossiter, M. W. (2012). *Women Scientists in America, vol. 3: Forging a New World since 1972* (Johns Hopkins University Press, Baltimore, MD and London).

Scerri, E. R. (2007). *The Periodic Table: Its Story and its Significance* (Oxford University Press, Oxford).

Sime, R. L. (1996). *Lise Meitner: A Life in Physics* (University of California Press, Berkeley, CA, Los Angeles, CA and London).

Simonton, D. (2006). *The Routledge History of Women in Europe Since 1700* (Routledge, London and New York, NY).

The Internet Database of Periodic Tables (conservator: Mark R. Leach) (https://www.meta-synthesis.com/webbook/35_pt/pt_database.php, last accessed 10 April 2019).

Van Spronsen, J. W. (1969). *The Periodic System of Chemical Elements: A History of the First Hundred Years* (Elsevier, Amsterdam, London and New York, NY).

Van Tiggelen, B. and Lykknes, A. (2019). Celebrate the Women Behind the Periodic Table, *Nature*, 565, pp. 559–561.

Watts, R. (2007). *Women in Science: A Social and Cultural History* (Routledge, London and New York, NY).

Whaley, L. A. (2003). *Women's History as Scientists: A Guide to the Debates* (ABC CLIO, Santa Barbara, CA, Denver, CO and Oxford).

Wikipedia Timeline of Chemical Element Discoveries (https://en.wikipedia.org/wiki/Timeline_of_chemical_element_discoveries, last accessed 26 March 2019).

Part 1

Old and New Understandings of the Elements

Dorothea Juliana Wallich (1657–1725) and Her Contributions to the Chymical Knowledge about the Element Cobalt

Alexander Kraft

Network Alchemy at the Gotha Research Centre of the
University of Erfurt, 99867 Gotha, Germany

Although cobalt compounds had already been in use for colouring glass and ceramics for centuries, the Swedish chemist Georg Brandt (1694–1768) is usually credited with discovering the element in 1735 [Weeks, 1960]. In the *Acta literaria et scientiarum Sveciae* of that year, Brandt published a detailed account in which he showed that cobalt was clearly distinct from bismuth after he had produced both metals from the same ore [Brandt, 1735]. But cobalt compounds were commercial products and objects of research long before Brandt's important publication. In 1705 and 1706 Dorothea Juliana Wallich published three chymical books under the pseudonym D.I.W. [Wallich, 1705a, 1705b, 1706]. These books were the result of her work with a cobalt-containing ore, and focussed on producing the philosophers' stone which was supposed to transmute metals such as lead and tin into silver and gold. Wallich also recorded some chymical experiments with cobalt compounds. By discovering the thermochromic reactions of certain cobalt compounds she contributed to the early understanding of the properties and behaviour of the element cobalt.

Dorothea Juliana Wallich and her Chymical Work

Dorothea Juliana was born in 1657 in the central German town of Weimar as Dorothea Juliana Fischer. Her father was Heinrich Fischer (1611–1665), the First Tax Collector for the Duke of Saxe-Weimar; her mother, Anna-Catharina Lippach, was the daughter of the well-known Protestant theologian David Lippach (1580–1653). On 1 March 1674, at the age of 16 years and seven months, Dorothea Juliana Fischer married the 34-year-old Johann Wallich, the Court Secretary of the Duke of Saxe-Weimar. The couple was wealthy, lived in Weimar and did not have any children [Kraft, 2017].

Around 1685, when Wallich was in her late twenties, she became interested in chymistry (the chemical practices of the time) including practical laboratory experiments.[1] Wallich engaged intensely with chymistry for the following 20 years and became a renowned expert in the area of Thuringia and Saxony.

Unfortunately, not much is known about her life prior to the publication of her three books in 1705/6. From this twenty-year period only small pieces of information survive, which Georg Ernst Stahl (1659–1734) published in a book [Stahl, 1718, pp. 249, 252–253] and put down in 1728 in a letter to the physician, chemist and mineralogist Johann Friedrich Henckel (1678–1744) in Freiberg in Saxony [Stahl, 1795, pp. 99–100]. Stahl was a prominent physician and chymist himself, a professor of medicine at the university of Halle between 1694 and 1715, and the First Court Physician to the king in Prussia in Berlin from 1715 until 1734. His development of the phlogiston theory[2] helped to pave the way from chymistry to modern chemistry [Beck, 1960].

Stahl lived in Weimar with his family between early 1687 and 1694. At the time he was the personal physician of the duke Johann Ernst III of

---

[1] The term "chymistry" was coined by Newman and Principe [Newman and Principe, 1998] to encompass both (in modern terms) alchemy and chemistry, and will be employed throughout the article.

[2] See Bret and Kawashima in this volume.

Saxe-Weimar (1664–1707, duke from 1683). Stahl knew Wallich very well: she was godmother to one of his daughters from 1690 onwards [Kraft, 2018]. According to Stahl, Wallich "had certainly more experience in chymical things ... than a large number of very scholarly and intellectual men" [Stahl, 1718, p. 249]. Stahl further reported that Wallich had been active for a while in the Saxon mining town of Schneeberg, successfully extracting more silver from cobalt ores than other chymists were able to do. Her three books were written in German and published in 1705 and 1706. They were the result of her chymical work, both in the laboratory and from studying the chymical literature of the time. Figure 1 shows the title page of Wallich's first book. Wallich wrote these books because she thought that she had found the secret "minera" that was considered necessary for producing the philosophers' stone.

Wallich called this "minera" or ore "Electrum minerale immaturum," "Magnesia" [Wallich, 1705a, p. 84] or "Markasitta plumbea" [Wallich, 1706, p. 272]. She also described exactly what the ore looked like. In their correspondence Wallich and those around her also called this ore "Wismuth." This ore may have been native bismuth mixed with safflorite ($CoAs_2$) and/or skutterudite ($CoAs_3$) from the Schneeberg region. Figure 2 shows a photo of such a mineral, bismuth with skutterudite from the Saxon Ore Mountains. In Wallich's time, this ore was used and processed as follows: first the bismuth metal was melted out and the remaining cobalt arsenide compounds used as a starting material for the production of smalt (blue cobalt glass) [Hill, 1748, p. 624].

Wallich and Cobalt

Cobalt compounds were used to colour glass or ceramics blue as early as in 2000 BC [Weeks, 1960]. However, the name "cobalt" dates from much later, and was originally a term for seemingly worthless ores found in the silver mining areas of Saxony and Bohemia. It came into use in the fifteenth century, and was derived from the German word "Kobold" for small mountain dwarfs or goblins. Miners thought that these vicious "Kobolde" had transformed silver ore into a useless material.

Around 1540, it was found that a blue coloured glass could be produced from cobalt ore. This cobalt oxide glass was called "smalt," and used for colouring glass and ceramics. Finely ground smalt could also be

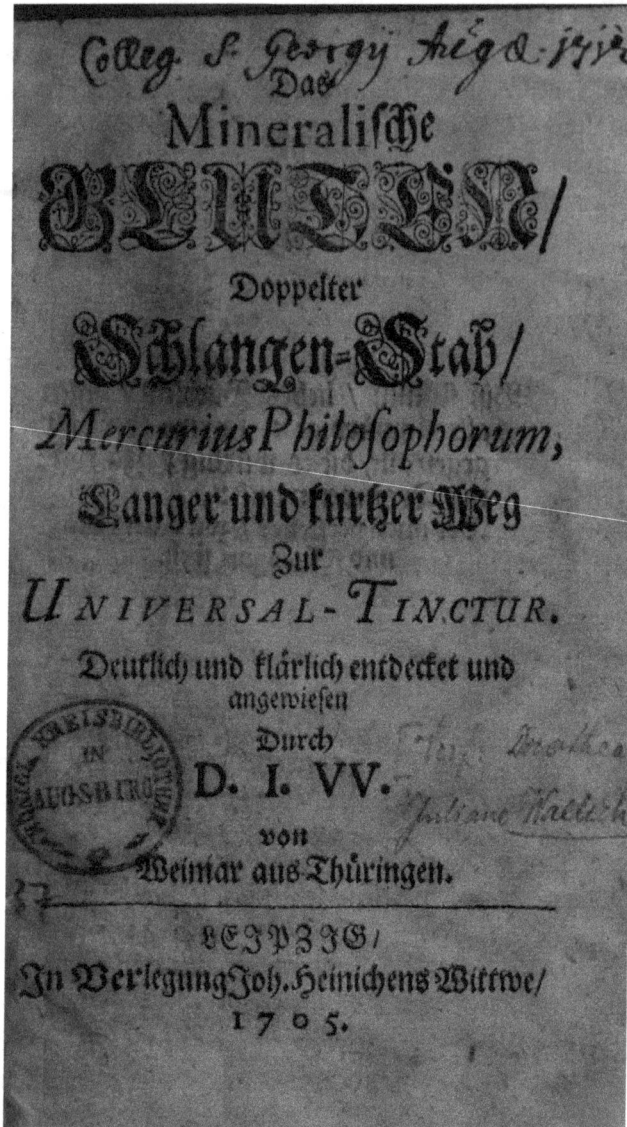

Figure 1.   Title page of Dorothea Juliana Wallich's first book, *Das mineralische Gluten, doppelter Schlangen-Stab, Mercurius Philosophorum, langer und kurtzer Weg zur Universal-Tinctur. Deutlich und klärlich entdecket und angewiesen*, written in 1704 and published in 1705 (Staats- und Stadtbibliothek Augsburg, Phys 1334, p. 1, urn:nbn:de:bvb: 12-bsb11275024-5).

Figure 2. Bismuth mixed with skutterudite from a uranium mining spoil tip near Schlema, Ore Mountains, Saxony, Germany (from the collection of the author, photo: Alexander Kraft, 2018).

used as a blue pigment in painting. Smalt production grew to be an important business in the area around the mining town of Schneeberg in Saxony [Bruchmüller, 1897]. In the mining region of Saxony and Bohemia there were different minerals which contained cobalt compounds, among other substances. In Wallich's time they bore designations like "marcasite," "wismuth," "bismuth," "glanzcobalt," "speiscobalt," "cobalt" and others [Bruchmüller, 1897, pp. V–VII]. No clear distinction was made between these minerals, and the same mineral could be referred to with different names. As mentioned above, the metal cobalt was still unknown.

As described above, Wallich's secret "minera" which is the main topic of her three books, was called "marcasite" or "bismuth," and in addition to bismuth metal also contained cobalt arsenide compounds. Her chymical experiments with this ore are described in detail in her third book [Wallich, 1706, pp. 238–277]. Figure 3 shows the title page of this book. Among other experiments, Wallich describes her treatment of this ore with nitric

Figure 3. Title page of Dorothea Juliana Wallich's third book, *Schlüssel zu dem Cabinet der geheimen Schatz-Cammer der Natur, zur Such- und Findung des Steins der Weisen, durch Fragen und Antwort gestellet. Verfertiget und der Welt gezeiget,* written in 1705 and published in 1706 (Staats- und Stadtbibliothek Augsburg, Phys 1334#(Beibd., p. 1, urn:nbn:de:bvb:12-bsb11275025-0).

acid.[3] To the rose-coloured solution which she obtained she added a sodium chloride solution. Wallich produced what we can today infer was impure cobalt(II) chloride solutions and impure solid cobalt chloride salt ($CoCl_2$). But most importantly, she discovered and reported the impressive thermochromic effects of these products: upon heating, the colour changed from rose-coloured via violet blue and sky blue to grass green. The colour faded back to pale rose-coloured upon cooling. It seems that Wallich was the first to describe the thermochromic effects of solid cobalt chloride and of cobalt chloride-containing aqueous solutions.[4]

However, in her books Wallich not only reported on the reaction of her cobalt-containing marcasite ore with nitric acid and on the results of the subsequent addition of sodium chloride solution. She also recorded several other chemical reactions in some detail, among them the reactions of cobalt-containing ore with citric acid, with ammonium chloride, with potassium nitrate, with sodium chloride and sodium sulphate, with antimony sulphide, mercury(II) chloride ("corrosive sublimate") and mercury. In addition, she reported reactions of her cobalt chloride product with ethanol-water mixtures and vinegar. Therefore, Dorothea Juliana Wallich

---

[3]To explain Wallich's chymical reactions with modern chemistry, I have used modern names of the substances involved.

[4]What exactly happens during colour change? At room temperature, in air with a usual humidity, solid cobalt(II) chloride exists as an hexahydrate: $CoCl_2 \cdot 6H_2O$. This is a rose-coloured compound. When heated the water is removed from the compound and blue water-free cobalt(II) chloride $CoCl_2$ is formed (dehydration). When the compound cools down cobalt chloride hexahydrate is formed again by uptake of water (hydration). This reversible dehydration/hydration reaction can be written as:

$$CoCl_2 \cdot 6H_2O \rightleftarrows CoCl_2 + 6\ H_2O \qquad (1)$$

Although the colour change of aqueous cobalt(II) chloride solutions is visually the same as in the solid, a different chemical reaction takes place. In an aqueous solution, there exists a concentration and temperature-dependent equilibrium between rose-coloured $Co(H_2O)_6^{2+}$ ions and blue coloured $CoCl_4^{2-}$ ions:

$$Co(H_2O)_6^{2+} + 4\ Cl^- \rightleftarrows CoCl_4^{2+} + 6\ H_2O \qquad (2)$$

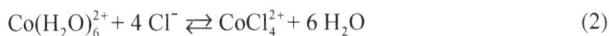

At constant concentrations an increase in temperature will shift the equilibrium to blue $CoCl_4^{2-}$ whereas lowering the temperature will favour the formation of rose-coloured $Co(H_2O)_6^{2+}$.

not only produced cobalt chloride, but also cobalt nitrate and cobalt oxide, and described some properties of these compounds. However, it should be noted that she never used the term "cobalt" in her books, preferring to refer to the ore as "Markasitta plumbea" or "Wismuth" and the cobalt chloride-containing product "Rosen-Farb Saltz," literally "rose-colour salt."

How did Wallich's publications influence other chemists? In 1739, Johann Heinrich Pott (1692–1777), a chemist from Berlin [Kraft, 2014], published a treatise about bismuth in which he also heavily referred to the work of Wallich [Pott, 1739, pp. 135–197]. However, because Brandt's 1735 article [Brandt, 1735], in which he described cobalt ("regulus cobalti") as a new metal, was not published until 1739 [Weeks, 1960, p. 157], Pott was not aware of these new research results or the distinction between the metals bismuth and cobalt.

Wallich's thermochromic research results formed the basis for the development of an invisible ink presented to the French Académie des sciences by Jean Hellot (1685–1766) in 1737 [Hellot, 1737]. Hellot reported that he learned to produce this solution in Paris, from an "Chymiste Allemand" from Stollberg [Hellot, 1737, p. 107]. This is a town located about 16 kilometres north-west of Schneeberg, in the Saxon Ore Mountains. In the same year, the German physician and chemist Hermann Friedrich Teichmeyer (1685–1744) claimed that he had already demonstrated this invisible ink in his lectures at the university of Jena *circa* six years earlier [Teichmeyer, 1737]. The rose-coloured writing with the cobalt chloride formulation on paper was almost invisible at room temperature. When the paper was heated the writing became visible in a blue-green colour. Information about this new invisible or sympathetic ink was soon included in chymistry textbooks such as Caspar Neumann's (1683–1737) posthumously published *Chymiae medicae dogmatico-experimentalis* [Neumann, 1754, pp. 625–627]. That this rose-coloured salt was a cobalt, and not a bismuth compound, was first mentioned by Johann Albrecht Gesner (1695–1760) in 1744 [Gesner, 1744, p. 20]. Later research found that pure aqueous solutions of cobalt chloride or pure solid cobalt chloride would change colour between rose-coloured and blue, not green, as described by Wallich and Hellot. In 1822, extracts from an 1804 letter from Georg Wilhelm Sigismund Beigel (1753–1837) to the chemist Martin Heinrich Klaproth (1743–1817) were published, as well as

Klaproth's answer. In the first letter, the green colour change at higher temperatures was explained as nickel and iron impurities in the processed cobalt compounds [Klaproth, 1822]. This is not surprising because naturally occurring cobalt arsenide minerals such as safflorite or skutterudite usually contain variable amounts of nickel and iron.

The thermochromic reactions of cobalt chloride which were discovered by Wallich are still in use today, e.g., in the form of experiments demonstrating a chemical equilibrium to students in schools and universities [Nguyen and Birdwhistell, 2014]. In technical contexts, the striking colour change of the dehydration/hydration reaction of solid cobalt(II) chloride is used as an indicator of water in desiccants, i.e. water absorbing solids. Cobalt chloride thermochromism has also been proposed as a basis for temperature sensors [Hao and Liu, 1990] and adaptive solar protection windows [Gadzuric *et al.*, 2012].

But to Wallich and her contemporaries, the impressive reversible colour change induced by changing temperature suggested the *cauda pavonis* (peacock's tail), a specific stage in the process of the great work (*opus magnum*), that is, the making the philosophers' stone. In the *cauda pavonis* stage of the *opus magnum* an array of iridescent colours appears, either all colours at the same time or one colour after the other. Witnessing what we now see as the thermochromism of cobalt chloride, Wallich interpreted the results as an indication that she was on the right path towards producing the philosophers' stone. Therefore, she subsequently devoted most of her time and resources to reaching the goal of *chrysopoeia*, i.e. the art of transmuting base metals into gold.

Dorothea Juliana Wallich as a Court Chymist

Following the publication of her books, Wallich became a sought-after chymist. For the first roughly five years after publication she worked for three German princes: Anton Günther II. of Schwarzburg-Sondershausen-Arnstadt (1653–1716, count from 1666), Johann Wilhelm Elector Palatine (1658–1716, duke from 1679), and Ernst Ludwig I of Saxe-Coburg-Meiningen (1672–1724, duke from 1706). In her contracts with these three clients, which were similar in nature, Wallich promised to produce gold by chymical operations from a certain amount of silver. She wanted

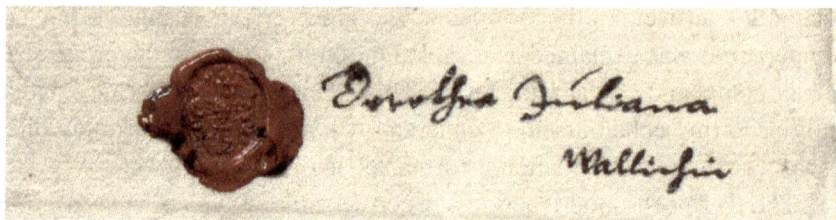

Figure 4. Seal and signature of Dorothea Juliana Wallich underneath the contract between her and duke Ernst Ludwig I. of Saxe-Coburg-Meiningen, dated 9 October 1708 (Landesarchiv Thüringen, Staatsarchiv Gotha, H. Ernst Ludwig zu S.-Meiningen Chymica vol. II; Geheimes Archiv E XI. Nr. 95, f. 144r, reproduced by kind permission of the Staatsarchiv Gotha).

to transmute *ca.* 10% of the silver into gold. Figure 4 shows Dorothea Juliana Wallich's signature under one of these contracts.

In all three cases, she had to prove first that her chymical process worked properly. After the proof, her contracting partner had to invest several thousand thaler to scale this chymical process to a large operation. The profit of this large-scale operation would be divided equally between Wallich and the corresponding duke. In Arnstadt, Düsseldorf and Coburg, Wallich tried to produce gold from silver according to her process. She was not successful in proving that her process worked with the requested three small samples, and lost nearly all of her money and her reputation between 1706 and 1710.

After this failure Wallich moved back to the small town of Arnstadt in Thuringia and lived there for the remaining 15 years of her life as a chymist, but without significant clients. The parish registers of the town church of Arnstadt state that she, the widow of a secretary, died a poor woman in February 1725 [Kraft, 2017]. She was 67 years old.

References

Beck, C. W. (1960). Georg Ernst Stahl, *J. Chem. Educ.*, 37, pp. 506–510.

Brandt, G. (1735). Dissertatio de semi-metallis, *Acta literaria et scientiarum Sveciae*, 4, pp. 1–12.

Bruchmüller, W. (1897). *Der Kobaltbergbau und die Blaufarbenwerke in Sachsen bis zum Jahre 1653* (Richard Zeidler, Crossen).

Gadzuric, S., Vranes, M. and Dozic, S. (2012). Thermochromic Cobalt(II) Chloro-Complexes in Different Media: Possible Application for Auto-Regulated Solar Protection, *Sol. Ener. Mater. Sol. Cells*, 105, pp. 309–316.

Gesner, J. A. (1744). *Historia cadmiae fossilis metallicae sive cobalti et ex illo praeparatorum* (Johann Andreas Rüdiger, Berlin).

Hao, T. and Liu, C.-C. (1990). An Optical Fiber Temperature Sensor Using a Thermochromic Solution, *Sens. Actuator A — Phys.*, 24, pp. 213–216.

Hellot, J. (1737). Sur un nouvelle encre simpatique I, II, *Mém. Acad. Roy. Sci.*, 39, 101–120, 228–247.

Hill, J. (1748). *A General Natural History or, New and Accurate Descriptions of the Animals, Vegetables, and Minerals, of the Different parts of the World,* (Osborne, London), Part 1.

Klaproth, M. H. (1822). Ein Brief an und von Klaproth aus dem J. 1804, über die Eigenschaften reinen Kobalts, *Ann. Phys.*, 71, pp. 109–111.

Kraft, A. (2014). Ein streitbarer Chemiker: Der Berliner Arzt Johann Heinrich Pott (1692–1777), *Mitteilungen der Fachgruppe Geschichte der Chemie der GDCh*, 24, pp. 3–32.

Kraft, A. (2017). Dorothea Juliana Wallich, geb. Fischer (1657–1725), eine Alchemistin aus Thüringen, *Genealogie: Deutsche Zeitschrift für Familienkunde*, 33, pp. 539–555.

Kraft, A. (2018). Maria Rosina Hilgund (1663–1691) aus Jena, die erste Ehefrau des Arztes und Chemikers Georg Ernst Stahl (1659–1734), *Zeitschrift für Mitteldeutsche Familiengeschichte*, 59, pp. 291–302.

Neumann, C. (1754). *Chymiae medicae dogmatico-experimentalis*, (Waisenhaus, Züllichau), 4. Band, 1. Teil.

Newman, W. R. and Principe, L. M. (1998). Alchemy vs. Chemistry: The Etymological Origins of a Historiographic Mistake, *Early Science and Medicine*, 3, pp. 32–65.

Nguyen, V. D. and Birdwhistell, K. R. (2014). Microwave Mapping Demonstration Using the Thermochromic Cobalt Chloride Equilibrium, *J. Chem. Educ.*, 91, pp. 880–882.

Pott, J. H. (1739). *Observationum et animadversionum chymicarum praecipue circa sal commune, acidum salis vinosum et wismuthum* (Johann Andreas Rüdiger, Berlin).

Stahl, G. E. (1718). *Zufällige Gedancken und nützliche Bedencken über den Streit vom sogenannten Sulphure* (Waisenhaus, Halle).

Stahl, G. E. (1795). Letter from Nov 27, 1728 in Berlin to J.F.Henckel in Freiberg, in *Mineralogische, Chemische und Alchymistische Briefe von reisenden und anderen Gelehrten an den ehemaligen Chursächsischen*

Bergrath J. F. Henkel, Dritter Theil, ed. n.n. (Walther, Dresden), pp. 91–101.

Teichmeyer, H. F. (1737). Anonymus quidam Jenensis sequentis nobis transmisit, philyris hisce inserenda, *Commercium litt. rei med. sci. natur.*, 7, pp. 91–93.

Wallich, D. J. (1705a). Das mineralische Gluten, doppelter Schlangen-Stab, Mercurius Philosophorum, langer und kurtzer Weg zur Universal-Tinctur. Deutlich und klärlich entdecket und angewiesen (Johann Heinichens Witwe, Leipzig).

Wallich, D. J. (1705b). Der philosophische Perl-Baum, das Gewächse der drey Principien, zu deutlicher Erklärung des Steins der Weisen, wie er mit seinen Wurtzeln in der äussern und finstern Welt, mit seiner Blüthe aber in der Paradiesischen und Licht-Welt und mit seiner reiffen Frucht in der englischen und himmlischen Welt stehet und wächset (Johann Heinichens Witwe, Leipzig).

Wallich, D. J. (1706). Schlüssel zu dem Cabinet der geheimen Schatz-Cammer der Natur, zur Such- und Findung des Steins der Weisen, durch Fragen und Antwort gestellet. Verfertiget und der Welt gezeiget (Johann Heinichens Witwe, Leipzig).

Weeks, M. E. (1960). Some Swedish Metals: Cobalt, in M. E. Weeks, *Discovery of the Elements*, 6th ed. (Journal of Chemical Education, Easton, PA), pp. 152–161.

Select Bibliography

Anders, J. (2016). *33 Alchemistinnen* (Vergangenheitsverlag, Berlin).

Kraft, A. (2017). Dorothea Juliana Wallich, geb. Fischer (1657–1725), eine Alchemistin aus Thüringen, *Genealogie: Deutsche Zeitschrift für Familienkunde*, 33, pp. 539–555.

Macrakis, K. (2015). *Prisoners, Lovers, and Spies: The Story of Invisible Ink from Herodotus to Al-Qaeda* (Yale University Press, New Haven, CT).

Newman, W. R. and Principe, L. M. (1998). Alchemy vs. Chemistry: The Etymological Origins of a Historiographic Mistake, *Early Science and Medicine*, 3, pp. 32–65.

Nummedal, T. (2007). *Alchemy and Authority in the Holy Roman Empire* (University of Chicago Press, Chicago, IL).

Nummedal, T. (2019). *Anna Zieglerin and the Lion's Blood: Alchemy and End Times in Reformation Germany* (University of Pennsylvania Press, Philadelphia, PA).

Principe, L. M. (2013). *The Secrets of Alchemy* (University of Chicago Press, Chicago, IL).

Sone, K. and Fukuda, Y. (1987). *Inorganic Thermochromism* (Springer, Heidelberg and Berlin).

Weeks, M. E. (1960). *Discovery of the Elements*, 6th ed. (Journal of Chemical Education, Easton, PA).

Emilie Du Châtelet and the Nature of Fire: *Dissertation sur la nature et la propagation du feu*

Brigitte Van Tiggelen

Mémosciences asbl, Voie du Vieux Quartier 18,
1348 Louvain-La-Neuve, Belgium
Science History Institute, 315 Chestnut Street,
Philadelphia, PA 19106, USA

In April 1738, about fifty years before Antoine-Laurent Lavoisier (1743–1794) reformed chemistry in France, the *Académie des sciences de Paris* (French Academy of Sciences) announced the winners of a competition devoted to the question of the nature and the propagation of fire. The Greeks considered fire, along with air, water and earth, to be an element. As further knowledge of matter was gained through its transformations such as metallurgy, another series of elements or principles introduced by the Arabs — sulphur, mercury and salt — were preferred, as they allowed for a better understanding of actual laboratory practice and were widely used in alchemy. The two systems did not exclude each other, and to chymists,[1] fire remained to be a mysterious phenomenon that was to be investigated.

Quite unusual was the fact that there were three equal winners in the French Academy's competition: Leonhard Euler (1707–1783), the Jesuit Louis-Antoine Lozeran du Fech (1691–1755) and Count Jean-Antoine de Créquy (1699–1762) all had their competition essays published [Académie Royale des Sciences, 1742]. But even more curious was the Academy's decision to publish two further essays from the eighteen pieces that had been submitted, in the same volume as the winning essays. One was written by François-Marie Arouet Voltaire (1694–1778), "one of our best poets" [Académie Royale des Sciences, 1742, p. 86],[2] and the other "by a young lady of high rank" [Académie Royale des Sciences, 1742, p. 85], which was the first piece ever sent by a woman, the famous Emilie Du Châtelet (1706–1742) (Figure 1).[3] In her *Dissertation sur la nature et la propagation du feu* ("Dissertation on the Nature and Propagation of Fire"), Madame Du Châtelet elaborated on physical experiments conducted at her castle of Cirey, during the summer and the autumn of 1737, while Voltaire, who had decided to participate in the competition, was her guest. While she encouraged Voltaire in his work, she came to the conclusion that she did not share his views and interpretations, and decided to secretly take part in the competition as well. Since participation was anonymous, this was something even a woman could contemplate at a time when women were still excluded from the Academy's meetings. Du Châtelet is the first female scientists whose work was published by the Academy [Du Châtelet, 1742]. Aware of this feat, and also being a perfectionist, she published a revised second edition in 1744 [Du Châtelet, 1744].[4]

---

[1] The term "chymistry" was coined by Newman and Principe [Newman and Principe, 1998] to encompass both (in modern terms) alchemy and chemistry, and will be employed throughout the article.

[2] All translations from the French are the author's unless otherwise indicated.

[3] Mention of the name of Du Châtelet appears on p. 85 [Académie Royale des Sciences, 1742, p. 85] and her work is printed on pp. 87–170. She also requested a correction (printed as an erratum on pp. 220–221), which was not the usual procedure, as essays were generally published in the version in which the Academy had first received and judged them.

[4] An English translation was published in 2009 [Du Châtelet, 2009, pp. 62–103].

Figure 1.   Gabrielle Émilie le Tonnelier de Breteuil, marquise Du Châtelet, French school after Maurice Quentin de La Tour, 18th century (private collection, Choisel, Château de Breteuil, public domain).

The Making of a *Femme Savante*

Born in 1706 into a family of lesser nobility, Gabrielle Émilie le Tonnelier de Breteuil was introduced to the learned circle of her time thanks to her father's appointment as "lecteur ordinaire du Roi," (official reader to the King) then "introducteur des ambassadeurs et étrangers" (master of protocol for ambassadors and foreigners) at Louis XIV's (1638–1715) court.[5] As the only daughter in the family, she received roughly the same

---

[5]For Du Châtelet's biography see [Badinter, 2006] and [Zinsser, 2006].

education as her brothers, and was instructed in both physical activity and scholarly subjects. Emilie learned ancient and modern languages, mathematics and science, as well as literature. But most of all she showed an ardent curiosity, which was encouraged by her father and his acquaintances, among them the permanent secretary of the Academy, Bernard Le Bovier de Fontenelle (1657–1757). Though it seems that Emilie de Breteuil was not sent to a convent, as was common for young women at the time, but educated at home by a governess and preceptors, she still acquired the typical feminine skills of hosting, dancing, singing, playing music, etc. Consequently, when she married the Marquis Florent-Claude Du Châtelet-Lomont (1695–1765) in 1725, she became the perfect mistress of the house as befitted her rank. Now the Marquise Du Châtelet, she had one daughter and two sons, the younger of which died as a child in 1734. As was the custom, Du Châtelet's children were looked after by wet nurses, which allowed her to resume her social activities quite soon after the births, that is, once the traditional six-week bedrest period for young mothers had passed.

Having thus performed her duties as a wife, that is, borne two sons, she was somewhat bored with the frivolities and amusements of her class, and returned to the passion of her life: mathematics. In order to follow the latest progress in the field, she hired the best mathematicians of the day as her tutors: Pierre Louis Moreau de Maupertuis (1698–1759), who was a member of the French Academy of Sciences, and Alexis Claude Clairaut (1713–1765).

Notwithstanding her devotion to acquiring knowledge, Du Châtelet was also a woman of her time, and it was not unusual for noble women to take lovers, often with the tacit consent of their husband. But the converse was, of course, also the case at times. Du Châtelet's choice was both daring and passionate: she embarked on a long-term affair and an even longer intellectual partnership with Voltaire. When they met in 1733, Voltaire had already achieved literary fame, suffered imprisonment in the Bastille prison, and fled to England. Upon his return to France in 1728 he was not allowed to enter Paris at first, and officially continued to work as a playwright, which progressively took him back to the French capital. Instead of keeping a low profile, he published a French version of the *Letters Concerning the English Nation* (London, 1733) in 1734. The *Lettres philosophiques ou Lettres anglaises* are considered one of the most important pamphlets from the French Enlightenment: Voltaire contrasted the French absolutist and the British constitutional monarchies, and presented the latter as more enlightened, more balanced in power and counterpower, and more respectful to human and religious rights. As was to be expected, Voltaire had to flee Paris once again, and he found refuge at the Cirey castle, a property of the Du Châtelets located in the region of Champagne, where he could not be apprehended. Except for a few trips to foreign countries, he stayed there with Du Châtelet for the following fifteen years.

Voltaire's political exile turned into a fruitful scientific retreat. With her social obligations restricted, Du Châtelet focussed even more on her intellectual endeavours. With Voltaire's financial support the property was restored, expanded and refurbished, and also provided with a theatre and a laboratory, or as it was then called a "cabinet de physique" ("a physical cabinet"). And far from being isolated from other scholars, Du Châtelet entertained many of them at Cirey, and corresponded and exchanged books with others [Gauvin, 2006]. The lovers shared a common interest in Newtonianism, which was yet another controversial topic in France at the time. Inspired by Isaac Newton's (1642/43–1727) methods of natural philosophy, the Newtonian worldview sought to uncover the laws that govern the universe; these laws were rational, best expressed in mathematical form, and they relied on forces that acted from a distance. The way to uncover them was through what was then called experimental philosophy, in which the scholar observed, experimented, and drew conclusions from the experiment, in a see-saw of inductive and deductive reasoning. Contemporary science still relies on this method. But in the first half of the eighteenth century, Newton's ideas were not welcomed by the French intellectual scene, which was still firmly rooted in Cartesianism.

Cartesianism, like Newtonianism, was a mechanical worldview, with corpuscles governed by laws of movement that could be expressed in the mathematical language, but it did not, for instance, accept the concept of forces acting from a distance (like gravitation as we now understand it), and favoured deduction from innate ideas over experimental induction. Of course, there was also the issue of "national" pride or sensitivity present in these debates. During his stay in England, Voltaire had acquired a definite taste for Newton's physics, and was intent on disseminating it in France — once more taking a provocative route. He started to experiment, especially with optics, in order to write a full account of Newton's ideas, and this was eventually published as the *Eléments de la philosophie de Neuton mis à la portée de tout le monde* ("Elements of the Philosophy of Newton for Everyone") in 1738.[6] Throughout this period his dear Emilie continued to collaborate with him, discussing the matter with expertise and a mathematical proficiency he could never have mastered. After a while, he encouraged her to write her own exposition of Newton's theory, a project she embarked on for the rest of her life. Depicted in the frontispiece of *Eléments de la philosophie de Neuton* as a muse who reflects the rays of the sun on Voltaire's desk (Figure 2), Du Châtelet was now to become a scientist in her own right. It was thus in a context of an advanced scientific environment and fuelled with the desire to disseminate Newton's ideas that Du Châtelet and Voltaire read the announcement of the academic competition in 1736.

A Burning Question: The Nature of Fire

When the Parisian Academy put the question of the nature of fire forward as an object of enquiry, this was a complex question — much more complex than a characterisation of the concept of fire as an Aristotelian element. Admittedly the four Aristotelian elements, earth, water, air and fire were still mentioned in the chemistry textbooks of the time, but the operational use of these elements in actual chymical practice was scarce. Fire was more than an element, as it had also been an analytical tool in the laboratory for centuries, was often held responsible for the decomposition of substances, and was active in well-established processes like cupellation or distillation. Based on everyday experience, fire was also associated with heat, light, and other phenomena including combustion and even electricity. To some chymists fire was just an element, but to others it was a principle, sometimes identified with the sulphur of the alchemists. At the very end of the seventeenth century, Georg Ernst Stahl (1659–1734) proposed the concept of phlogiston, i.e. the principle of combustibility, which was relinquished during combustion and calcina-

---

[6]Voltaire expanded his work in 1745, and both editions were translated into English under the title *The Elements of Sir Isaac Newton's Philosophy* (1738; 2nd expanded ed. 1745).

Figure 2. Frontispiece of Voltaire's *Elémens de la philosophie de Neuton: Contenant la métaphysique, la théorie de la lumière, et celle du monde* (Prault, Paris, 1738) (The Science History Institute).

tion (rusting). Phlogiston was a chemical body that combined with other substances, and via chemical reactions it could be fixed or released. It was invisible yet material.

Discussions of fire, with a mechanistic approach, were also at the heart of natural philosophy. René Descartes (1596–1650) described the particles of fire as the most subtle corpuscles: so subtle that they could infiltrate every interstice like a liquid and move faster than every other body. According to Newtonian theory these corpuscles ought to obey the law of gravitation and carry weight. But precisely here emerged an experimental fact that both chymists and natural philosophers had to face: when metals were calcined, they gained weight despite the release of phlogiston.

The Academy had Cartesian leanings, but also knew about the competing theories of fire, and wanted to leave the investigation open to other theoretical frameworks. The announcement of the competition in the April 1736 issue of the gazette *Mercure de France* ("The Mercury of France") [N. N., 1736, p. 766][7] stressed that the competition would select the best explanation(s) provided, but not in any event be an endorsement of the principles underlying these explanation(s). In other words, one could submit an essay based on Newtonian principles, and even have a chance of being selected as the best piece, but this would not result in the Academy adopting Newton's doctrine from that moment on.

Though both Voltaire and Du Châtelet were engaged in writing a treatise on physics, Voltaire saw an opportunity to advance his cause, and decided to compete for the prize; he had Du Châtelet's full support. They believed that their recent studies on the nature of light — a topic on which Newton's method had proved promising — would also have laid the groundwork. They conducted a series of experiments melting metals, some of them on a large scale in order to study the phenomenon from all angles, which was possible thanks to the fact that nearby forges belonged to the Du Châtelets. For other experiments they equipped themselves thoroughly with thermometers, prisms and lenses, a vacuum pump and burning lenses. They also read everything they could put their hands on: works by the chymists Herman Boerhaave (1668–1738) and Wilhelm Homberg (1652–1715), the Newtonian experimental philosopher Pieter van Musschenbroek (1692–1761), Descartes, and, of course, Newton [Joly, 1995; De Gandt, 2001].

Voltaire's plan for the competition essay was to devise a Newtonian explanation: he hoped to demonstrate that fire was, indeed, made of

---

[7]The competition is announced on p. 765.

particles, which followed the rule of universal attraction. The predominant chemist of the time, Boerhaave, for his part, stated that fire had no weight. Boerhaave thought of fire as an agent that kept particles in motion and apart from each other, and as the cause of elasticity and electricity (which was a recently discovered phenomenon at the time). As time passed, Voltaire proved to be unable to draw the correct conclusions from the experiments they were both witnessing. Blinded by his Newtonian perspective, he concluded that all the results supported the interpretation of a gravitational force at play and fire as a weighty corpuscle. Du Châtelet felt she could not subscribe to this reasoning, as many of the experiments were, in fact, inconclusive as to which particle of fire was bearing weight, and thus, submitted to the law of gravitation. In the best of cases, changes in weight during processes could be explained attributing weight not to fire, but to other causes in which fire was not instrumental. She furthermore believed that Voltaire's interpretations of Boerhaave were wrong. As they argued with each other, and Du Châtelet realised that she was unable to convince Voltaire of what she felt was true, she decided, only a few weeks before the deadline, to write an essay and compete herself, to represent her own conclusions. She kept her participation secret from Voltaire. The only person who knew about it was her husband, who likely needed to give his consent, since the potential selection of her piece would make this a public issue [Kawashima, 2013].

In her essay [Du Châtelet, 1744] (Figure 3), Du Châtelet followed Boerhaave's strategy of studying fire via its manifestations, in which she included light, colours and heat, because these made it possible to experiment with fire. Like Boerhaave, she did not focus on combustion, as the impact of combustion could vary with the substances. By contrast, dilation — she and Boerhaave called it "rarefaction" — was a direct consequence of fire. Therefore, to gain a complete view of the phenomenon, she also studied the absence of fire, that is, in her terms, cold, and confirmed the universality of rarefaction through the constant condensation that can be observed when substances are cooled down.

Not only were her interpretations different and more nuanced than Voltaire's, but Du Châtelet also had a striking ability to record experiments faithfully as well as to make original observations. Among other things she used a crystal to diffract light, observed that the different

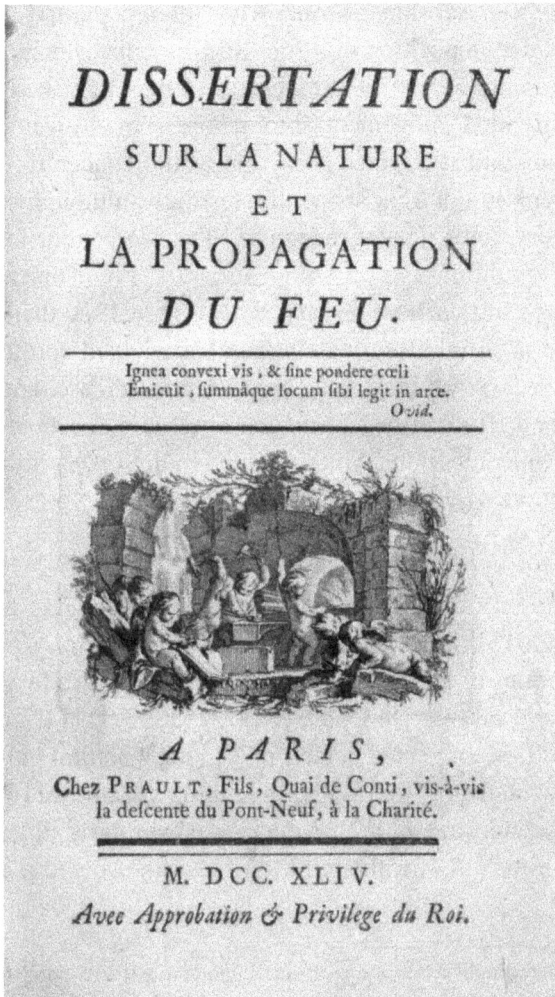

Figure 3.   Title page of Gabrielle Emilie le Tonnelier de Breteuil Du Châtelet, marquise, *Dissertation sur la nature et la propagation du feu* (Prault fils, Paris, 1744) (The Science History Institute).

colours of the spectrum yielded different levels of warmth, and even that warmth prevailed in a zone close to the end of the spectrum, since the candlewax melted beyond the red end of the visible spectrum, we now call infra-red. Also following Boerhaave, and contrary to Voltaire, Du Châtelet considered the distinctive property of fire to be the increase of the volume of the substances, and when continued for a long time, eventually the decomposition of a substance into its principles. She even saw fire as a counter-force to attraction, and therefore the natural force that brings life and movement into nature, which would otherwise remain compact and still, since attraction would concentrate and immobilize everything into a tight space. Interestingly, although she preferred Newton's theory, Du Châtelet recognised that the conclusions she drew from the observations agreed with Descartes' philosophy. Her open-mindedness was furthermore obvious when she realised that experiments on the change in weight after calcination had yielded confusing results: sometimes there was an increase, sometimes a decrease, and sometimes no change at all. By contrast, consumed by his defence of Newton, or perhaps his own convictions, Voltaire failed to deliver an informed scientific investigation.

***

The *Dissertation* on fire started Du Châtelet's career as a writer [Kawashima, 2013]. In 1740 she published her *Institutions de physique* ("The Foundations of Physics")[8] that were presented as a textbook for her son Louis Marie Florent Du Châtelet (1727–1793) [Du Châtelet, 1740]. While she had first immersed herself in Newton's natural philosophy, she had now discovered Gottfried Wilhelm Leibniz's (1646–1716) doctrine of monades [Hagengruber, 2012], which was similarly little known in France at the time.[9] As for the work on fire, she avoided a close attach-

---

[8]The first complete English translation of this important text was published in 2018 by [Brading *et al.*, 2018] as a result of the recent efforts to rediscover female philosophers and their work. These translations are partly based on [Bour and Zinsser, 2009, pp. 105–200] and [Patton, 2014, pp. 332–342].

[9]Like atoms, monads are basic individual substances, indivisible, and considered by Leibniz the fundamental existing things that form the universe; unlike atoms, that have a finite extension and are, in fact, the smallest unit of extension, monads are non-extended. By contrast, according to the corpuscular worldview, monads are soul-like entities, and their properties are within them, and do not depend on inter-relational actions like forces. This approach not only allows for the inclusion of the human individual as a monad, but also accommodates the presence of God, who maintains the pre-existing harmony that explains the apparent inter-relation of monads [Burnham, s.d.].

ment to one or the other system, and included and discussed both of them in her treatise. Unbiased and always eager to explore new ideas, Du Châtelet's thought was also in constant flux [Zinsser and Hayes, 2006], and she was unafraid of upsetting her peers who were, of course, all men. She engaged in a public discussion with the French scholar Jean-Jacques Dortous de Mairan (1678–1771) [Du Châtelet, 1741], in a published correspondence on the question of the so-called "vis viva" which, interestingly, also involved energy, a concept that was not clearly defined until the mid-nineteenth century. Being a perfectionist, Du Châtelet used this opportunity to reissue her dissertation in an amended form, since she had found too many typographical errors in the version published by the Academy.

In spring 1749, when Du Châtelet found out she was expecting a child from the young poet Jean François de Saint-Lambert (1716–1803), another of her lovers, the completion of her exposition of Newton's ideas in French was a race against time. Since she had passed the age of 40 at that time, she worried that she would not survive her pregnancy, or the delivery. Indeed, a few days after giving birth to a daughter, she died from pulmonary embolism, probably caused by a blood clot that had developed after the delivery. She was surrounded by her husband, Saint-Lambert and Voltaire, who were all equally affected by her death. The baby girl did not survive, either. Voltaire, who had publicly acknowledged Du Châtelet's expertise on many occasions, was responsible for the publication of her master piece, on which she worked until the very last days of her life: the *Principes mathématiques de la philosophie naturelle, Par feue Madame la Marquise Du Chastellet* in two volumes ("Mathematical Principles of the Natural Philosophy by late the Madame la Marquise Du Châtelet") [Du Châtelet, 1759]. Less well known is her *Dissertation sur la nature du feu*, and yet at the very beginning of her scientific career, Du Châtelet had already demonstrated a commitment to the scientific method and a sense for chymistry that many of her contemporary scholars of natural philosophy did not match. Of course we now know fire is not an element, and studying its nature and its propagation belongs to the field of thermodynamics. Du Châtelet was actually dealing with concepts which scientists were not able to define until more than a century later. It is a pity that her work on the question of the nature of fire was not understood and followed up on, unlike her exposition of Newton, which gained immediate popularity in France and built her reputation as a female scientist in her own right.

References

Académie Royale des Sciences, ed. (1742). *Recueil des pièces qui ont Remporté le Prix de l'Académie Royale des Sciences Depuis leur Fondation jusqu'à Présent, avec les Pièces qui y sont Concouru, Depuis 1738 jusqu'en 1740: Tome Quatrième, Contenant les Pièces Depuis 1738 jusqu'en 1740* (Imprimerie Royale, Paris).

Badinter, E. (2006). *Émilie, ou l'ambition féminine au XVIIIe siècle*, 2nd ed. (Flammarion, Paris).

Bour, I. and Zinsser, J. (2009). *Emilie Du Châtelet: Selected Philosophical Writings* (Chicago University Press, Chicago).

Brading, K., Elder, J., Green, A., Hanson, J., Montes. L., Seul, A., Sloan, P. R., Solomon, M., Steeger, J., Wells, A. (2018). *Foundations of Physics* (https://www.kbrading.org/translations, last accessed 8 February 2019).

Burnham, D. (s.d.). Gottfried Leibniz: Metaphysics, *The Internet Encyclopedia of Philosophy* (https://www.iep.utm.edu/leib-met/#H9, last accessed 20 January 2019).

De Gandt, F., ed. (2001). *Cirey dans la vie intellectuelle: La réception de Newton en France* (Studies on Voltaire and the Eighteenth Century, 11) (Voltaire Foundation, Oxford).

Du Châtelet, E. (1740). *Institutions de physique* (Prault, Paris).

Du Châtelet, E. (1741). *Réponse de Mme*** à la lettre que M. de Mairan, secrétaire perpétuel de l'Académie royale des sciences, lui a écrite, le 18 février 1741, sur la question des forces vives* (Foppens, Bruxelles).

Du Châtelet, E. (1742). Dissertation sur la nature et la propagation du feu, in Académie Royale des Sciences, ed. (1742), *Recueil des Pièces qui ont Remporté le Prix de l'Académie Royale des Sciences Depuis leur Fondation jusqu'à Présent, avec les Pièces qui y sont Concouru, Depuis 1738 jusqu'en 1740: Tome Quatrième, Contenant les Pièces Depuis 1738 jusqu'en 1740*, (Imprimerie Royale, Paris), pp. 85–168.

Du Châtelet, E. (1744). *Dissertation sur la nature et la propagation du feu* (Prault, Paris).

Du Châtelet, E. (1759). *Principes mathématiques de la philosophie naturelle, Par feue Madame la Marquise Du Chastellet*, 2 vols. (Desaint et Saillant, Paris).

Du Châtelet, E. (2009). *Selected Philosophical and Scientific Writings,* ed. Zinsser, J. P., transl. Bour, I. and Zinsser, J. P. (University of Chicago Press, Chicago, IL).

Gauvin, J.-F. (2006). Le cabinet de physique du château de Cirey et la philosophie naturelle de Mme Du Châtelet et Voltaire, in Zinsser, J. P. and Hayes, J. C., eds., *Émilie Du Châtelet: Rewriting Enlightenment Philosophy and Science* (Studies on Voltaire and the Eighteenth, 1) (Voltaire Foundation, Oxford), pp. 163–202.

Hagengruber, R., ed. (2012). *Émilie Du Châtelet: Between Leibniz and Newton* (Springer, Dordrecht).

Joly, B. (1995). Voltaire chimiste: L'influence des théories de Boerhaave sur sa doctrine du feu, *Revue du Nord*, 312, pp. 817–843.

Kawashima, K. (2013). *Emilie du Châtelet et Marie-Anne Lavoisier: Science et genre au XVIIIe siècle* (Honore Champion, Paris).

N. N. (1736). Prix proposé par l'Académie Royale des Sciences pour l'année 1738, *Mercure de France*, avril 1736, pp. 765–766.

Newman, W. R. and Principe, L. M. (1998). *Alchemy vs. Chemistry: The Etymological Origins of a Historiographic Mistake*, Early Science and Medicine, 3, pp. 32–65.

Patton, L. (2014). *Philosophy, Science, and History: A Reader* (Routledge, 2014, New York).

Zinsser, J. P. (2006). *La dame d'esprit: A Biography of the Marquise Du Châtelet* (Viking, New York, NY).

Zinsser, J. P. and Hayes, J. C., eds. (2006). *Émilie Du Châtelet: Rewriting Enlightenment Philosophy and Science* (Studies on Voltaire and the Eighteenth, 1) (Voltaire Foundation, Oxford).

Select Bibliography

Badinter, E. (2006). *Émilie, ou l'ambition féminine au XVIIIe siècle*, 2nd ed. (Flammarion, Paris).

De Gandt, F., ed. (2001). *Cirey dans la vie intellectuelle: La réception de Newton en France* (Studies on Voltaire and the Eighteenth Century, 11) (Voltaire Foundation, Oxford).

Detlefsen, K. (2018). Émilie Du Châtelet, *The Stanford Encyclopedia of Philosophy*, ed. Zalta, E. N., winter 2018 (https://plato.stanford.edu/archives/win2018/entries/emilie-du-chatelet/, last accessed 8 February 2019).

Du Châtelet, E. (1744). *Dissertation sur la nature et la propagation du feu* (Prault, Paris).

Hagengruber, R., ed. (2012). *Émilie Du Châtelet: Between Leibniz and Newton* (Springer, Dordrecht).

Joly, B. (1995). Voltaire chimiste: L'influence des théories de Boerhaave sur sa doctrine du feu, *Revue du Nord*, 312, pp. 817–843.

Kawashima, K. (2013). *Emilie du Châtelet et Marie-Anne Lavoisier: Science et genre au XVIIIe siècle* (Honore Champion, Paris).

N. N. (2019). Project Vox, Duke University (http://projectvox.library.duke.edu/content/du-ch%C3%A2telet-1706-1749, last accessed 8 February 2019).

Terrall, M. (1995). Émilie Du Châtelet and the Gendering of Science, *History of Science*, 33, pp. 283–310.

Zinsser, J. P., ed. (2005). *Men, Women, and the Birthing of Modern Science* (Northern Illinois University Press, DeKalb, IL).

Zinsser, J. P. (2006). *La dame d'esprit: A Biography of the Marquise Du Châtelet* (Viking, New York, NY).

Madame Lavoisier's Diffusion and Defence of Oxygen Against Phlogiston: Her Translations of Richard Kirwan's Essays

Patrice Bret* and Keiko Kawashima[†]

*Centre Alexandre Koyré & Comité Lavoisier,
Académie des Sciences, 23 Quai de Conti, 75006 Paris, France
[†]Department of General Studies, Nagoya Institute of Technology,
Gokiso-cho, Showa-ku, Nagoya, 466-8555, Japan

Antoine-Laurent Lavoisier (1743–1794) is often referred to as the "father of the Chemical Revolution," which included the radical notion of a chemical element as simple substance and a new theory of combustion. Above all, Lavoisier was the individual who named oxygen and demonstrated the crucial role of that element in calcination (heating to high temperatures in air) and all kinds of combustion, including respiration [Bret, 2018].

Oxygen was the core of Lavoisier's new "pneumatic chemistry," also known as the "theory of the French chemists," because, to some extent, his research method prefigured modern collaborative research. Indeed, Lavoisier often worked at his laboratory in Paris Arsenal with other

Figure 1. Jacques-Louis David, portrait of the Lavoisiers (1787) (Metropolitan Museum of Art, New York; public domain).

chemists, such as Louis-Bernard Guyton de Morveau (1737–1816), Claude-Louis Berthollet (1746–1822), Antoine-François de Fourcroy (1755–1809) and several younger chemists, along with mathematicians, mainly Pierre-Simon Laplace (1749–1827) and Gaspard Monge (1746–1818). Furthermore, a key member of Lavoisier's group was his wife, Marie-Anne Pierrette Paulze-Lavoisier (1758–1836) (Figure 1).

Unlike the male savants of the Arsenal, Mme Lavoisier never signed her contributions, except for her etchings of her husband's chemical instruments. Yet she was an active collaborator across the full range of his encyclopaedic activities, and particularly helpful in the success and widespread diffusion of the Chemical Revolution, which led to the victory of oxygen over phlogiston [Kawashima 2005, 2011, 2013]. In particular, she was central in the translation and refutation of two pamphlets authored by the phlogistician Richard Kirwan (1733–1812).

Madame Lavoisier: Her Husband's Collaborator and a Member of the Arsenal Group

Marie-Anne was born into a wealthy bourgeois family. She was not yet fourteen in 1771, when she married Lavoisier, a twenty-eight year-old general farmer (tax collector) and junior partner of her father, Jacques-Alexis Paulze (1723–1794). With that union, Lavoisier's patron shielded his daughter from the marriage arrangements — expected by her grandun-cle, Abbé Terray, the Controller General of Finance — with a penniless fifty-year-old aristocrat. Paulze was also a man of culture, who taught his daughter "to value serious study and to respect deserving people"[1] [Guizot, 1841, p. 74]. The young girl, who had only received a traditional education in a convent school, saw in her husband one of those deserving people.

To be worthy of her husband, young Mme Lavoisier applied herself to serious study. She gained extensive knowledge, mastering Latin, English and Italian, and learning chemistry systematically from Lavoisier's collaborators, Jean-Baptiste Michel Bucquet (1746–1780), and then Philippe-Joachim Gengembre (1764–1838) [Grimaux, 1896; Duveen, 1953; Scheler, 1985; Goupil, 1993b; Bret, 2003; Poirier, 2004; Kawashima, 2008, 2015]. She was also trained in drawing and painting by the famous artist Jacques Louis David (1740–1825), and acquired engraving skills [Pinault-Sørensen, 1994].

Like other Paris *salonnières* (enlightened ladies who regularly assembled authors for literary and philosophical discussions in their *salons*), Mme Lavoisier received her husband's collaborators and French or foreign scholarly visitors for lunch or English tea, including Benjamin Franklin (1706–1790) and the English agronomist Arthur Young (1741–1820), who told of her participation in chemical discussion in his trave-

---

[1] All translations from the French are the authors' unless otherwise indicated.

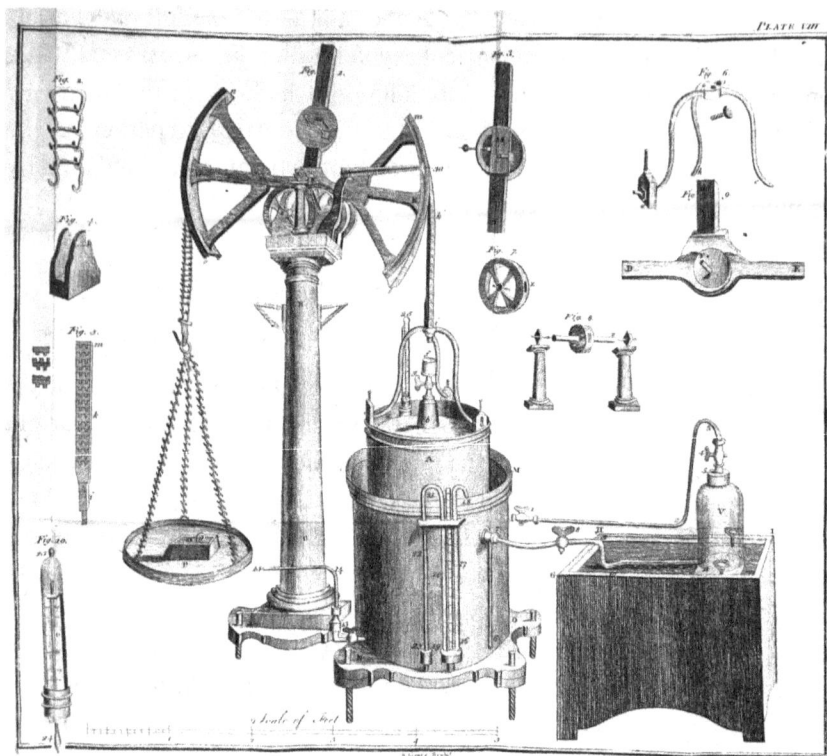

Figure 2. Lavoisier's gasometer, drawn and engraved by Mme Lavoisier for her husband's *Traité élémentaire de chimie* (1789). From the English translation *Elements of chemistry in a new systematic order* (1790), plate VIII. (The Natural Science Library, NTNU University Library.)

logue. In addition to an oil portrait of Franklin, the thirteen copperplate illustrations of chemical instruments attached to Lavoisier's *Traité élémentaire de chimie* [Lavoisier, 1789] provide evidence for Mme Lavoisier's drawing and etching skills (Figure 2). Moreover, in three of her four drawings of laboratory experiments at the Arsenal [Beretta, 2012], she portrayed herself as a record keeper of experiments amidst a group of male experimenters (Figure 3). As a testimony to this record-keeping role, her handwriting appears, among the hands of her husband and his collaborators, in Lavoisier's laboratory notebooks [Lavoisier, 1773–1789]. Both those invaluable historical materials convey the

Figure 3. Lavoisier's experiments on respiration on a man at rest. Sepia by Mme Lavoisier (1791) (Private collection; photo J.-L. Charmet; reproduced by kind permission of the Académie des sciences).

appearance of Lavoisier's modern laboratory as well as the concrete role Mme Lavoisier played as mediator of the experimental practice.

For her husband's benefit, since he was not skilled in foreign languages, Mme Lavoisier also translated English and Italian papers and letters into French. In 1787 she was even encouraged by Lavoisier's group to follow the example of Mme Picardet (1735–1820) in Dijon, who had published dozens of chemical translations in learned journals, alongside a collection of chemical memoirs by the German-Swedish apothecary Carl Wilhelm Scheele (1742–1786), the first discoverer of "fire air" (oxygen) [Bret, 2008a; 2008b, 2016]. And indeed, Mme Lavoisier published the translations of *An Essay on Phlogiston* (1787) and *Of the Strength of Acids* (1791) by the renowned Irish chemist Richard Kirwan, Fellow of the Royal Society of London and Fellow of the Royal Irish Academy, who opposed Lavoisier's theory of oxygen. Her first published translation, *Essai sur le phlogistique* (1788), occupied an important role in the oxygen *versus* phlogiston debate, as it contained comments and refutation arguments, leading to the victory of the oxygen theory — and thus Lavoisier's theory — and the subsequent abandonment of the latter theory [Kawashima, 2000].

Lavoisier and the Chemical Revolution *vs.* Phlogiston

The phlogiston theory had entered mainstream science in Lavoisier's youth. Based on the idea of ancient elements, it had been developed in the early decades of the eighteenth century by Georg Ernst Stahl (1659–1734) after Johann Joachim Becher (1635–1682). The phlogiston was supposed to be contained in any combustible body and to be released during combustion. Despite that first rather coherent chemical theory, a conflicting issue occurred in the calcination of metal: the metal was heavier after calcination than before, although it was supposed to have lost its phlogiston. During Stahl's era this phenomenon was viewed as unimportant, for the change in quality was more important than the change in quantity. Furthermore, not all combustion yielded an increase in weight (for instance combustion with sulphur and phosphorus did not). By the mid-eighteenth century this troubled the phlogisticians. They now saw phlogiston either as a mere "principle" without material substance, or even with a negative weight. Some of them, like Kirwan, thought it was nothing but the "inflammable air" that Lavoisier recognized as a constituent element of water (hydrogen).

In the 1770s, Lavoisier found that common air diminished during the calcination of metal, and that the increased weight of the calcined material coincided with the decreased weight of the air. Thus, he realized that during combustion a part of the air was combined with burning bodies. He was therefore convinced that the phlogiston theory was wrong. He conducted thorough quantitative experiments using large, expensive and precise experimental equipment that was later engraved by his wife (Figure 2). Then, he argued that the "vital air" — recently discovered independently as "fire air" by Carl Wilhelm Scheele in Sweden, and as "dephlogisticated air" by Joseph Priestley (1733–1804) in England — was the cause of both combustion and acidification, and he proved that it was also a constituent element of water. Lavoisier opposed all phlogisticians. Even though they acknowledged his experimental data, there was a complete incompatibility between the ideas of combustion caused by the burning or calcining body itself, *versus* combustion related to the surrounding environment, i.e. the presence or absence of that gas essential for life and fire.

In addition to Lavoisier's pneumatic theory of oxydo-reduction, the naming of chemical substances was another achievement of the Chemical Revolution. The *Méthode de nomenclature chimique* (*Method of Chemical Nomenclature*) designed and published by Guyton de Morveau, Lavoisier, Berthollet and Fourcroy in 1787 replaced the various puzzling and inconsistent terms often inherited from alchemy with systematic ones: alongside the simple substances or elements, the names of compounds directly indicated the elements mixed in them. Subsequently, the concept of the element and the nomenclature were linked inseparably for the antiphlogistonists. One of Mme Lavoisier's challenges was to pay attention to this aspect in her translations.

Madame Lavoisier's Defence of Oxygen Against Phlogiston

The *Essai sur le phlogistique* was not just a translation of Kirwan's defence of phlogiston: it was also a counter-attack by Lavoisier's party, and each section was followed by an extensive rebuttal written by a member of the group. To some extent, as much as translating the text, Mme Lavoisier also had to help refute the author's theory.

In the "Translator's Preface" that preceded Kirwan's introduction, Mme Lavoisier summarized the history of the phlogiston theory and presented the essence of the Lavoisier theory against it. Though emphasizing Kirwan's efforts to adapt phlogiston to modern discoveries, she considered it a hypothesis. Kirwan's introduction was then followed by Lavoisier's note presenting the principles of his antiphlogiston theory. In short, right from the start of the French version, even before the text of the book itself, the Lavoisiers attacked Kirwan's defence of phlogiston on either side of the original introduction, before all chemists from their group took turns in systematically providing their criticisms. This culminates in the concluding note by Guyton de Morveau, following Kirwan's conclusion, and providing the actual conclusion to the whole work.

As for the terminology, some interesting strategies occur: Mme Lavoisier faithfully translated Kirwan's traditional terminology into

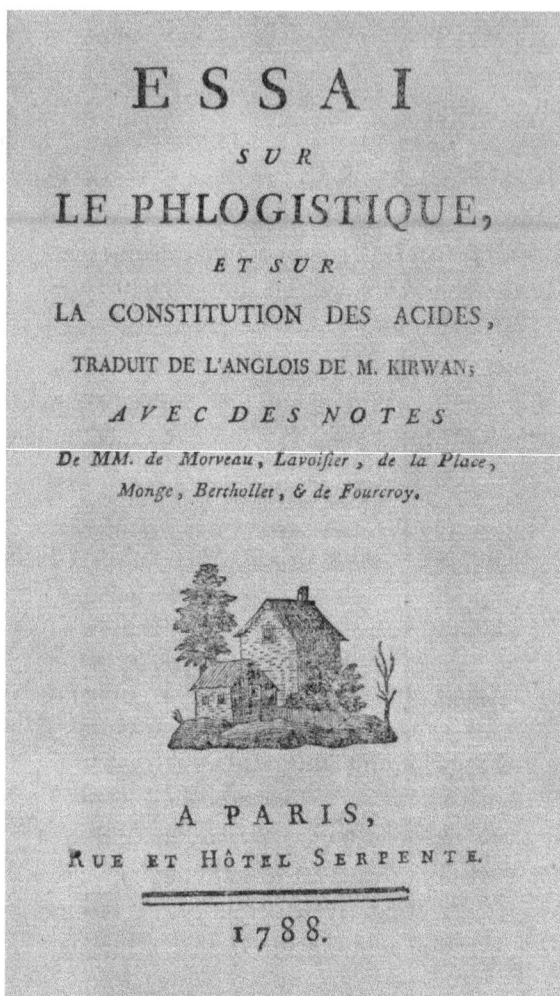

Figure 4.   Title page of Kirwan's *Essai sur le phlogistique*, translated by Mme Lavoisier (1788) (The Science History Institute).

French in the titles of each chapter, using the old nomenclature. Lavoisier and his colleagues, by contrast, used the new nomenclature in their counter-argumentation, and by looking at the table of contents, readers could compare old and new nomenclatures. Moreover, the new terms were added in parentheses to Mme Lavoisier's translation of Kirwan's original text, for instance "l'acide marin déphlogistiqué (acide muriatique oxigéné)" for "dephlogisticated marine acid" — later identified as an element and named "chlorine" by Humphry Davy. Such tactics can be seen as an evidence of the entire group's collaborative work, and their in-depth discussions during numerous meetings. In her correspondence and in the manuscript version of her preface [Paulze-Lavoisier, 1788], Mme Lavoisier actually acknowledged the help she had received from Guyton de Morveau, in both chemical and language matters, and from Président de Virly, the only one who had recently spent sixteen months in Britain, had visited Kirwan and was practiced in speaking English [Bret 2016].

Mme Lavoisier's second published translation of *De la force des acides,* appeared four years later (1792) in the *Annales de chimie,* a journal edited by Lavoisier's group [Kirwan, 1792]. It provides an additional proof of the couple's collaboration. While her name is not mentioned on the title page of *Essai sur le phlogistique*, in this second publication "Madame L." appears as translator. However, her original manuscript reveals that her husband corrected it: even the critical footnotes are in his handwriting, although he gallantly attributed them to her, by carefully adding "translator's notes [Paulze-Lavoisier, 1792]." Together, they removed the old concept of the element and turned it into a tool for the dissemination of Lavoisier's scientific theory, built upon oxygen.

Gender-Related Contradictions of Madame Lavoisier, 18th-Century *femme savante*

As mentioned above, the name of the translator was not revealed on the first page of the *Essai sur le phlogistique* (Figure 3). Yet, the letters to the Lavoisiers and even learned journals show that savants throughout Europe knew that it was the work of Mme Lavoisier. She also fought phlogiston theory as a salonnière. Using her charm and social skills, Mme Lavoisier planned various events for advertising Lavoisier's oxygen chemistry skilfully. The most curious was a somewhat allegoric theatre performance at the Arsenal in 1788, proposed by Jean-Henri Hassenfratz (1755–1827), another member of the group, to show "Oxygen fighting three-quarters-defeated Phlogiston" or "the Genie of the new chemistry slaying the Hypothesis" [Goupil, 1993a, No. 945].

European scholars celebrated this talented woman. The Swiss naturalist Horace Bénédict de Saussure (1740–1799) was converted to a supporter of oxygen theory while reading the *Essai sur le phlogistique*. He highly esteemed Mme Lavoisier's contribution to the new chemistry, and wrote that her stunning and accurate translation of the book was fatal to Kirwan's reputation [Goupil, 1993a, No. 1089]. Such praises pleased Mme Lavoisier, of course, but she knew the limits of her position, as her husband's assistant rather than a savant in her own right. At the age of thirty she confessed to Saussure that she felt herself a helpless young girl when in the presence of her husband and his colleagues [Bret, 1997–2012, VI, No. 1106]. Such a self-deprecating comment may be considered a consequence of the gender bias and social convention of the time.

Convent schools educated girls only to become dutiful and religious wives and mothers. Women's lives, even in public, were devoted to helping men, and no public achievements were expected from them. Marie Geneviève Charlotte Thiroux d'Arconville (1720–1805), a writer also active in the field of chemistry in an earlier period, condemned such a gender norm that did not allow women to become authors, noting "[i]f their works are bad, they are whistled at; if they are good, they are ignored" [Bret, 2011, pp. 84–85; Bret and Van Tiggelen, 2011].

While Mme Lavoisier certainly gained more knowledge than most women, and was praised by the scholars of the "Republic of Letters" — the late-seventeenth and eighteenth-century intellectual network of learned individuals centuries in Europe and the Americas —, her position continued to be separate from other members'. Indeed, in the abovementioned drawings of the laboratory (Figure 3), she visually expressed this lonely position for herself: though participating to the work of the group, as sole record keeper, she is the only woman represented in the picture, somewhat apart from the men at the centre of the action.

Interestingly, her feminine role served the greater cause in her special and socially accepted skills as a salonnière, a translator, a painter and an etcher, as a secretary and record keeper [Kawashima, 2015]. Once a widow, she further kept the memory of her late husband, publishing the works he had revised for a new edition [Lavoisier, 1805; Duveen and

Scheler, 1959; Beretta, 2001], and keeping his scientific instruments and papers. As late as in 1818 she drafted the life of her late husband for Georges Cuvier's entry "Lavoisier" in Michaud's *Biographie universelle* [Gillispie, 1956]. Meanwhile, she had married Benjamin Thompson, Count Rumford (1753–1814) in 1805, probably hoping to revive the collaboration she had enjoyed with Lavoisier, but the relationship soon ended in divorce. Mme Lavoisier's achievements were firmly connected with her marriage to Lavoisier.

Although a woman of the eighteenth century, in which most female collaborators were rendered invisible to history behind the fame of male scholars, Mme Lavoisier left her mark as an assistant to the development and dissemination of the Chemical Revolution. This was a significant step for women's history and the elements' history alike.

References

Works Written, Translated or Edited by Mme Lavoisier

Kirwan, R. (1788). *Essai sur le phlogistique et sur la constitution des acides, traduit de l'anglais, avec des notes de MM. de Morveau, Lavoisier, de la Place, Monge, Berthollet et de Fourcroy* ("rue et hôtel Serpente," Paris) [translation; preface, pp. v–xj; three translator's notes, pp. 27–28, 45, 58–59].

Kirwan, R. (1792). De la force des acides et de la proportion des substances qui composent les sels neutres, *Annales de Chimie*, XIV, juillet, pp. 152–211, pp. 238–286 [translation; four translator's notes, pp. 191, 204, 206, 246].

Lavoisier, A.-L. (1773–1789). Laboratory notebooks [manuscripts], 14 vols. [including notes by Mme Lavoisier], Fonds Lavoisier, Archives de l'Académie des Sciences, Paris, 5 J 23.

Lavoisier, A.-L. (1789). *Traité élémentaire de chimie*. 2 vols. (Cuchet, Paris) [the 13 copperplates drawn and engraved by Mme Lavoisier].

Lavoisier, A.-L. (1805). *Mémoires de physique et de chimie*. 2 vols. (Du Pont, Paris) [edititorial work; two-page unpaginated preface].

Paulze-Lavoisier, M.-A. P. (1788). "Préface du traducteur" [manuscript draft for *Essai sur le phlogistique*], Fonds Lavoisier, Archives de l'Académie des Sciences, Paris, 5 J 3, No.118.

Paulze-Lavoisier, M.-A. P. (1792). "De la force des acides" [manuscript draft translation], Fonds Lavoisier, Archives de l'Académie des Sciences, 5 J 5, No. 340.

Other Sources

Beretta, M. (2001). Lavoisier and his Last Printed Work: The *Mémoires de physique et de chimie* (1805), *Annals of Science*, 58, pp. 327–356.

Beretta, M. (2012). Imaging the Experiments on Respiration and Transpiration of Lavoisier and Séguin: Two Unknown Drawings by Madame Lavoisier, *Nuncius*, 27, pp. 163–171.

Bret, P., ed. (1997–2012). *Œuvres de Lavoisier: Correspondance*, vols. VI–VII (Académie des sciences, Editions Hermann, Paris).

Bret, P. (2003). Lavoisier (Marie-Anne-Pierrette Paulze, épouse), 1758–1836, in *Dictionnaire de biographie française* (Letouzey et Ané, Paris), vol. XIX, col. 112–113.

Bret, P. (2008a). Picardet, Claudine, in *New Dictionary of Scientific Biography*, ed. Koertge, N. (Charles Scribner's Sons and Thomson Gale, Farmington Hills, MI), vol. VI, pp. 97–98.

Bret, P. (2008b). Les promenades littéraires de Madame Picardet. La traduction comme pratique sociale de la science au XVIIIe siècle, in *Traduire la science, hier et aujourd'hui,* ed. Duris, P. (Publications de la Maison des sciences de l'homme d'Aquitaine, Bordeaux), pp. 125–152.

Bret, P. (2011), Mme d'Arconville, femme de sciences au temps des Lumières, *Pour la science*, 402:April, pp. 84–87.

Bret, P. (2016). The Letter, the Dictionary and the Laboratory: Translating Chemistry and Mineralogy in Eighteenth-Century France, *Annals of Science*, 73:2, pp. 122–142.

Bret, P. (2018). Oxygène: Le premier élément, in *L'Europe: Encyclopédie historique*, ed. Charles, C. and Roche, D. (Actes Sud, Arles), pp. 1491–1493.

Bret, P. and Van Tiggelen, B., eds. (2011). *Madame d'Arconville (1720–1805): Une femme de lettres et de sciences au siècle des Lumières* (Hermann, Paris).

Duveen, D. I. (1953). Madame Lavoisier, *Chymia*, 4, pp. 13–29.

Duveen, D. I. and Scheler, L. (1959). Des illustrations inédites pour les *Mémoires de chimie*, ouvrage posthume de Lavoisier, *Revue d'histoire des sciences*, 12, pp. 345–353.

Gillispie, C. C. (1956). Notice biographique de Lavoisier par Madame Lavoisier, *Revue d'histoire des sciences*, 9, pp. 52–61.

Goupil, M., ed. (1993a). *Œuvres de Lavoisier: Correspondance*, vol. V (Académie des sciences, Paris).

Goupil, M. (1993b). Madame Lavoisier, in *Œuvres de Lavoisier: Correspondance*, vol. V, ed. Goupil, M. (Académie des sciences, Paris), pp. 273–275.

Grimaux, E. (1896). *Lavoisier 1743–1794*, 2nd ed. (Félix Alcan, Paris).

Guizot, F. (1841). *Madame de Rumford* (Crapelet, Paris).

Kawashima, K. (1998). Madame du Châtelet et Madame Lavoisier, deux femmes de science, *La Revue. Musée des arts et métiers*, Mars, pp. 22–29.

Kawashima, K. (2000). Madame Lavoisier et la traduction française de l'*Essay on phlogiston* de Kirwan, *Revue d'histoire des sciences*, 53:2, pp. 253–263.

Kawashima, K. (2005). Madame Lavoisier: Participation of a salonnière in the Chemical Revolution, in *Lavoisier in Perspective*, ed. Beretta, M. (Deutsches Museum, Munich).

Kawashima, K. (2008). Paulze-Lavoisier, Marie-Anne-Pierrette, in *New Dictionary of Scientific Biography*, ed. Koertge, N. (Charles Scribner's Sons and Thomson Gale, Farmington Hills, MI), vol. VI, pp. 44–45.

Kawashima, K. (2011). Women's Translations of Scientific Texts in the 18th Century: A Case Study of Marie-Anne Lavoisier, *Historia Scientiarum*, 21:2, pp. 123–137.

Kawashima, K. (2013). *Émilie du Châtelet et Marie-Anne Lavoisier* (Honoré Champion, Paris).

Kawashima, K. (2015). Lavoisier, Marie-Anne-Pierrette Paulze, in *Dictionnaire des femmes des Lumières*, 2 vols., eds. Krief, H., and André, V. (Honoré Champion, Paris), vol. 2, pp. 688–692.

Perrin, C. (1989). The Lavoisier–Bucquet Collaboration: A Conjecture, *Ambix*, 36:1, pp. 5–13.

Pinault-Sørensen, M. (1994). Madame Lavoisier, dessinatrice et peintre, *La Revue, Musée des arts et métiers*, 6, pp. 23–25.

Poirier, J.-P. (2004). *La science et l'amour: Madame Lavoisier* (Pygmalion, Paris).

Scheler, L. (1985). Deux lettres inédites de Madame Lavoisier, *Revue d'histoire des sciences*, 38, pp. 121–130.

Select Bibliography

Beretta, M., ed. (2005). *Lavoisier in Perspective* (Deutsches Museum, Munich).

Bret, P. and Moerman, E. (2014). Sciences et arts, in *Histoire des traductions en langue française: XVIIe–XVIIIe siècle, 1610–1815*, eds. Chevrel, Y. *et al.*, (Éd. Verdier, Lagrasse), pp. 595–722.

Gargam, A. and Bret. P., eds. (2014). *Femmes de sciences de l'Antiquité au XIXe siècle: Réalités et représentations* (Éditions universitaires de Dijon, Dijon).

Kawashima, K. (2013). *Émilie du Châtelet et Marie-Anne Lavoisier* (Honoré Champion, Paris).

Poirier, J.-P. (1998). Lavoisier: *Chemist, Biologist, Economist* (University of Pennsylvania Press, Philadelphia, PA).

Poirier, J.-P. (2004). *La science et l'amour, Madame Lavoisier* (Pygmalion, Paris).

Rayner-Canham, M. F. and Rayner-Canham, G. W. (2001). *Women in Chemistry: Their Changing Roles from Alchemical Times to the Mid-Twentieth Century* (Chemical Heritage Foundation, Philadelphia, PA).

Schiebinger, L. (1989). *The Mind Has No Sex? Women in the Origins of Modern Science* (Harvard University Press, Cambridge, MA).

Part 2

The Glory of Analytical Chemistry: The Elements Multiply

Elementary Chemistry: Mrs Jane Marcet and the Popularisation of Chemistry

Peter E. Childs

Emeritus Senior Lecturer, Department of Chemical Sciences, University of Limerick, Limerick, Ireland

A Literary Lady

Mrs Jane Marcet (née Haldimand, 1769–1858) (Figure 1) is an important figure in chemical education and in the popularisation of chemistry in the nineteenth century [Armstrong, 1938; Derrick, 1985; Childs, 2018; Meyer and Patterson, 2018]. She was born to Swiss parents, then living in London, and she later married a Swiss scientist, Alexander Marcet (1770–1822), in 1799.

In addition to having contacts in Geneva, Jane Marcet was part of an influential intellectual circle in London, of scientists, writers and economists, who met in her house. She was acquainted, amongst others, with the English natural philosophers Humphry Davy (1778–1829), William Wollaston (1766–1828), John Dalton (1766–1844), Smithson Tennant (1761–1815), Charles Hatchett (1765–1847) and Michael Faraday (1791–1867), as well as foreign authorities in the chemical arts, such as

Figure 1.   Jane Marcet (1769–1858). From Bette Polkinghorn, *Jane Marcet: An Uncommon Woman* (Forestwood, 1993). Reproduced by kind permission of the Pasteur family.

the Swede Jöns Jacob Berzelius (1779–1848). These contacts, as well as those with her husband, helped her keep in touch with current scientific developments. Around 1801, Marcet came up with the idea of presenting science to the public, particularly women, in the form of a dialogue between an older woman (Mrs B; Mrs. Bryan in later editions) and two teenage girls, Caroline and Emily [Dreifuss and Sigrist, 2012]. The book, *Conversations on Chemistry, in which Elements of that Science are Familiarly Explained and Illustrated by Experiments* [Marcet, 1806] (Figure 2), was published anonymously in 1806 and went through many

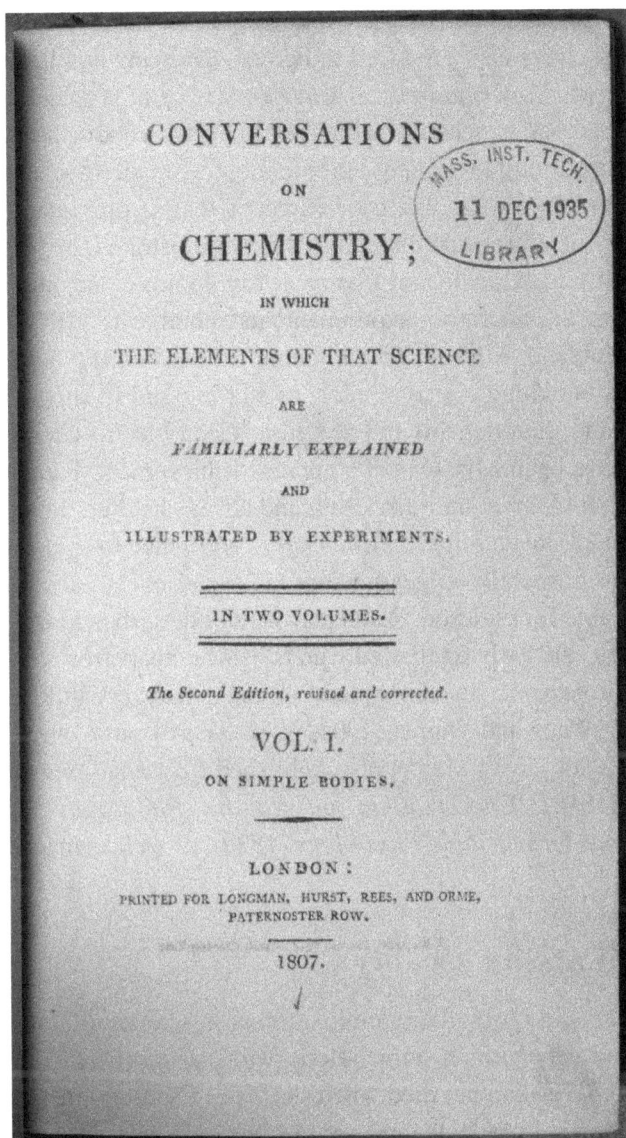

Figure 2. *Conversations on Chemistry*, title page, second edition, 1807 (MIT Libraries, Institute Archives and Special Collections, Cambridge, Massachusetts, USA reproduced by kind permission of MIT Libraries).

editions, in England and America, and tens of thousands of copies were sold. Editions were also produced in France, Germany and Italy.

In America *Conversations on Chemistry* became a widely used and influential textbook in schools, particularly in girls' schools (academies) — despite competition from many other texts. And academy chemistry was no picnic. In the schools that used Marcet's text, it provided "an up-to-date review of European chemical theory, illustrated by experiment, requiring an understanding of chemical terminology and facility in the manipulation of laboratory equipment and chemicals" [Lindee, 1991, p. 23]. In England, Michael Faraday credited reading the book in 1810, while still a bookbinder's apprentice, as starting and feeding his interest in science and directing his interest from electricity to chemistry. This may well have been Mrs Marcet's greatest achievement. Faraday kept in touch with Mrs Marcet until her death and always held her in high regard. Jane Marcet's *Conversations in Chemistry* contributed thus in establishing chemistry as a socially engaging discipline and eventually as a school subject suitable for children. Watts and Weiner describes Marcet's efforts as "breasting relatively uncharted waters" since, apart from the very few university professors, "most people of all ranks had very little knowledge of science" [Watts and Weiner, 2004, p. 3]. Apart from *Conversations in Chemistry* (1806), Mrs Marcet also published *Conversations on Political Economy* (1816), *Conversations on Natural Philosophy* (1819) and *Conversations on Vegetable Physiology* (1829), as well as many books for children.

Conversing About Chemistry

Mrs Marcet had no formal academic training in science although she was given a good education at home, along with her brothers, following the Swiss custom. Her mother died when she was 15 so that Jane had to manage her father's household and act as hostess. After her marriage to Alexander Marcet in 1799 she continued to host frequent gatherings of the great and the good, the scholars and the scientists. Her husband encouraged her to attend Humphry Davy's popular lectures at the Royal Institution, and this sparked her interest in chemistry and in communicating chemistry. She and her husband exemplify scientific collaboration between couples. Dreifuss and Sigrist [2012, p. 27] comment on their fruitful partnership:

> More decisive than either partner's single contribution, the complementarity between his chemical knowledge and her popular-science writing, energized within a supportive scientific and literary circle of friends, culminated in the making of a bestseller.

Conversations in Chemistry consists of 26 Conversations. In no. 1 Marcet introduces the concept and definition of elements and their classification, then follow conversations on light and caloric (the substance of heat), which appeared in The French chemist Antoine-Laurent Lavoisier's (1743–1794) table of elements from 1789 among the elements. Conversation no. 6 deals with oxygen and nitrogen, no. 7 with hydrogen, no. 8 with sulphur and phosphorus, no. 9 with carbon and no. 10 with metals and their reactions. The remaining conversations discuss "compound bodies," as they were called at the time.

In her opening conversation Mrs B. clearly explains the nature of the chemical elements in the form of a dialogue with her young listeners. She starts from the popular view of the elements (air, earth, fire and water) and clearly explains Lavoisier's definition of an element: that an element is a body that cannot be decomposed further. As demonstrated in the extract in Box 1, Marcet explains the difference between decomposition and division by referring to a loaf of bread: separating it into its ingredients is very different from cutting it into pieces. When she first wrote *Conversations in Chemistry*, John Dalton's ideas on atoms were still unpublished, but she included atoms in later editions. Mrs Marcet's clear explanations laid a good foundation of the concept of an element for two generations of students.

What were the main sources of Mrs Marcet's scientific knowledge, which she repackaged so effectively in *Conversations in Chemistry*? Firstly, her contact with London's leading scientists, as well as overseas scientists, at the dinner parties given at home and those she attended in other houses, e.g. at The Scottish science writer and polymath Mary Sommerville (1780–1872) house. Secondly, her attentive presence and notetaking at public science lectures in the Royal Institution. Thirdly, her conversations with Alexander and their work in a home laboratory, where

Box 1. Jane Marcet's exposition of the nature of chemical elements [Marcet, 1817, Conversation no. 1].

MRS. B.

If every individual substance were formed of different materials, the study of chemistry would, indeed, be endless; but you must observe that the various bodies in nature are composed of certain elementary principles, which are not very numerous.

CAROLINE.

Yes; I know that all bodies are composed of fire, air, earth, and water; I learnt that many years ago.

MRS. B.

But you must now endeavour to forget it I have already informed you what a great change chemistry has undergone since it has become a regular science. Within these thirty years especially, it has experienced an entire revolution, and it is now proved, that neither fire, air, earth, nor water, can be called elementary bodies. For an elementary body is one that has never been decomposed, that is to say, separated into other substances; and fire, air, earth, and water, are, all of them susceptible of decomposition.

EMILY.

I thought that decomposing a body was dividing it into its minutest parts. And if so, I do not understand why an elementary substance is not capable of being decomposed, as well as any other.

MRS. B.

You have misconceived the idea of decomposition; it is very different from mere division. The latter simply reduces a body into parts, but the former separates it into the various ingredients, or materials, of which it is composed. If we were to take a loaf of bread, and separate the several ingredients of which it is made, the flour, the yeast, the salt, and the water, it would be very different from cutting or crumbling the loaf into pieces.

EMILY.

I understand you now very well. To decompose a body is to separate from each other the various elementary substances of which it consists.

Table 1. Elements discovered by Mrs Marcet's friends and contacts.

Name	Elements discovered
Humphry Davy	Na, K, B, Ca, Ba
J.J. Berzelius	Ce, Th, Se, Si, Zr
Students of Berzelius	Li, La, Er, Tb, V
William Wollaston	Rh, Pd
Smithson Tennant	Os, Ir
Charles Hatchett	Nb

she repeated experiments seen at the Royal Institution. Fourthly, her access, due to her wealth and social position, to books and journals, and through Alexander, vicariously into London's scientific elite. Fifthly, the contacts and friendships she made were kept up throughout her long life through correspondence, as she tried to keep up-to-date with the rapid growth of science and incorporate it into later editions [Hilton, 2007].

One of the interesting things about several of the chemists Mrs Marcet met socially is that they discovered new chemical elements (Table 1). Mrs Marcet was thus in a position to hear first-hand of these new discoveries and to include them in subsequent editions of her book.

Davy's discovery of new elements by electrolysis starting in 1807 (potassium and sodium) was controversial at the time, but Mrs Marcet quickly incorporated them into the third edition in 1809 and onwards (see extract from the eighth edition in Box 2). In the 1818 US edition, Thomas Cooper wrote: "In this publication, she has explained and explicitly adopted all Sir Humphrey [*sic*] Davy's discoveries and opinions even when his other contemporaries in chemical investigations have not yet dared to follow him" [quoted in Rayner-Canham and Rayner-Canham, 2001, p. 34].

This extract shows how Mrs Marcet was up-to-date with the latest discoveries and quickly incorporated them into the revised editions of her book. She made sure her readers knew the value of scientific discovery and the excitement of finding new elements previously unknown, and the dialogue clearly insists on the beauty of pursing knowledge for the sake of knowledge, independently from any foreseeable utility.

The period from 1800 to 1858, when Mrs Marcet was writing and revising her book, was a fertile time for the discovery of new elements. Of

Box 2. Extract from the 8th edition of *Conversations in Chemistry*, 1822.

MRS. BRYAN.

Before we part, I must introduce to your acquaintance the curious metals which Sir Humphry Davy has recently discovered. The history of these extraordinary bodies is yet so much in its infancy that I shall confine myself to a very short account of them; it is more important to point out to you the vast, and apparently inexhaustible, field of research which has been thrown open to our view by Sir Humphry Davy's memorable discoveries, than to enter into a minute account of particular bodies or experiments.

CAROLINE.

But I have heard that these discoveries, however splendid and extraordinary, are not very likely to prove of any great benefit to the world, as they are rather objects of curiosity than of use.

MRS. BRYAN.

Such may be the illiberal conclusions of the ignorant and narrow minded; but those who can duly estimate the advantages of enlarging the sphere of science, must be convinced that the acquisition of every new fact, however unconnected it may at first appear with practical utility, must ultimately prove beneficial to mankind.

CAROLINE.

What a satisfaction Sir Humphry Davy must have felt, when by an effort of genius he succeeded in bringing to light, and actually giving existence to these curious bodies, which without him might perhaps have ever remained concealed from our view.

the 25 elements identified in this period, 20 were found by chemists known to Mrs Marcet (Table 1). Five more elements were discovered before 1869, bringing the total to 63, when Dimitri Mendeleev (1834–1907) (simultaneously with Lothar Meyer, 1830–1895) proposed his new classification of the elements in his periodic law. Mendeleev could only devise his table because enough elements had been discovered by then to allow trends and patterns to be discerned, which were convincing enough to be accepted. Even before Mendeleev's breakthrough, other attempts had been made to systematise the chemical elements e.g. the German chemist Johann Döbereiner's (1780–1849) triads (1817). Mrs Marcet may have been familiar with these new ideas, but she drew on the chemical ideas and theories of her time, especially the classification and the definition of Lavoisier. She was twenty years old when Lavoisier's *Traité élémentaire de chimie* appeared in 1789, and she died eleven years before Mendeleev's first periodic table was presented, at the age of 89. As Rosenfeld [2001, p. 790] commented:

> Mrs. Marcet followed Lavoisier's scheme of classification of the elementary substances as set out in the 1796 English translation and considered light, electricity, and heat or caloric as "imponderable agents."

***

Jane Marcet's great achievement was to popularise chemistry as a subject, especially for women, making the rapidly expanding science intelligible and interesting to the middle and upper classes. She was the best-selling author of her day and her books became very influential in the United States as textbooks, especially in girls' schools. By clearly clarifying the nature of the chemical elements and their properties and reactions, and by keeping her readers up-to-date with the latest discoveries, Mrs Marcet fully played the role of a mediatrix of science [Neeley, 1992, p. 213].

It would be hard to over-emphasise the value of this role played by Mrs Marcet and her writings in the first half of the nineteenth century. Many more people read her books on elementary chemistry than read a scientific journal or attended a scientific lecture, and so she helped to create a positive climate for scientific education, which blossomed in the second half of the nineteenth century. This developed chemical literacy among the general public, at a time when chemistry was becoming the dominant science of the century. If she had lived, Mrs Marcet would no doubt have played a role in disseminating and popularising the periodic law.

References

Armstrong, E. V. (1938). Jane Marcet and her 'Conversations on Chemistry', *J. Chem. Educ.*, 15:2, pp. 53–57.

Childs, P. E. (2018). Pioneers of Science Education, 3: Jane Marcet (née Haldimand) (1 January 1769–28 June 1858): 'Populariser of Chemistry',

Science, 53:2, pp. 23–26 (http://www.ista.ie/wp-content/uploads/2018/04/Science-March-2018.pdf, last accessed 21 August 2018).

Derrick, M. E. (1985). What Can a Nineteenth Century Chemistry Textbook Teach Twentieth Century Chemists? *J. Chem. Educ.*, 62:9, pp. 749–751.

Dreifuss, J.-J. and Sigrist, N. T. (2012). The Making of a Bestseller: Alexander and Jane Marcet's Conversations on Chemistry, in *For Better or Worse: Collaborative Couples in the Sciences*, eds. Lykknes, A., Opitz, D. L. and Van Tiggelen, B. (Springer, Heidelberg), pp. 19–32.

Hilton, M. (2007). *Women and the Shaping of the Nation's Young: Education and Public Doctrine in Britain 1750–1850* (Routledge, London).

Lindee, S. M. (1991). The American Career of Jane Marcet's *Conversations on Chemistry*, 1806–1853, *Isis,* 82:1, pp. 8–23.

Marcet, J. (1806). *Conversations in Chemistry, in which Elements of that Science are Familiarly Explained and Illustrated by Experiments*, 2 vols. (Longman, London).

Marcet, J. (1817). *Conversations in Chemistry*, vol. I, 5th ed. (Longman, Hurst, Bees, Orme and Brown, London), pp. 7–12 (https://archive.org/details/conversationsonc01marcuoft/page/n3, last accessed 19 October 2018).

Meyer, M. and Patterson, G. (2018). Mrs Jane Marcet and 'Conversation in Chemistry', in *Preceptors in Chemistry*, ACS Symposium Series, ed. Patterson, G. (American Chemical Society, Washington, DC), pp. 83–109.

Neeley, K. A. (1992). Woman as mediatrix: Women as Writers on Science and Technology in the Eighteenth and Nineteenth Centuries, *IEEE Transactions on Professional Communication*, 35:4, pp. 208–216.

Rayner-Canham, M. F. and Rayner-Canham, G. (2001). *Women in Chemistry: Their Changing Roles from Alchemical Times to the Mid-Twentieth Century* (Chemical Heritage Foundation, Philadelphia, PA).

Rosenfeld, L. (2001). The Chemical Work of Alexander and Jane Marcet, *Clinical Chemistry*, 47:4, pp. 784–792.

Watts, R. and Weiner, G. (2004). Women, Wealth and Power: Women and Knowledge Production. Producers and Consumers: Women Enter the Knowledge Market. Presented at the annual conference of the Women's History Network, Hull, England, 3–5 September 2004 (http://www.gabyweiner.co.uk/wp-content/uploads/2014/04/WHN.pdf, last accessed 15 August 2018).

Select Bibliography

Dreifuss, J.-J. and Sigrist, N. T. (2012). The Making of a Bestseller: Alexander and Jane Marcet's Conversations on Chemistry, in *For Better or Worse: Collaborative Couples in the Sciences*, eds. Lykknes, A., Opitz, D. L. and Van Tiggelen, B. (Springer, Heidelberg), pp. 19–32.

Marcet, J. (1817). *Conversations in Chemistry*, vol. I, 5th ed. (Longman, Hurst, Bees, Orme and Brown, London), pp. 7–12 (https://archive.org/details/conversationsonc01marcuoft/page/n3, last accessed 19 October 2018).

Meyer, M. and Patterson, G. (2018). Mrs Jane Marcet and 'Conversation in Chemistry', in *Preceptors in Chemistry*, ACS Symposium Series, ed. Patterson, G. (American Chemical Society, Washington, DC), pp. 83–109.

Rossotti, H. (2006). *Chemistry in the Schoolroom: 1806. Selections from Mrs Marcet's Conversations on Chemistry* (AuthorHouse, Bloomington).

Ordering the Platinum Metals — The Contribution of Julia V. Lermontova (1846/47–1919)

Gisela Boeck

Institute of Chemistry, University of Rostock,
D-18051 Rostock, Germany

The six platinum-group metals are ruthenium (Ru), rhodium (Rh), palladium (Pd), osmium (Os), iridium (Ir) and platinum (Pt). They have fairly similar physical and chemical properties. They occur together in the same mineral deposits. So it is difficult to separate them from each other chemically. Separating them was necessary in order to determine their atomic weights[1] and to find their place in the periodic table, which initially was ordered by increasing atomic weight [Griffith, 2008]. In 1871, a woman named Julia Vsevolodovna Lermontova worked on the separation process of the platinum group, a topic which had been discussed in the chemical community for a long time.

Today, we know the atomic weight (mass) of each member of this group and are able to record their electronic configuration, which allows

---

[1] For historical reasons the term "atomic weight" is used in connection with data from the nineteenth century.

Table 1. Data known for the platinum metals today [IUPAC, 2017].

	Group 8	Group 9	Group 10
period	5	5	5
atomic number	44	45	46
symbol	Ru	Rh	Pd
atomic mass	101,07(2)	102,90549(2)	106,42(1)
electron configuration	[Kr] $4d^7 5s^2$	[Kr] $4d^8 5s^1$	[Kr] $4d^{10}$
period	6	6	6
atomic number	76	77	78
symbol	Os	Ir	Pt
atomic mass	190,23(3)	192,217(2)	195,084(9)
electron configuration	[Xe] $4f^{14} 5d^6 6s^2$	[Xe] $4f^{14} 5d^7 6s^2$	[Xe] $4f^{14} 5d^9 6s^1$

us to explain their place in the periodic table in a straightforward manner (Table 1). The elements from the fifth period of the table are referred to as the light platinum metals, and those from the sixth period as the heavy platinum metals [Riedel, 2002, pp. 846–860].

Palladium and rhodium were discovered by William Hyde Wollaston (1766–1828), and osmium and iridium by Smithson Tennant (1761–1815) at the beginning of the nineteenth century. Ruthenium was isolated in 1844 by Carl Claus (1796–1864).

As early as in 1812, Johann Wolfgang Döbereiner (1780–1849) had extracted and isolated platinum metals from American platinum ores. In 1823 he observed the ignition, by finely divided platinum powder, of a stream of hydrogen directed at it. This discovery led him to the construction of a pneumatic gas lighter, the so-called "Döbereiner Feuerzeug," in which, in modern terms, zinc metal reacts with dilute sulfuric acid to produce hydrogen gas. The platinum sponge catalyzes the reaction of hydrogen with atmospheric oxygen, which heats the catalyst and ignites the hydrogen, producing a gentle flame.

Döbereiner also tried to order the elements. For some of them he had shown that the atomic weight of the middle element lay roughly half-way between those of the other two, for instance, the atomic weight

Table 2.   The arrangement of the platinum metals by Carl Claus [Claus, 1860, p. 289].

positive end of the horizontal row	principal series	**osmium**	**iridium**	**platinum**	negative end of the horizontal row
	secondary series	**ruthenium**	**rhodium**	**palladium**	

of the second element of the triad, sodium, is the arithmetic mean of the atomic weights of the first element, lithium, and the third, potassium. These groups of three elements were later called "triads." Döbereiner also applied this idea to the "metallic igniters" platinum, iridium and osmium, as well as to palladium, pluran[2] and rhodium [Döbereiner, 1829; 1895].

In 1864 William Odling (1829–1921) stated for rhodium and ruthenium the weight 104 and for palladium 106,5 — in his table one finds the symbol Pt twice, but the first must be Pd — , for platinum and iridium the weight 197, and for osmium 199 [Odling, 1864, p. 643].

Carl Claus placed the platinum metals as in Table 2. The three most abundant ones form the principal series ("Hauptreihe") — these were referred to above as the heavy platinum metals. The second horizontal row — Claus called it "Nebenreihe" (the light platinum metals) — also includes platinum metals of similar atomic weights. But the weights are about the half of the elements from the main row. Claus also checked the pairs Os-Ru, Ir-Rh and Pt-Pd, and noted their marked similarities in chemical reactions [Claus, 1860].

The above-mentioned search for a classification of elements was initially based upon chemical and physical properties. But Döbereiner already used the atomic weight as a criterion. John Dalton's (1766–1844) concept of atoms opened the way to the determination of atomic weights at the beginning of the nineteenth century. But there were several problems with this, e.g. the distinction between an atom and a molecule, or the relative atomic weight reference. In 1860, the first international chemistry congress, held in Karlsruhe, Germany, provided some answers to these questions.

---

[2]Pluran, or pluranium, was mentioned in 1827 by Gottfried Osann (1796–1866) [Osann, 1829, p. 133], but it was later shown not to be an element.

The German chemist Lothar Meyer (1830–1895), who, like the Russian chemist Dmitri I. Mendeleev (1834–1907), attended this congress, published a first classification of elements in 1864 [Meyer, 1864, pp. 135–139]. There he used the following atomic weights: ruthenium 104,3; rhodium 104,3; palladium 106; platinum 197,1; iridium 197,1 and osmium 199. In 1868, in an unpublished manuscript [Seubert, 1913, pp. 6–7] Meyer grouped the metals as follows: Ru and Pt belong to group 2, Rh and Ir to group 3 and Pd and Os to group 4 in that table. Finally, in 1870 he used the atomic weights which are presented in Table 3. On the basis of this data he located the platinum metals in the periodic table at the place which they occupy today [Meyer, 1870 and Seubert, 1913, pp. 9–17].

Meyer argued:

> To obtain this arrangement, some few of the elements whose atomic weights have been found to be nearly equal, and which have probably not been very carefully determined, must be rearranged somewhat, Os before Ir and Pt, and these before Au. Whether this reversal of the series corresponds to the properly determined atomic weights must be shown by later research [Meyer, 1870].[3]

The question mark for the atomic weight of osmium marks that Meyer was not sure of it. In the first version of his periodic table Mendeleev placed the platinum metals in the eighth group [Mendeleev, 1869, p. 405] and provided the atomic weights, as shown in Table 4. The similarities in atomic weights made it hard to order the metals in the periodic table.

Later Mendeleev [Mendeleev, 1871 and Seubert, 1913, p. 54] changed the order of the weights: Os 195?, Ir 197, Pt 198? Again, the question

Table 3. Meyer's arrangement of the platinum metals in 1870 [adopted by the author from Seubert, 1913, p. 11].

symbol	atomic weight	symbol	atomic weight	symbol	atomic weight
Mn	54.8	Ru	103.5	Os	198.6?
Fe	55.9	Rh	104.1	Ir	196.7
Co, Ni	58.6	Pd	106.2	Pt	196.7

---

[3]Translation from [McDonald and Hunt, 1982].

Table 4. Atomic weights of the platinum metals used by Mendeleev in 1869 [adopted by the author from [Mendeleev, 1869, p. 405].

symbol	atomic weight	symbol	atomic weight
Rh	104.4	Pt	197.4
Ru	104.4	Ir	198
Pd	106.6	Os	199

Table 5. Atomic weights published by Karl Seubert [adopted from Seubert, 1891, p. 279].

element	atomic weight	element	atomic weight	element	atomic weight
Ru	101,4	Rh	102,7	Pd	106,35
Os	190,3	Ir	192,5	Pt	194,3

marks indicate that he was not sure that these were their exact values. Like Meyer, Mendeleev found the several methods of determination and calculation of the atomic weights problematic. One very important prerequisite for exact results was the investigation of pure substances. These uncertainties led to intensive work to improve them in several laboratories.

Lothar Meyer's disciple Karl Seubert (1851–1921) conducted extensive research on a new determination of the atomic weight of the platinum metals, published in 1891 (Table 5).

Seubert explicitly mentioned the necessity for these corrections which would clarify the place of these metals in the periodic table [Seubert, 1891]. But one problem remained: the analysis of the platinum ores which were the main source for the platinum metals. Today, several separation processes exist for this group of elements [Proske *et al.*, 1953]. The industrial process first yields platinum, and the expensive rhodium and iridium are not separated until the end. This is a long process; even at the end of the twentieth century patents were issued for the invention of better methods for separating of rhodium and iridium [Patent, 1999].

Julia Vsevolodovna Lermontova — Her Biography

The search for a separation process for platinum metals was already pursued in the 1870s by a female chemist: Julia Vsevolodovna Lermontova (1846/47–1919).

Figure 1.   A. Karvain, photo of Julia V. Lermontova (1846/47–1919), undated (Institute Mittag-Leffler, Djursholm, special collection; reproduced by kind permission of the Institute Mittag-Leffler).

Lermontova (Figure 1) — known as the second woman worldwide to receive a doctoral degree in chemistry in 1874[4] — was born on 21 December 1846 (Julian calendar)/2 January 1847 (Gregorian calendar) in St. Petersburg. As a child, she moved with her parents to Moscow, where she was educated by private teachers. She was very interested in science,

---

[4] Apparently, the first woman was Lydia Sesemann (1845–1925) who earned her degree six months before Lermontova, in Zurich, Switzerland [Creese, 1998, pp. 23–24].

especially in chemistry. At that time in the Russian Empire it was impossible for a woman to study chemistry at a university. Thanks to the support of her best friend Sofya Vasilyevna Kovalevskaya (1850–1891), who became the first female teaching professor of mathematics worldwide in 1884, Julia Lermontova was able to accompany her to Heidelberg in 1869 and attended lectures delivered by Robert Bunsen (1811–1899) [Lermontova, 1951]. Many other Russian scientists also travelled to Heidelberg for varying durations. Mendeleev was working there in 1860/61, and returned to Heidelberg quite often. It was probably there during one of his visits that he first met Julia Lermontova. Later Lermontova and Kovalevskaja went on to Berlin. Julia conducted research on organic chemistry under August Wilhelm Hofmann (1818–1892; ennobled from 1890) in his private laboratory in Berlin. The work on di-iodium methane was used for her dissertation, and she received her doctorate in 1874 at the University of Göttingen. In the same year Julia Lermontova returned to Russia, where she became the second female member of the Russian Chemical Society in 1875 (the first was Anna Fëdorovna Volkova (d. 1876)).

Lermontova received the opportunity to conduct research in the field of organic chemistry in the laboratory of Vladimir Vasilyevich Markovnikov (1837–1904) at the University of Moscow, and worked on the synthesis of 1,2-dibromine propane and glutaric acid. After an extended illness she moved to St. Petersburg to work in the laboratory of Aleksandr Mikhaylovich Butlerov (1828–1886). But in 1880 she decided to go back to Moscow. Markovnikov offered her a place in his laboratory, and she started working on oil research in Markovnikov's laboratory, assisting him in his investigations into the composition of Caucasian petroleum. She was probably the first female chemist working in this field [Creese, 1998, p. 21]. Not fully satisfied with this work, Lermontova looked for another position. She was invited to supervise the practical courses in chemistry at the Higher Women's Courses in St. Petersburg in 1880, but did not accept the invitation, because she was afraid that this offer would be denied by the Ministry of Education on account of her not holding a Russian scientific degree. In August 1880, she mentioned that she was not able to move to St. Petersburg for family reasons. In fact, Sofya Kovalevskaya asked Julia Lermontova to look after her daughter Fufa during her own scientific travels. From 1883 onwards, Fufa permanently lived with her godmother Julia. Sofya wanted to start her lectureship in Stockholm undisturbed, and in 1891, she died at the young age of 41, leaving full custody of Fufa to Julia.

In the 1880s Julia Lermontova abandoned pure chemical research. She tried to establish a school for women together with her sister, however without success. With the help of an inheritance she moved to the village Semenkovo near Moscow, where she became interested in agricultural sciences, and died there on 16 December 1919.[5]

Lermontova's "Work on Platinum Residues"

This title was given by Julia Lermontova to her results of using a separation method of platinum metals in an unpublished manuscript. It is likely that she conducted this experimental work in Bunsen's laboratory. The manuscript recording these unpublished results is held at the Mendeleev-Archive in St. Petersburg, and was first discussed by Musabekov [1967].

As already mentioned Lermontova and Mendeleev met prior to the summer of 1871. She was recommended to Mendeleev by Leon Nikolaevich Shishkov (1830–1907/08), who had also studied under Bunsen in Heidelberg and mentioned that Julia Lermontova was looking for a platinum ruble for her work [Musabekov, 1967, p. 53]. It is not clear whether the proposal for the analyses of platinum metals was made by Mendeleev. Letters indicate that Mendeleev and Julia Lermontova had discussed the problem of the separation of the platinum group. Perhaps Mendeleev encouraged her work in this direction. We know indeed from Lermontova's letter that Mendeleev made known to her what type of research he wished her to pursue. And as pointed out above, Mendeleev was interested in precise atomic weights, especially for this group. For this purpose, he required pure substances.

Julia Lermontova likely learned the most important facts about the platinum metals in Bunsen's lectures in Heidelberg. In his course on experimental chemistry Bunsen discussed the separation of platinum from the other platinum metals. Bunsen also talked about Russian coins, and about the separation of Ir, Pd and Rh [Curtius, 1908, p. 22]. Furthermore, it can be assumed that Julia Lermontova gathered some experience when working with Bunsen's new pump for analysis.

---

[5]For more biographical details see [Lermontova, 1951; Musabekov, 1967; Tollmien, 1997; Creese, 1998; Roussanova, 2003; Offereins, 2011].

Julia Lermontova analysed two samples. One was rich in platinum. It also contained Pd, Ir and Rh, but no Os and Ru. The second sample contained all platinum metals except Os. The general procedure for analysis was to find compounds which were insoluble in water or other solvents, and therefore could be separated step by step. At first Ru was separated as $Ba[RuCl_6]$. The other metals were transferred to the chlorides and later precipitated in the sequence Pt, Pd, Rh, Ir [Musabekov, 1967, pp. 53–61].

Julia Lermontova described the procedure quite extensively, but neither quantitative information nor any specification concerning the colour of the precipitates were available. This is unusual, because the different colours were important for the identification of the several metals. Lermontova wanted to analyse a third sample but ran out of time: with her move to Berlin she left the field of platinum metals behind.

During her years in St. Petersburg Julia Lermontova had some contact with Mendeleev: her cousin (likely the physicist and photochemist Vladimir Vladimirovich Lermantov (1845–1919) [Musabekov, 1967, p. 72]) was working with him. But she never continued her initial work for Mendeleev, and instead translated a chapter from Mendeleev's book about the oil industry in Pennsylvania and in the Caucasus [Mendeleev, 1877] — the book connected with his obligations as a consultant to the Russian government in oil-related matters.

***

Julia Lermontova's mostly forgotten contribution to the determination of the right atomic weights may have been small, but the separation of the platinum metals was an important prerequisite for exact values. Her description roughly agrees with the separation process as it is used still today. It is also a testimony to the forgotten contribution of chemists who provided the required robust chemical analysis and precise atomic weights to those who were elaborating the "big picture" in devising the periodic table.

References

Claus, C. (1860). Neue Beiträge zur Chemie der Platinmetalle, *Journal für Praktische Chemie*, 80:2, pp. 282–317.

Creese, M. R. S. (1998). Early Women Chemists in Russia: Anna Volkova, Iuliia Lermontova, and Nadezhda Ziber-Shumova, *Bulletin for the History of Chemistry*, 21, pp. 19–24.

Curtius, T. and Rissom, J., eds. (1908). *Geschichte des Chemischen Universitäts-Laboratoriums zu Heidelberg seit der Gründung durch Bunsen: Zur Feier der Enthüllung des Bunsendenkmals in Heidelberg* (F. W. Rochow, Heidelberg).

Döbereiner, J. W. (1829). Versuch zu einer Gruppirung der elementaren Stoffe nach ihrer Analogie, *Poggendorfs Annalen der Physik und Chemie*, 15. Reprint in Meyer. L., ed. (1895). *Die Anfänge des natürlichen Systems der Elemente* (Engelmann, Leipzig), pp. 3–8.

Griffith, W. P. (2008). The Periodic Table and the Platinum Group Metals, *Platinum Metals Review*, 52:2, pp. 114–119.

IUPAC 2017. Atomic Weights (http://www.sbcs.qmul.ac.uk/iupac/AtWt/, last accessed 16 January 2019).

Lermontova, J. V. (1951). Vospominanija o Sofie Kovalevskoi, in Sofya V. Kovalevskaya, *Vospominanija i pisma*, ed. Strajch, S. J. (Izdatelstvo Akademii Nauk SSSR, Moskow).

McDonald, D. and Hunt, L. B. (1982). *A History of Platinum and its Allied Metals* (Johnson Matthey, London).

Mendeleev, D. I. (1869). Über die Beziehung der Eigenschaften zu den Atomgewichten der Elemente, *Zeitschrift für Chemie*, 12, pp. 405–406.

Mendeleev, D. I. (1871). Die periodische Gesetzmäßigkeit der chemischen Elemente, *Annalen der Chemie und Pharmacie*, VIIIth supplement, pp. 133–229.

Mendeleev, D. I. (1877). *Neftjanaja promyshlennost v Severo-Amerikanskom Shtate Pensilvanija i na Kavkaze* (Obshchestvennaja polza, St. Petersburg).

Meyer, L. (1864). *Die modernen Theorien der Chemie und ihre Bedeutung für die chemische Statik* (Maruschke & Berendt, Wrocław).

Meyer, L. (1870). Die Natur der chemischen Elemente als Funktion ihrer Atomgewichte, *Annalen der Chemie und Pharmacie*, VIIth supplement, pp. 354–364.

Musabekov, J. S. (1967). *Julia Vsevolodovna Lermontova 1846–1919* (Nauka, Moskow).

Odling, W. (1864). On the Proportional Numbers of the Elements, *The Quarterly Journal of Science*, 1, pp. 642–648.

Offereins, M. (2011). Julia Lermontova (1846–1919), in *European Women in Chemistry*, eds. Apotheker, J. and Sarkadi, L. S. (Wiley, Weinheim), pp. 27–30.

Osann, G. (1829). Ueber uralisches Platin, *Archiv für die gesammte Naturlehre*, 16, pp. 129–137.

Patent DE19746134C1 25.03.1999. Verfahren zur Trennung der Platingruppenmetalle (http://www.patent-de.com/19990325/DE19746134C1.html, last accessed 4 November 2018).

Proske, O., Blumenthal, H. and Ensslin, F. (1953). Analyse der Metalle (Springer, Berlin *et al.*).

Riedel, E. (2002). *Anorganische Chemie*, 5th ed. (Walter de Gruyter, Berlin *et al.*).

Roussanova, E. (2003). *Julia Lermontowa (1846–1919): Die erste promovierte Chemikerin* (Books on Demand, Norderstedt).

Seubert, K. (1891). Die Atomgewichte der Platinmetalle, *Justus Liebig's Annalen der Chemie*, 261, pp. 272–279.

Seubert, K. (1913). Anmerkungen, in *Das natürliche System der chemischen Elemente*, Ostwald's Klassiker der exakten Wissenschaften No. 68, ed. Seubert, K. (Wilhelm Engelmann, Leipzig), pp. 119–134.

Tollmien, C. (1997). Zwei erste Promotionen: Die Mathematikerin Sofja Kowalewskaja und die Chemikerin Julia Lermontowa, in *"Aller Männerkultur zum Trotz": Frauen in Mathematik und Naturwissenschaften*, ed. Tobies, R. (Campus Verlag, Frankfurt am Main), pp. 83–109.

Select Bibliography

Creese, M. R. S. (1998). Early Women Chemists in Russia: Anna Volkova, Iuliia Lermontova, and Nadezhda Ziber-Shumova, *Bulletin for the History of Chemistry*, 21, pp. 19–24.

Curtius, T. and Rissom, J., eds. (1908). *Geschichte des Chemischen Universitäts-Laboratoriums zu Heidelberg seit der Gründung durch Bunsen: Zur Feier der Enthüllung des Bunsendenkmals in Heidelberg* (F. W. Rochow, Heidelberg).

Griffith, W. P. (2008). The Periodic Table and the Platinum Group Metals, *Platinum Metals Review*, 52:2, pp. 114–119.

Hoß-Hitzel, S. B. (2003). "Es lebt sich himmlisch in Heidelberg": Robert Wilhelm Bunsen und seine Korrespondenz (Universität Heidelberg, Heidelberg).

Lermontova, J. V. (1951). Vospominanija o Sofie Kovalevskoi, in *Sofija V. Kovalevskaja, Vospominanija i pisma*, ed. Strajch, S. J. (Izdatelstvo Akademii Nauk SSSR, Moskow).

Mendeleev, D. (1877). L'origine du pétrole, *Revue scientifique*, 18, pp. 409–416 (translated by Lermontova).

Musabekov, J. S. (1967). *Julia Vsevolodovna Lermontova 1846–1919* (Nauka, Moskow).

Offereins, M. (2011). Julia Lermontova (1846–1919), in *European Women in Chemistry*, eds. Apotheker, J. and Sarkadi, L. S. (Wiley, Weinheim), pp. 27–30.

Pietsch, E. and Haller, E. (1971). Nachweis der Platinmetalle, in *Gmelins Handbuch der anorganischen Chemie,* ed. Gmelin-Institut für anorganische Chemie und Grenzgebiete (Springer, Berlin and Heidelberg).

Proske, O., Blumenthal, H. and Ensslin, F. (1953). Analyse der Metalle (Springer, Berlin *et al.*).

Roussanova, E. (2003). *Julia Lermontowa (1846–1919): Die erste promovierte Chemikerin* (Books on Demand, Norderstedt).

Tollmien, C. (1997). Zwei erste Promotionen: Die Mathematikerin Sofja Kowalewskaja und die Chemikerin Julia Lermontowa, in *"Aller Männerkultur zum Trotz": Frauen in Mathematik und Naturwissenschaften*, ed. Tobies, R. (Campus Verlag, Frankfurt am Main), pp. 83–109.

Zajceva, E. A. (2013). Pervye shagi russkich zhenshchin na puti k chimicheskomy obrazovaniju (XIX–nachalo XX vv.), in Zh*enshchiny-chimiki*, ed. Lunin, V. V. (Janus-K, Moskow).

Women and Analytical Chemistry: Reflections on the Chemical Skill Needed for Investigating the Elements

Anders Lundgren

Department for History of Science and Ideas, Uppsala University
Box 629, 751 26 Uppsala, Sweden

Because of the perceived need for chemistry in the mining industry, chemical analysis has always been a prominent part of chemistry in Sweden [Lundgren, 2008]. A good chemist was a good analyst — and a good analyst was a prerequisite for discovering and investigating new elements. In Sweden analysis remained to be the dominant way of discovering elements during the nineteenth century. The discovery of erbium, terbium, vanadium, thulium, holmium and a host of other elements happened in this country, and all these found their way into the periodic table, as well as some others — norium, wasium, etc. — which were later shown to be nonexistent. This essay will reflect on the role of women chemists in analytical chemistry with examples from Sweden. I will do so with the help of the common metaphor that equates laboratory work with work in the kitchen. This metaphor is appropriate, not only because many chemists often jokingly refer to their chemistry in cookery terms, but more importantly because chemistry was, up to the end of the nineteenth century, often carried out in homes, in physical proximity to the kitchen, and sometimes in it. There existed hardly any boundary between household, kitchen and laboratory. Jöns Jacob Berzelius' (1879–1848) laboratory was an integrated part of his household and adjacent to kitchen and bedroom [Söderbaum, 1929, p. 310]. Equipment for chemical experiments was often the same as that used in kitchens, as were techniques necessary for analysis, such as grinding with mortar and pestle, decanting, boiling, etc. Kitchen ovens were regularly used for sand baths. Wives, daughters, housemaids and other members of the household therefore naturally contributed to the work, and often acted as laboratory assistants. Antoine Lavoisier's wife Marie-Anne Paulze (1758–1836)[1] was not unique in this regard. Two examples from Sweden are Sara Pohl (1751–1793), Carl Wilhelm Scheele's housekeeper, and Anna Sundström (1785–1871) [Trofast, 2018] Berzelius' housekeeper — known among his pupils as "die strenge Anna" ("the strict Anna"), who, as Berzelius stated, "took care of my person and of my laboratory glasses" [quoted in Trofast, 1979, p. 54], and as Berzelius' friend Carl Palmstedt (1785–1870) wrote to him in 1818, she knew "where everything stands and perhaps knows that better than you [Berzelius]" [Trofast, 1979, p. 63].[2] She took active part in moving laboratory utensils to Berzelius' new home in 1818 [Söderbaum, 1929, p. 185; Trofast, 1979, pp. 79, 90, 93–95,148]. Palmstedt wrote to Berzelius in 1819 that she "lives in the laboratory, where she arranged all glass vessels in neat groups, so that nothing broke" [Trofast, 1979, p.114] (Figure 1).

Kitchen and Laboratory: A Longstanding Tradition

Pohl's and Sundström's main tasks included the washing of glassware — without clean vessels any analytical operation will fail — and the cleaning of the lab. But daily work in the laboratory must have increased their chemical knowledge, especially its practical part. Sundström was thus so "accustomed to all my apparatuses and their names, that I actually think

---

[1] See Bret and Kawashima in this volume.
[2] All translations from the Swedish are the author's, unless otherwise indicated.

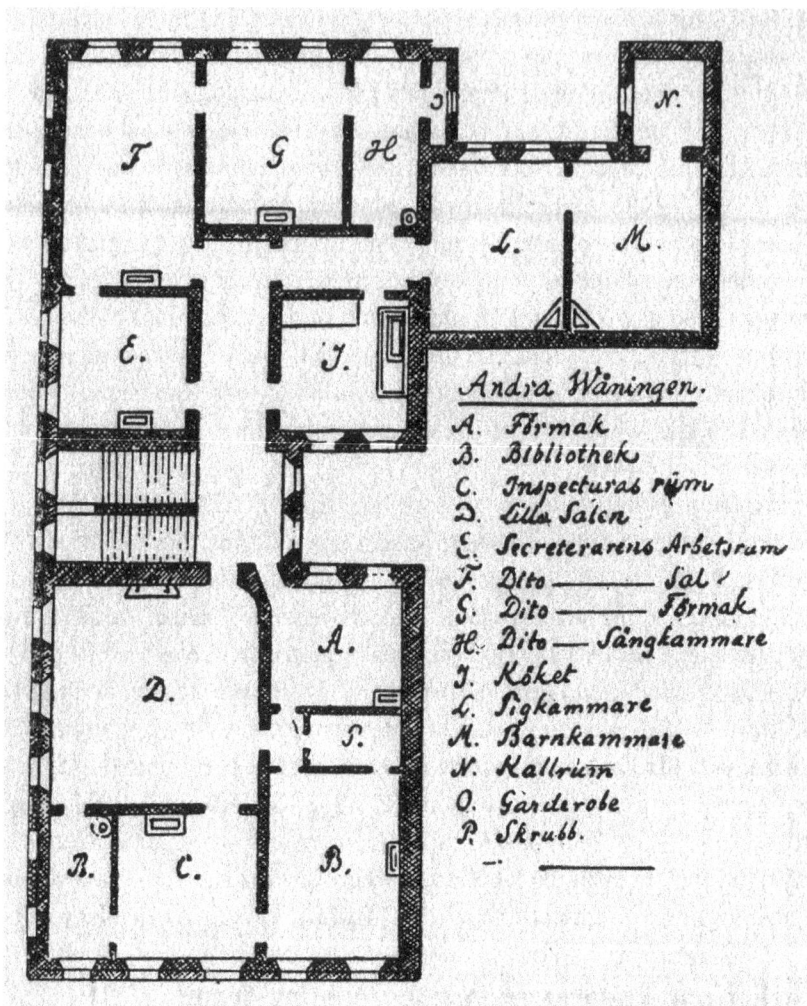

Figure 1. "Berzelius' flat and laboratory in the house of the Royal Academy of Science from 1819." H, G, F were rebuilt as living rooms, L and M as laboratories, and E as Anna Sundström's chamber. J is the kitchen [from Söderbaum, 1929, p. 183].

one could directly ask her to distil hydrochloric acid" [Söderbaum, 1929, p. 145], and Scheele could hardly have discovered all the elements he is said to have discovered, without any help. Does this make Pohl and Sundström analytical chemists? Hardly, but their, and their equals', knowledge of everyday chemistry must have been considerable. Sundström was working at Berzelius' laboratory when lithium, selenium and vanadium were isolated, she is today known as the "first woman chemist in Sweden," and a scholarship has been named after her. But what Sundström, Pohl and others specifically did is difficult to establish; they are never mentioned in scientific publications, only in correspondences and memoirs, generally with words indicating an active role. That they and their forgotten fellows worked both in kitchens and in laboratories at the same time, and sometimes with the same operations, points to the close links between the kitchen and the laboratory.

The connections between chemistry, kitchen and family continued in the nineteenth century. The professor's home was often housed in the institutional building itself, to enable him to follow the experiments closely. The boundaries between household and laboratory were therefore easy to cross for family members. Astrid Cleve (1875–1968), the daughter of the chemistry professor Per Cleve (1840–1905), recalls that in the 1890s the noises and smells from the laboratory influenced the everyday life of the family, and that her father often took her to the laboratory. She was part of both an academic and a family sphere. These two spheres interacted in her choice of research areas, one of which was rare earth metals, and in the context of which she published on the element ytterbium [Espmark, 2012].[3]

Throughout the late nineteenth century, women wishing to study natural sciences often chose chemistry as their subject. Their number is difficult to assess, since women and men were not listed separately in student statistics [Frängsmyr, 2010]. But especially at Stockholm University many female students, while still a minority of all students, appear to have attended chemistry courses — even though only few of them achieved a scientific career. Most became secondary school teachers or disappeared from public life upon marriage. Some, like Astrid Cleve, continued as researchers or assistants.

Two Examples

Some light might be thrown on the question why chemistry was so popular among women by the kitchen metaphor, and Louise Hammarström (1849–1917)

---

[3] See also Espmark and Nordlund in this volume.

and Naima Sahlbom (1871–1957), two female chemists in Sweden, offer interesting case studies.[4] Neither of them discovered a new element, but their work with elements and their chemical practice raise some questions about the relation between women and analytical chemistry.

Hammarström did not receive any formal education but learned chemistry as a private pupil at A. W. Cronquist's (1846–1910) chemical-technical bureau in Stockholm, where she worked as assistant. In 1882 she moved to a mining district in Sweden to work at various iron works, and then opened her own analytical-metallurgical bureau in 1893.

We know more about Naima Sahlbom [Lundgren, 2000; Bergwik, 2016]. From 1884 she studied chemistry at Uppsala University (under Per Cleve), where her abilities as an analytical chemist were put to use by other chemists. For a while she worked at an analytical metallurgical bureau, but thanks to the support of some professors she secured work in geological surveys, and received the opportunity to study radioactivity and mineralogy in Germany. In 1910 she earned a PhD in chemistry at Neuchâtel University in Switzerland with a thesis on colloid chemistry [Sahlbom, 1910]. In 1914 she settled in Stockholm and opened her own analytical metallurgical bureau, specializing in precision analysis for geological purposes and in radioactive substances. Sahlbom is also, and better, known for her work in international women peace organisations, especially in the fight against the use of war gases.

Sahlbom's and Hammarström's careers were quite different, and yet they had much in common: they were women, they were chemists, they were analysts, they opened their own analytical metallurgical bureau — and both were unmarried. Sahlbom wrote for publication, if not extensively, while Hammarström never did, although it is probable that much of her work was embedded in Cronquist's analysis of arsenic compounds [Hillmo, 1994], later known to be one of Hammarström's specialities.

Louise Hammarström

Idun, a Swedish weekly journal intended for women and for home, but also celebrating successful women, published two articles on Hammarström

---

[4]On the use of the kitchen in teaching the periodic system, see Opitz in this volume.

[-r-r, 1895; N. N., 1909]. They reflect common attitudes towards women chemists and how they were supposed to behave and act. This attitude, or rather cliché, was certainly also embraced by the women themselves. From the articles we learn that in 1903 Hammarström bought a house for herself to live in, which she "furnished [with] a laboratory and adjacent localities in 5 rooms" [N. N., 1909, p. 271]. It was said that Hammarström's laboratory saw more than 29,000 analyses in 16 years, and throughout this period her kitchen and laboratory were located close to each other in her home. She also employed two assistants, both women: Elsa Cronquist, probably the daughter of A. W. Cronquist, in 1899, and the yet unknown Maria Lejdström in 1902. *Idun*, perhaps innocently, also relates some interesting motives for women to engage with analytical chemistry: their love of order and their ability to tolerate long and monotonous tasks.

> The always easy hand of a woman, which rarely "stumbles around", but puts everything in its right place, and which day after day doesn't get tired in the seemingly monotonous work, offers conditions for uniformity in the work, exactly where uniformity is absolutely necessary — in chemical analysis [-r-r, 1895, p. 369].

Help from men was only needed "when crushing and grinding ores," that is, for their physical strength [N. N., 1909, p. 271]. The fact that such attitudes existed helps explain a connection between the experiences of being an analytical chemist and being a woman. Training as women at home and in the kitchen taught these women a skill which trained them for work in analytical chemistry. Hammarström's laboratory was a laboratory in which routine analyses were perfected, not a scientific laboratory. Her iron and arsenic analyses were not meant to give new knowledge about these elements in themselves, but to establish how much of an element certain samples contained.

Naima Sahlbom

Sahlbom's position was different. She was academically educated, but did not enter an academic career, but concentrated on her analytical metallurgical bureau instead. Her laboratory must also be considered a laboratory for routine analysis. In fact, she published very little after opening the bureau, but she analysed Swedish spring water for radioactivity and tested geologically interesting substances for other scientists and governmental authorities. Her geological analyses had the same purpose as Hammarström's iron and arsenic analyses: to establish the composition and the amount of different elements in certain substances, not to increase our scientific understanding of certain elements. The analyses of spring water, published together with the professor of geology Hjalmar Sjögren (1856–1922), were not regular chemical analyses, but measured radioactivity with an Elster-Geitel electroscope. Sahlbom found that most of the radioactivity originated from the element radium [Sjögren and Sahlbom, 1907].

There are also connections between kitchen, chemistry and laboratory in Sahlbom's work. In a popular article on the history of cookery she mentions the "complicated techniques of the modern kitchen" in terms reminiscent of laboratory work [Sahlbom, 1908, p. 263]. At the same time she wanted to "theoretically understand the many problems of the kitchen," which would "create a more lasting interest to domestic labour than only a purely practical, often mechanical work is able to do" [Sahlbom, 1908, p. 269]. She also published on the use of chemistry in the kitchen when propagating the making of wine from local berries and fruit [Sahlbom, 1907–1908]. Even if her academic background and formal interest in theory distinguishes her from Hammarström, her view of a perfectly ordered and clean laboratory would certainly have gained support both from Hammarström and from "die strenge Anna" (Figure 2).

***

From these short observations some preliminary thoughts on the connection between women, chemistry and analytical chemistry can be made. Most elements in the periodic table up to the end of the nineteenth century were discovered by traditional chemical analysis. Many female chemists became analytical chemists. None of the women mentioned above showed any immediate interest in the periodic table, but their lack of interest in this subject was shared by the chemical community at large [Kaji *et al.*, 2015]. With their chemical analysis, however, they helped to confirm and

Figure 2.   A modern kitchen according to a chemist [Sahlbom 1908, p. 658].

continue the mapping of the different elements onto the periodic table. These skills coincided with the skills needed to do good work in the kitchen. A chemist should be clean, have a sense for order, and know the place of everything; all of which were also virtues prized in the kitchen. Since the kitchen and the laboratory were often located physically very close to each other, the step from one to the other was easy to take. And in the laboratory, women helpers working as assistants, whether relatives or housekeepers, should remain invisible. Just as in the kitchen, food on the table should not refer to she who prepared it.

Of course male chemists also analysed chemical substances under difficult conditions and with technical skill, and the majority of analytical chemist certainly were men. According to *Idun*, however, men strove further, not the least for a higher income, and therefore worked as chemical analysts for ten years at the most [-r-r, 1895, p. 369]. This is an exaggeration, but be that as it may, women generally stayed in the roles of analysts, partly because that was expected from them, and partly because career possibilities did not open up for them due to social and cultural reasons. In order to learn more about the role of women in chemistry, and especially in analytical chemistry, it would be more rewarding to study those who did not deviate from the normal pattern than to study those who did. The significance of their work for the periodic table would then also become more obvious. Such a study would need to look behind the published works, and behind the official stories. That this is not an easy job should not prevent us from trying.

References

Bergwik, S. (2016). *Kunskapens osynliga scener: Vetenskapshistorier 1900–1950* (Makadam, Göteborg).

Espmark, K. (2012). *Utanför gränserna: En vetenskapshistorisk biografi om Astrid Cleve von Euler*, Historiska studier. Skrifter från Umeå universitet, vol. 5 (Umeå universitet, Umeå).

Frängsmyr, C. (2010). *Uppsala universitet 1852–1916* 2:1, Skrifter rörande Uppsala universitet. C, Organisation och historia (Uppsala universitet, Uppsala).

Hillmo, T. (1994). *Arsenikprocessen: Debatt och problemperspektiv kring ett hälso- och miljöfarligt ämne i Sverige 1850–1919*, Linköping Studies in Arts and Science (Linköpings universitet, Linköping).

Kaji, M., Kragh, H. and Palló, G., eds. (2015). *Early Responses to the Periodic System* (Oxford University Press, Oxford).

Lundgren, A. (2000). Naima Sahlbom, *Svensk biografiskt lexikon*, 31, pp. 225–228.

Lundgren, A. (2008). Gruvor och kemi under 1700-talet i Sverige: Nytta och vetenskap, *Lychnos*, pp. 7–43.

N. N. (1909). 60-årig kvinnlig kemist: Louise Hammarström och hennes laboratorium, *Idun*, 22:22, p. 271.

-r-r. (1895). Louise Hammarström, *Idun*, 8:47, pp. 369–370.

Sahlbom, N. (1907–08). Bär- och fruktvinsberedning i hemmet, *Tidskrift för hemmet*, 3–4, pp. 46–55.

Sahlbom, N. (1908). Ströftåg inom köksvetenskapen: 1. Ur kokkonstens historia, *Varia*, 11, pp. 263–269.

Sahlbom, N. (1910). *Kapillaranalyse kolloider Lösungen* (Theodor Steinkopff, Verlagsbuchhandlung, Dresden).

Sjögren, H. and Sahlbom, H. (1907). Undersökningar af radioaktiviteten hos svenska källvatten, *Arkiv för kemi, mineralogi och geologi*, 3:2.

Söderbaum, H. G. (1929). *Jac. Berzelius: Levnadsteckning* vol. 2 (Kungl. Svenska vetenskapsakademien, Stockholm).

Trofast, J., ed. (1979). *Brevväxlingen mellan Jöns Jacob Berzelius och Carl Palmstedt* vol. 1 (Kungl. Svenska vetenskapsakademien, Stockholm).

Trofast, J. (2018). Anna Sundström: Berzelii hushållerska och hjälpreda (http://kemisamfundet.se/wp-content/uploads/2016/04/Anna_Sundstr%C3%B6m.pdf, last accessed 25 November 2018).

Select Bibliography

Abir-Am, P. G. and Outram, D. (1989). *Uneasy Careers and Intimate Lives: Women in Science, 1789–1979* (Rutgers University Press, New Brunswick, NJ).

Berg, A., Florin, C. and Wisselgren, P., eds. (2011). *Par i vetenskap och politik* (Boréa bokförlag, Umeå).

Bergwik, S. (2016). *Kunskapens osynliga scener: Vetenskapshistorier 1900–1950* (Makadam, Göteborg).

Espmark, K. (2012). *Utanför gränserna: En vetenskapshistorisk biografi om Astrid Cleve von Euler*, Historiska studier. Skrifter från Umeå universitet, vol. 5 (Umeå universitet, Umeå).

Klein, U. and Spary, E. C., eds. (2010). *Materials and Expertise in Early Modern Europe: Between Market and Laboratory* (University of Chicago Press, Chicago, IL).

Long, P. O. (2011). *Artisan/Practitioners and the Rise of the New Science* (Oregon State University Press, Corvallis, OR).

Rossiter, M. (1982–2012). *Women Scientists in America* (Johns Hopkins University Press, Baltimore, MD).

Astrid Cleve Von Euler on Ytterbium and Selenium

Kristina Espmark and Christer Nordlund

Department of Historical, Philosophical and Religious Studies,
Umeå University, 901 87 Umeå, Sweden

In the course of her lifetime the Swedish scientist Astrid Cleve
(1875–1968) was active in a number of different fields of research (Figure 1).
Her initial interest was in botany, the subject in which she earned her
doctorate at Uppsala University in 1898. In doing so, she became the first
woman in Sweden with a PhD in science. Another early interest was the
analysis and systematisation of diatoms. This was a life-long pursuit,
crowned with the monograph *Die Diatomeen von Schweden und Finnland*
("The Diatoms of Sweden and Finland"), for which she was awarded the
honorary title of professor by the Swedish government in 1955. After
World War I she mainly conducted research in geology and archaeology,
as well as, to some extent, theology.

However, the only scientific subject in which Cleve was professionally active in academic employment was chemistry, which she pursued at
the end of the nineteenth century and the beginning of the twentieth century, first as a teaching and research assistant at Stockholm University
College, and then as a director of an industrial wood chemistry laboratory,

Figure 1. Astrid Cleve, 1948, photographer Gunnar Sundgren (Uppsala University Library, public domain).

located in Skoghall. Later on she also worked temporary as a science writer [Espmark, 2012].

Cleve did much of her research work in chemistry with others, and in particular, she published several articles together with the chemist Hans von Euler-Chelpin (1873–1964), to whom she was married from 1902 to 1912; but she also published some work under her own name. In this essay, we would like to make particular note of two of Cleve's single-handed publications, which in various ways address the elements ytterbium and selenium. The first related to her own basic research, conducted during her time at Stockholm University College, and the second concerned popular science, with the aim of contributing to the general education of Swedish society.

Contributions to Knowledge Regarding Ytterbium

The step into the world of chemistry was not a big one for Astrid Cleve. Her father, Per T. Cleve (1840–1905), was a professor of chemistry at Uppsala University and she literally grew up at the University's Department of Chemistry, where the family had their home [von Euler, 1906]. There she studied chemistry alongside botany, and early on began to collaborate with her father, as well as with other chemists present in this setting. She distinguished herself as a skilled experimentalist and, even before completing her doctorate, she was employed as a teaching assistant at Stockholm University College's chemistry laboratory. Her work included lectures on organic and inorganic chemistry and the supervision of practical student exercises. Besides teaching, Cleve had the opportunity to conduct her own research in the lab in Stockholm, an opportunity she made use of during her spare time.

In her years as a teaching assistant in Stockholm, Cleve published a few short papers on metallic calcium [Cleve, 1900] and organic super oxides [Cleve, 1902b]. But above all, she followed in the footsteps of her father and a number of other Swedish chemists, notably Carl Gustaf Mosander (1797–1858) and Lars Fredrik Nilson (1840–1899). This work concerned the rare earth metals. Per T. Cleve was interested in this group of elements, as well as their relationship to the periodic system, even before he was appointed professor at Uppsala [Cleve, 1874 a–d], and his research had contributed to the discovery of the elements holmium (named after Stockholm) and thulium (named after "thule," the Latin name for the Nordic region). Like the substances yttrium, terbium, ytterbium, erbium and scandium, they had been found in the Ytterby mine in the Stockholm archipelago. Per T. Cleve continued with these studies until the mid-1880s. He then placed his knowledge and his collection of materials at the disposal of his students. His daughter Astrid was one among them [von Euler, 1906].

Astrid Cleve focused particularly on investigating the properties of the rare earth metal ytterbium, which had been discovered in 1878 by the

series, and to translating books — such as the Norwegian chemist Ellen Gleditsch's (1879–1968) text *Radioaktivitet og grunnstofforvandling* ("Radioactivity and Elementary Transformation") (1924) and the American physicist Robert Andrews Millikan's (1868–1953) *Science and Life* (1924) — she also published a book under her own name in the series. The title was *Det underbara grundämnet selen* ("The Wonderful Element Selenium," 1925) (see Figure 2). Here, Cleve once more focussed on an element, but this time with the general Swedish public as her audience rather than her colleagues in chemistry.

The discovery of selenium is associated with the Swedish chemist Jöns Jacob Berzelius (1779–1848), who in 1818 extracted the substance from pyrite that originated from the Falu mine. It was given the name "selenium" from the Greek name for the moon ("selene"), to mark its chemical relationship with the element tellurium, which was named after the Latin name for the earth ("tellus"). In contrast to ytterbium, selenium was a non-metallic element of which Cleve herself had no in-depth scientific experience. But she found the substance interesting and of importance to the Swedish public. In her preamble, she writes that selenium, "a relatively rare neighbour of the well-known sulphur," had recently "come forth from its obscurity" and "without doubt will be even more spoken of in the future" thanks to its "unique ability to electrically react to changes of light" [Cleve-Euler, 1925, p. 5].

Det underbara grundämnet selen begins with a historical account of the discovery of the substance. Then, selenium and its properties are described — primarily its ability to conduct electricity upon illumination — and the basic principles of physics and chemistry needed to generally understand them. Simple metaphors with the aim of making the abstract knowledge comprehensible play an important role in the descriptions. The second half of the book is about selenium's practical areas of application, such as signalling systems, picture telegraphy, television and sound film — inventions that were based on the substance's special sensitivity to light and which made it clear that it was related to the breakthrough of electricity in modern society.

Altogether, the book provides a broad and easy-to-read introduction to the remarkable element selenium and its practical potential. It can be described as a manifestation of a popular science ideal that Cleve herself

Figure 2.   Upper cover of Astrid Cleve's book on selenium, *Det underbara ämnet selen* (Uppsala, 1925).

formulated: to make the serious and complicated pleasant by weaving the scientific knowledge into a historical and individually oriented account where theoretical knowledge is related to people and practical applications [Cleve-Euler, 1926].

When the series "Vad vi veta" was launched, *Det underbara grundämnet selen* was marketed as a Swedish original work written by Astrid Cleve-Euler. The book received good reviews. One enthusiastic reviewer wrote: "In this book, the author tells about the virtually unnoticed element selenium in a truly captivating and interesting way ... The book can without doubt be counted as one of our most original and pleasant works of popular science" [Bolin, 1926, p. 83]. It also proved to be authoritative, e.g. it was used as a reference for the entry on selenium in the *Nordisk familjebok* [Vesterberg, 1926], the foremost Swedish encyclopaedia of the time.

But *Det underbara grundämnet selen* was actually not an original work. Closer inspection shows that it extensively builds on the book *The Moon-Element: An Introduction to the Wonders of Selenium*, which the Irish physicist and inventor Edmund Edward Fournier d'Albe (1868–1933) had published just one year previously [1924]. Cleve indeed makes no secret of the fact that she used *The Moon-Element* when it comes to the account of the selenium-based inventions. Placing the texts side by side, it is also apparent that some changes were made in the Swedish version. For example, Cleve skipped over a section where electrons and protons are compared with the help of the stereotypical gender roles of the times; the former as "an active and mobile male sex" and the latter as "a passive, receptive, and more or less stationary female sex" [Fournier d'Albe, 1924, p. 15]. At the same time, Cleve added some of her own reflections about the social dimensions of science, such as the difficulties to get scientific results published if the results challenged established truths, and especially if the scientist behind the results was not an established male professor. These reflections might well have been related to her own experiences as an amateur geologist; her papers were often refused, and when she wrote the book on selenium she was, in fact, subject to a public scientific debate regarding a controversial geological theory put forward by her. But the descriptions of the inventions and the science, and even what seem to be her own thoughts in *Det underbara grundämnet selen*, are mere translations [Espmark, 2012, pp. 186–188].

Translations were not uncommon in the Swedish popular science text series of the time. Of the first eight books published in the series "Vad vi veta," no fewer than six were translations. In another series, "Tro och forskning" ("Belief and Research"), published by the same publisher, only two out of seven books were original works. One may then ask why the publishing company claimed that *Det underbara grundämnet selen* was Cleve's own work when she was named as the translator of two other works at the same time. It would have definitely been more accurate to describe the book as a translated adaptation.

***

The elements ytterbium and selenium were among the substances which Astrid Cleve researched and on which she published during her time as a chemist. As an employee at Stockholm University College, she conducted basic research on ytterbium with the aim to deepen the knowledge of the rare earth metals' properties and relationships. In doing so, she walked in her father's analytical footsteps, but her research was independent and of significance to the continued research on the element, even if not all of her results withstood the test of time. Both the article about ytterbium and the popular book about selenium were well received when they were published; one received a scientific award and the other positive reviews. These successes did not, however, significantly affect Cleve's situation. After her years as a teaching assistant at Stockholm University College she never again found employment at a university, and with the book about selenium — which actually was a translated adaptation — she ultimately said farewell to chemistry altogether. As one of few multidisciplinary women in a male-dominated and discipline-oriented scientific community, Cleve remained an academic outsider for the rest of her life, albeit ultimately with the honorary title of professor, bestowed on her in 1955, the year she turned 80 years old.

References

Bolin, I. (1926). Astrid Cleve-Euler, Det underbara grundämnet selen, *Biblioteksbladet*, 11, p. 83.

Cleve, A. (1900). Om metalliskt kalcium, dess framställning och egenskaper, *Svensk Kemisk Tidskrift*, 12, pp. 53–58.

Cleve, A. (1901). Bidrag till kännedomen om ytterbium, *Öfversigt af Kongl. Vetenskapsakademiens Förhandlingar*, 58, pp. 573–618.

Cleve, A. (1902a). Beiträge zur Kenntnis des Ytterbiums, *Zeitschrift für Anorganische Chemie*, 32, pp. 129–169.

6ssf.

Cleve, A. (1902b). Om organiska superoxider, *Svensk Kemisk Tidskrift*, 14, pp. 94–97.

Cleve, P. T. (1874a). Bidrag till jordmetallernas kemi. I. Torium, *Bihang till Kongl. Svenska Vetenskaps-Akademiens Handlingar*, 2:6, pp. 1–26.

Cleve, P. T. (1874b). Bidrag till jordmetallernas kemi. II. Lantan, *Bihang till Kongl. Svenska Vetenskaps-Akademiens Handlingar*, 2:7, pp. 1–23.

Cleve, P. T. (1874c). Bidrag till jordmetallernas kemi. III. Didym, *Bihang till Kongl. Svenska Vetenskaps-Akademiens Handlingar*, 2:8, pp. 1–24.

Cleve, P. T. (1874d). Bidrag till jordmetallernas kemi. IV. Yttrium och Erbium, *Bihang till Kongl. Svenska Vetenskaps-Akademiens Handlingar*, 2:12, pp. 1–11.

Cleve-Euler, A. (1925). *Det underbara grundämnet selen* (J. A. Lindblads, Uppsala).

Cleve-Euler, A. (1926). Populärvetenskap, *Tidskrift för hemmet*, 22, pp. 29–32.

Espmark, K. (2012). *Utanför gränserna: En vetenskapshistorisk biografi om Astrid Cleve von Euler* (Umeå University, Umeå).

Espmark, K. and Nordlund, C. (2012). Married for Science, Divorced for Love: Success and Failure in the Collaboration Between Astrid Cleve and Hans von Euler-Chelpin, in *For Better or For Worse? Collaborative Couples in the Sciences*, eds. Lykknes, A., Opitz, D. L. and Van Tiggelen, B. (Springer Birkhäuser, Heidelberg), pp. 81–102.

Fournier d'Albe, E. E. (1924). *The Moon-Element: An Introduction to the Wonders of Selenium* (D. Appleton and Company, New York, NY).

Vesterberg, K. A. (1922). Ytterbium, *Nordisk familjebok*, 33, pp. 625–626.

Vesterberg, K. A. (1926). Selen, *Nordisk familjebok*, 38, p. 263.

Von Euler, H. (1906). Per Teodor Cleve, *Kungl. Svenska Vetenskapsakademiens Årsbok 1906*, 4, pp. 187–217.

Thyssen, P. and Binnemans, K. (2011). Accommodations of the Rare Earths in the Periodic Table: A Historical Analysis, in *Handbook of the Physics and Chemistry of Rare Earths*, vol. 41, eds. Gschneidner, K. A. Jr., Bünzli, J.-C. G. and Pecharsky, V. K. (Elsevier North-Holland, Amsterdam), pp. 1–93.

Weeks, M. E. (1932). The Discovery of the Elements XVI: The Rare Earth Elements, *Journal of Chemical Education*, 9, pp. 1751–1773.

Select Bibliography

Creese, M. R. S. and Creese, T. M. (2004). *Ladies in the Laboratory II: West European Women in Science 1800–1900. A Survey of Their Contributions to Research* (The Scarecrow Press, Inc., Lanham, MD and Oxford).

Espmark, K. (2012). *Utanför gränserna: En vetenskapshistorisk biografi om Astrid Cleve von Euler* (Umeå University, Umeå).

Espmark, K. and Nordlund, C. (2012). Married for Science, Divorced for Love: Success and Failure in the Collaboration Between Astrid Cleve and Hans von Euler-Chelpin, in *For Better or For Worse? Collaborative Couples in the Sciences*, eds. Lykknes, A., Opitz, D. L. and Van Tiggelen, B. (Springer Birkhäuser, Heidelberg), pp. 81–102.

Frängsmyr, C. (2010). *Uppsala universitet 1852–1916*, 2 vols. (Uppsala University, Uppsala).

Thyssen, P. and Binnemans, K. (2011). Accommodations of the Rare Earths in the Periodic Table: A Historical Analysis, in *Handbook of the Physics and Chemistry of Rare Earths*, vol. 41, eds. Gschneidner, K. A. Jr., Bünzli, J.-C. G. and Pecharsky, V. K. (Elsevier North-Holland, Amsterdam), pp. 1–93.

From Miss Swallow's "Insoluble Residue" to the Discovery of Samarium and Gadolinium

Joanne A. Charbonneau and Richard E. Rice

P.O. Box 1210, Florence, MT 59833 USA

During the past century Ellen Henrietta Swallow Richards (1842–1911) has been recognised as the founder or a pioneer in such diverse fields as home economics, ecology, sanitary engineering, public health, nutrition, and women's education [Hunt, 1912; Clarke, 1973; Musil, 2014; Swallow, 2014]. As a teacher and chemist, she adeptly applied the principles and techniques of her scientific training in all her work throughout her long career at the Massachusetts Institute of Technology (MIT). Less well known is that she was also considered "an eminent mineralogist in her day" [Schneiderman, 2018, p. 4] and was highly regarded for her careful and accurate chemical analyses of minerals, as well as of water samples, at a time when most women were not yet allowed in college classrooms.

A Passion for Learning

Born on the family farm in rural Massachusetts, Ellen (Figure 1) was fortunate in that both her parents had attended the New Ipswich (New Hampshire)

Figure 1.   Ellen Swallow as a young woman (courtesy of the MIT Museum).

Academy, and thus her home schooling was undoubtedly superior to that of the local one-room schoolhouse. The Swallow family left the farm when Ellen was sixteen, one of the reasons for the move being that she was able to obtain some formal education at the Westford (Massachusetts) Academy. She excelled in all her studies, but was especially drawn to science and mathematics. After finishing at the Academy in 1862, Ellen was thwarted for several years in her desire for further education. She busied herself with teaching, working in her father's general store and at the post office, attending public lectures, caring for her mother, and earning money from tutoring, sewing, cleaning, and any other work she could find. With incredible luck, she read in *Godey's Lady's Book* about a college specifically for young women that had recently opened in Poughkeepsie, New York. Vassar College offered genuine college-level courses in various subjects, including science and mathematics. To Ellen, this seemed providential. After several years of saving for her first-year expenses, she entered in 1868 as a special student[1] [Clarke, 1973, pp. 5–19; Swallow, 2014, pp. 1–24].

---

[1]Ellen was initially accepted at Vassar as a non-degree student, presumably because of her limited formal education, but she quickly proved herself an exceptional student and became a member of the senior class the following year.

At Vassar, Ellen Swallow studied astronomy with Maria Mitchell (1818–1889), the first professor hired at the new college, and chemistry with Charles S. Farrar (1825–1903), the first chair of chemistry and physics. Although both disciplines appealed to her, as did almost any field of study, Ellen was greatly influenced by Farrar's "intellectual power" and his emphasis on the importance of science in solving practical problems. He urged his students to work with great care in the chemistry laboratory "as the profession of an analytical chemist is very profitable and means very nice and delicate work fitted for ladies' hands." Ellen took nearly every science course offered at Vassar, even extra electives such as the one in which she and two other students carried out chemical analyses on "everything … from 'shoe-blacking to baking powder'" [Hunt, 1912, pp. 38–42, 60].

After graduating from Vassar in 1870, Ellen Swallow had intended to teach in Argentina, but her plan fell through, and her subsequent attempts to find work as a chemist closer to home were unsuccessful. None of the companies she applied to were willing to hire a woman, but one of them, in rejecting her application, suggested that she inquire whether MIT, then known as Boston Tech (Figure 2), would accept her for further study in chemistry. In December 1870 she enrolled as MIT's first female student — again as a special student[2] [Clarke, 1973, pp. 20–25]. In spite of considerable faculty opposition to her enrolment, the young woman quickly won over several professors: she studied mineralogy with Professor Robert Richards (1844–1945), assisted Professor William Ripley Nichols (1847–1886) with his survey of public water bodies for the Massachusetts State Board of Health, and collaborated with Professor John Ordway (1823–1909) on problems related to industrial chemistry.

Swallow's student years at MIT coincided with the tenure of Professor James Mason Crafts (1839–1917), who encouraged the school's few chemistry students "to undertake 'the investigation of unsolved problems'" [Cross, 1919, p. 160]. Her first independent project involved the

---

[2]This time, however, her status as a special student masked an ulterior motive on the part of the MIT faculty. While she would "have any and all advantages which the Institute has to offer without charge of any kind," the administration could claim that she was not enrolled as a student if her presence at the Institute was called into question [Clarke, 1973, p. 24].

THE INSTITUTE OF TECHNOLOGY, BOSTON.

Figure 2. Rogers Building, Back Bay, Boston — the first building built for the Massachusetts Institute of Technology (from *Harper's Magazine*, February 1879, p. 472).

chemical analysis of silver-bearing minerals from mines in Colorado. Professor Richards gave her several samples of these minerals, which had never been analysed, and asked her to determine their appropriate classification and comparative composition. In addition to answering these questions, Swallow undertook an original investigation into analysing arsenic and antimony in the presence of other metals in the samples. In 1873 she described this work in her thesis for the Bachelor of Science degree at MIT [Swallow, 1873a] and at the age of thirty became its first female graduate.

Professor Richards, whom she married in 1875, also gave Swallow an iron ore sample "reported to contain Vanadium" though he was unsure of its composition [Clarke, 1973, pp. 42–43]. Although vanadium was first identified in 1801 and rediscovered in 1830, it was not isolated as a pure metal until 1868 by Henry Roscoe (1833–1915) [Fontani *et al.*, 2015, pp. 14–16], who observed, "there is, I believe, no metal more difficult to obtain than vanadium" [Roscoe, 1869, p. 688]. After several months of examining the sample, as well as studying the scientific literature on vanadium and corresponding with authorities around the world, she found that the ore contained 0.02 per cent vanadium. For this work she was awarded a Master of Arts from Vassar [Swallow, 1873b], her third degree in three years.

Miss Swallow's "Insoluble Residue"

Ellen Swallow's first published article details her chemical analysis of a small sample (about 7 g) of samarskite [Swallow, 1875a], a black orthorhombic mineral now known to be a complex mixture of the oxides of rare earth elements and other metals. Originally discovered in the southern Ural Mountains in Russia about 1840, samarskite was first found in North America some ten years later in Rutherford County in western North Carolina and again in 1874 in nearby Mitchell County. A sample from this last location was given to Swallow by Thomas Bouvé (1815–1896), then president of the Boston Society of Natural History. In his prefatory remarks to Swallow's paper, he described how he had obtained three small samples of the mineral and given one of them to "Miss Swallow, who appears to have done herself great credit by the thoroughness with which she performed the analysis, and by the full account given of her method" [Bouvé, 1875].

In her analysis, Swallow used three different methods: (1) that of Rudolph Hermann (1805–1879) [1869], then the most recent analysis of samarskite, but of a Russian sample, (2) that of T. Sterry Hunt (1826–1892) [1852], an earlier analysis, but of the first samarskite found in North America, and (3) essentially that of Hermann, but with modifications from other sources of analytical methods, as well as from her own observations of her previous analyses. Her results, as well as those of Hunt and Hermann, are given in Table 1.

Swallow detected all the same metals as in the two other analyses, but in addition, she found a small amount of material precipitated by

Table 1. Results of analyses of the composition of samarskite by Swallow on a sample from Mitchell County, North Carolina, by Hunt [1852] on a sample from Rutherford County, North Carolina, and by Hermann [1869] on a sample from Russia, near Miask (now Miass) [Swallow, 1875a, p. 426]. None of the analyses listed in the table add up to exactly 100 per cent because laboratory measurements always contain some experimental error.

		Mitchell Co., N. Carolina, 1875.	Rutherford Co., N. Carolina, Dr. Hunt, 1852	Miask, Urals, Hermanu, 1869
The metallic acids of the Tantalic Group		54.96	54.81	56.36
Oxide of Tin	SnO_2	0.16		
Oxide of Uranium	UO	9.91	UO_3 17.03	16.63
Oxide of Iron	FeO	14.02	14.07	8.87
Oxide of Manganese	MnO	0.91		1.2
Oxide of Cerium (La, Di)	CeO	5.17	3.95	2.85
Yttria	YO	12.84	11.11	13.29
Magnesia	MgO	0.52		0.50
Insoluble residue from the oxalate of Cerium		1.25		
Loss on ignition		0.66[1]	0.24	0.33
		100.40	101.21	100.03

ammonium oxalate that was insoluble in both water and hydrochloric acid (labelled as "Insoluble residue from the oxalate of Cerium" in Table 1). This was only slightly more than 1 per cent of the sample and had not been reported in either of the other analyses, but it was too small an amount for further investigation. Swallow repeated her analysis of the mineral's composition twice more to confirm her initial results, and she reportedly mentioned to other people in her laboratory that this residue might contain elements that were then unknown [Richards, 1936, p. 160]. Despite this tantalising prospect — and the fact that it was eventually shown to contain the then unknown elements samarium and gadolinium — she carried out no further studies on samarskite.

Although done with samarskite, Swallow was not quite finished with investigating minerals and ores. A month after her paper on samarskite, she published another paper on the chemical analysis of minerals associated with lead ore from the North Shore of Massachusetts [Swallow, 1875b]. Two years later, together with Margaret S. Cheney (1855–1882) [Atwood, 2005], a student in the Women's Laboratory[3] [MIT Archives, 1999; Durant, 2006], she worked out a method for estimating nickel in pyrrhotites and smelting mattes [Cheney and Richards, 1877]. As a result of her work on minerals during this period, Ellen Swallow Richards was inducted into the American Institute of Mining Engineers in 1875 as its first female member [Hunt, 1912, p. 245].

The Continuing Challenge of Samarskite

Other chemists who had read Swallow's paper on samarskite realised that the mineral warranted further investigation. Soon J. Lawrence Smith (1818–1883) [1877a] and Oscar D. Allen (1836–1913) [1877] each published studies of samarskite, comparing Swallow's analysis with their own. Although neither reported the presence of an insoluble oxalate residue, they both noted that it was unknown how much of the material reported as cerium was actually cerium, and whether it might also contain other elements.

In 1877 only six of the rare earth elements were known (lanthanum, cerium, erbium, terbium, yttrium, didymium[4]), but eight more were discovered during the late 1870s and 1880s,[5] a time of intense investigation into this group of elements. In his table of "verified and unverified" elements, Baskerville [1904, pp. 163, 170–171] listed ten elements from samarskite that were proposed during the decade beginning in 1877 though only a few of them were ultimately recognised as new elements and accepted into the chemical pantheon with the name chosen at the time. The first of these proposed elements was mosandrum, which Smith [1877b] tentatively announced in May 1877. The following year the

---

[3]Ellen Richards was instrumental in establishing this space in 1876 for the instruction of women in the chemical sciences. It closed in 1883 when a new laboratory was built for both men and women at MIT.

[4]Later found to consist of praseodymium and neodymium.

[5]Nine, if both praseodymium and neodymium are counted as newly discovered elements.

contents of a sealed packet, which he had deposited with the French Academy of Sciences in order to establish his priority, were revealed. The material contained correspondence from 1877 with the chemist Marc Delafontaine (1837–1911), who was also seeking new elements from samarskite, and Smith's claim for the new element was published [Smith, 1878]. According to Smith, experiments of Jacques-Louis Soret (1827–1890) and "conclusions" of Jean-Charles Galissard de Marignac (1817–1894) made it "evident" that the element recently proposed by Delafontaine [1878a] was identical to the one that Smith had discovered. Marignac [1878] and Delafontaine [1878b], both of whom were attempting to establish their own priorities with new elements, quickly disputed Smith's claim and argued that mosandrum was nothing but terbium, a rare earth that had been known for many years. The lively exchange among these three men did not completely settle the matter though mosandrum was not accepted as a new element.

The controversy over mosandrum was definitively settled several years after the death of J. Lawrence Smith in 1883. The French chemist Paul-Émile Lecoq de Boisbaudran (1838–1912) obtained a sample of Smith's own samarskite preparation from Dr Joseph Benson Marvin (1852–1913), a friend and colleague of Smith at the University of Louisville. Using spectroscopic techniques, Lecoq de Boisbaudran [1886] demonstrated that Smith's "impure mosandrum" sample contained at least four rare earths: samarium and gadolinium, as well as didymium and terbium. The latter two rare earths had been known since the 1840s, but samarium and gadolinium were not identified until 1879 [Lecoq de Boisbaudran, 1879a, 1879b] and 1880 [Marignac, 1880], respectively. Thus, samarium and gadolinium were not known at the time that Smith proposed the element mosandrum, but they were definitely present in his samarskite preparation, just as they had been in Ellen Swallow's "insoluble residue."

***

Swallow's 1875 publication on samarskite was the first to note the presence of an insoluble residue, which suggested to her the possibility of unknown elements in the mineral. The further analysis of samarskite, as well as of many other minerals, fell to chemists such as Smith, Marignac, and Delafontaine, each of them competing for the priority of discovery of new elements. Swallow's seminal work on samarskite pointed the way for other investigators to discover those unknown elements that she thought might be there. According to her husband Robert Richards, "[s]he came near to being one of those immortals who have identified new elements in the earth's crust" [Richards, 1936, p. 160].

Meanwhile, there were numerous other projects at MIT to which Ellen Swallow Richards could apply her knowledge, talents, and abilities. Moving beyond science for its own sake, she devoted herself to broader issues with social implications. For example, she performed most of the analytical work in a wide-ranging study of industrial and domestic pollution in rivers and lakes for the state of Massachusetts by Professor Nichols [1874]; she headed the science section of the Society to Encourage Studies at Home, intended primarily for the self-education of women outside academia [Hunt, 1912, pp. 155–157]; she was instrumental in obtaining funding for the Women's Laboratory, in which she taught and which provided scientific training to about 500 women during the years 1876–1883 [MIT Archives, 1999] — mostly as an unpaid associate. Ellen Richards also investigated the purity of foods sold in stores to the public [Richards, 1880]. Her strong belief in science for the common good led her far beyond the chemical analysis of minerals to significant and far-reaching work as an advocate for women's education and as an activist for clean air, water, and food during the Progressive Era in the U.S. [Richardson, 2002].

References

Allen, O. D. (1877). Chemical Constitution of Hatchettolite and Samarskite from Mitchell County, North Carolina, *Am. J. Sci.*, 14, 3rd ser., pp. 128–131.

Atwood, S. (2005). A Haven for Women: One Alumna's Legacy, *MIT Technol. Rev.* (https://www.technologyreview.com/s/404619/a-haven-for-women-one-alumnas-legacy/, last accessed 15 January 2019).

Baskerville, C. (1904). The Elements: Verified and Unverified, *Chem. News*, 89, pp. 109–110, 121–123, 135–137, 150–151, 162–163, 170–171, 186–187, 194–195, 210.

Bouvé, T. T. (1875). [Introductory remarks], *Proc. Boston Soc. Nat. Hist.*, 17, p. 424.

Cheney, M. S. and Richards, E. S. (1877). A New and Ready Method for the Estimation of Nickel in Pyrrhotites and Mattes, *Am. J. Sci.*, 14, 3rd ser., pp. 178–181.

Clarke, R. (1973). *Ellen Swallow: The Woman Who Founded Ecology* (Follett Publishing, Chicago, IL).

Cross, C. R. (1919). Biographical Memoir of James Mason Crafts 1839–1917, *Biogr. Mem. Nat. Acad. Sci. USA*, 9, pp. 159–177 (http://www.nasonline.org/publications/biographical-memoirs/memoir-pdfs/crafts-james.pdf, last accessed 15 January 2019).

Delafontaine, M. (1878a). Sur un nouveau métal, le philippium, *C. R. Hebd. Seances Acad. Sci.*, 87, pp. 559–561.

Delafontaine, M. (1878b). Sur le Mosandrum, de M. Lawrence Smith, *C.R. Hebd. Seances Acad. Sci.*, 87, pp. 600–602.

Durant, E. (2006). A Lab of their Own: The Women's Laboratory Was an Experiment, *MIT Technol. Rev.* (https://www.technologyreview.com/s/405825/a-lab-of-their-own/, last accessed 15 January 2019).

Fontani, M., Costa, M. and Orna, M. V. (2015). *The Lost Elements: The Periodic Table's Shadow Side* (Oxford University Press, New York, NY).

Hermann, R. (1869). Fortgesetzte Untersuchungen über die Zusammensetzung des Samarskits, sowie Bemerkungen über die chemische Constitution der Verbindungen der Niobmetalle, *J. prakt. Chem.*, 107, pp. 139–159.

Hunt, C. L. (1912). *The Life of Ellen H. Richards* (Whitcomb & Barrows, Boston, MA).

Hunt, T. S. (1852). Examination of Some American Minerals, *Am. J. Sci.*, 14, 2nd ser., pp. 340–346.

Lecoq de Boisbaudran, P. E. (1879a). Nouvelles raies spectrales observées dans des substances extraites de la samarskite, *C. R. Hebd. Seances Acad. Sci.*, 88, pp. 322–324.

Lecoq de Boisbaudran, P. E. (1879b). Recherches sur le samarium, radical d'une terre nouvelle extraite de la samarskite, *C. R. Hebd. Seances Acad. Sci.*, 89, pp. 212–214.

Lecoq de Boisbaudran, P. E. (1886). Sur la mosandrine de Lawrence Smith, *C. R. Hebd. Seances Acad. Sci.*, 102, pp. 647–648.

Marignac, C. (1878). Observation sur la découverte, annoncée par M. L. Smith, d'une nouvelle terre appartenant au groupe du cérium, *C. R. Hebd. Seances Acad. Sci.*, 87, pp. 281–283.

Marignac, C. (1880). Sur les terres de la samarskite, *Arch. Sci. Phys. Nat.*, 3, 3rd ser., pp. 413–438.

MIT Archives. (1999). Women's Laboratory, 1876–1883 (https://libraries.mit. edu/archives/exhibits/esr/esr-womenslab.html, last accessed 15 January 2019).

Musil, R. K. (2014). Don't Harm the People: Ellen Swallow Richards, Dr. Alice Hamilton, and their Heirs take on Polluting Industries, *in Rachel Carson and Her Sisters: Extraordinary Women Who Have Shaped America's Environment* (Rutgers University Press, New Brunswick, NJ), pp. 53–88.

Nichols, W. R. (1874). On the Present Condition of Certain Rivers of Massachusetts, *Fifth Annual Report of the State Board of Health of Massachusetts* (Wright & Potter, Boston, MA), pp. 61–152.

Richards, E. H. (1880). The Adulterations of Some Staple Groceries, in *First Annual Report of the State Board of Health, Lunacy, and Charity of Massachusetts, 1879* (Rand, Aberg, Boston, MA), pp. 55–65.

Richards, R. H. (1936). *Robert Hallowell Richards: His Mark* (Little, Brown, Boston, MA).

Richardson, B. (2002). Ellen Swallow Richards: "Humanistic Oekologist," "Applied Sociologist," and the Founding of Sociology, *Am. Sociol.*, 33, pp. 21–57.

Roscoe, H. E. (1869). Researches on Vanadium – Part II, *Phil. Trans. Roy. Soc.*, 159, pp. 679–692.

Schneiderman, J. S. (2018). Beginning with Mineralogy: Ellen Swallow Richards and Earth System Science, in *Women and Geology: Who Are We, Where Have We Come From, and Where Are We Going?* Geological Society of America Memoir 214, ed. Johnson, B. A. (The Geological Society of America, Boulder, CO), pp. 1–9.

Smith, J. L. (1877a). Examination of American Minerals. No. 6 – Description of Columbic Acid Minerals from New Localities in the United States, Embracing a Reclamation for the Restoration of the Name Columbium to the Element Now Called Niobium. Description and Analyses of Columbite, Samarskite, Euxenite, and Fergusonite, and the New Species Hatchettolite, and Rogersite, *Am. J. Sci.*, 13, 3rd ser., pp. 359–369.

Smith, J. L. (1877b). On the Earthy Oxides of Samarskite, *Proc. Acad. Nat. Sci. Philadelphia*, 29, p. 194.

Smith, J. L. (1878). Le Mosandrum: un nouvel élément, *C. R. Hebd. Seances Acad. Sci.*, 87, pp. 148–151.

Swallow, E. H. (1873a). Notes on Some Sulpharsenites and Sulphantimonites from Colorado, B.S. thesis, MIT.

Swallow, E. H. (1873b). Notes on the Estimation of Vanadium in an Iron Ore from Cold Spring, N.Y., M.A. thesis, Vassar College.

Swallow, E. H. (1875a). Analysis of Samarskite from a New Locality, *Proc. Boston Soc. Nat. Hist.*, 17, pp. 424–428.

Swallow, E. H. (1875b). Notes on the Chemical Composition of Some of the Mineral Species Accompanying the Lead Ore of Newburyport, *Proc. Boston Soc. Nat. Hist.*, 17, pp. 462–465.

Swallow, P. C. (2014). *The Remarkable Life and Career of Ellen Swallow Richards: Pioneer in Science and Technology* (John Wiley & Sons, Hoboken, NJ).

Select Bibliography of Works by Ellen Swallow Richards[6]

Richards, E. H. (1882). *The Chemistry of Cooking and Cleaning: A Manual for Housekeepers* (Estes & Lauriat, Boston, MA).

Richards, E. H. (1884). *First Lessons on Minerals* (Ginn, Heath, Boston, MA).

Richards, E. H. (1886). *Food Materials and Their Adulterations* (Estes and Lauriat, Boston, MA).

Richards, E. H. (1887). *Home Sanitation: A Manual for Housekeepers* (Ticknor, Boston, MA).

Richards, E. H. (1889). *Domestic Economy as a Factor in Public Education* (New York College for the Training of Teachers, New York, NY).

Richards, E. H. (1899). *The Cost of Living as Modified by Sanitary Science* (Wiley, New York, NY).

Richards, E. H. (1899). *Plain Words About Food*, Rumford Kitchen Leaflet (Rockwell and Churchill, Boston, MA).

Richards, E. H. (1901). *The Cost of Food: A Study in Dietaries* (Wiley, New York, NY).

Richards, E. H. (1904). *The Art of Right Living* (Whitcomb & Barrows, Boston, MA).

Richards, E. H. (1904). *First Lessons in Food and Diet* (Whitcomb & Barrows, Boston, MA).

Richards, E. H. (1905). *The Cost of Shelter* (Wiley, New York, NY).

Richards, E. H. (1907). *Sanitation in Daily Life* (Whitcomb & Barrows, Boston, MA).

Richards, E. H. (1908). *The Cost of Cleanness* (Wiley, New York, NY).

Richards, E. H. (1908). *Laboratory Notes on Industrial Water Analysis: A Survey Course for Engineers* (Wiley, New York, NY).

Richards, E. H. (1910). *Euthenics, the Science of Controllable Environment: A Plea for Better Living Conditions as a First Step Toward Higher Human Efficiency* (Whitcomb & Barrows, Boston, MA).

---

[6]A listing of many other publications by and about Ellen Richards may be found at https://libraries.mit.edu/archives/exhibits/esr/esr-bibliography.html (last accessed 15 January 2019).

Richards, E. H. (1911). *Conservation by Sanitation; Air and Water Supply; Disposal of Waste (Including a Laboratory Guide for Sanitary Engineers)* (Wiley, New York, NY).

Richards, E. H. and Talbot, M. (1894). *Food as a Factor in Student Life: A Contribution to the Study of Student Diet* (University of Chicago Press, Chicago, IL).

Richards, E. H. and Woodman, A. G. (1900). *Air, Water, and Food from a Sanitary Standpoint* (Wiley, New York, NY).

Fluorine and the Research of Vicenta Arnal, María Del Carmen Brugger, and Trinidad Salinas

Ignacio Suay–Matallana

Instituto Interuniversitario López Piñero-Universidad Miguel Hernández

Fluorine is a very pale yellow-green gas, the lightest halogen element, with atomic number 9. It is the most electronegative element of all, reacting with almost all other elements, including some noble gases. It was not isolated until late nineteenth century, though its compounds have been known for several centuries.

Because of its toxicity and reactivity (fluorine immediately reacts with water, forming hydrogen fluoride), it is known as "the chemists' killer element." It has caused serious injuries, blinded or killed many scientists throughout its history. It caused the respiratory problems suffered by Joseph Louis Gay–Lussac (1778–1850) and Louis Jacques Thénard (1777–1857), Humphry Davy's (1778–1829) laboratory accident, the explosion experienced by the British chemist George Gore (1826–1908), and the deaths of the Belgian chemist Paulin Louyet (1818–1850) and the French chemist Jerome Nicklès (1820–1869) [McDowell, 2017]. Fluorine also causes some environmental problems due to the expansion of the aluminium industry and the intense use of fertilizers. Its use in some countries' drinking water supply is also controversial.

Traditional histories of science consider the French chemist and pharmacist Henri Moissan (1852–1907) as the scientific hero who discovered and isolated fluorine in 1886, and, in turn, was awarded with the Nobel Prize in Chemistry in 1906. Moissan's success of isolating fluorine consisted of the electrolysis of potassium bifluoride (KHF_2) dissolved in liquid hydrogen fluorine. Thanks to the use of platinum, a very inert metal, and the application of low temperatures to reduce fluorine's activity, this was possible. Later work was done by Otto Ruff (1871–1939), who became the leading author on fluorine during the early decades of the twentieth century, and was awarded the German Liebig Medal for his studies of many fluorides of metals and non–metals [Partington, 1989, 355].

Despite such a variety of scientists from different periods and countries, contributions by women to the study, analysis, and application of fluorine are not generally mentioned in many traditional historiographies. The following sections consider work in fluorine chemistry by Spanish women scientists during the 1920s and the 1930s.

Collaboration and Exchanges: A Research Group on Analyses of Water and Fluorine

The chemist, director of the Spanish customs laboratory, and professor of chemical analyses at the Faculty of Pharmacy at Madrid José Casares Gil (1866–1961) was an expert on chemical analyses. He initiated research on fluorine determination in water because he had found high concentrations of fluorine in some mineral waters of the North of Spain. In 1902 he visited the laboratory of Henri Moissan in Paris to test different gravimetric, volumetric and gasometrical determinations of fluorine [Suay-Matallana, 2017]. During the 1920s and 1930s his laboratory was open to many students, some of them women, who researched fluorine chemistry and other chemical analyses. Fourteen women obtained government fellowships to study and research chemistry with Casares [Magallón, 1997]. These practical courses had two main goals: to offer practical training for primary and secondary school teachers interested in didactics, and to improve the laboratory skills of science graduates before they embarked on research abroad.

One of the women scientists travelling abroad at the time was Jenara Vicenta Arnal Yarza (1902–1960), who was born into a family of farmers, and initially worked as primary school teacher. A few years later, in 1931, she became the first woman to receive a PhD in chemistry in Spain [Delgado *et al.*, 2004]. From 1930 onwards, Arnal was also the second woman working as a director of a department of physics and chemistry in a Spanish secondary school.[1] Because of her scientific degree, she did not require additional experimental training at the Casares' lab prior to her travel of learning. Arnal specialised in electrochemistry, and in 1930, Casares approved her application to go to the University of Basel for study with Friedrich Fichter (1869–1952), professor of inorganic chemistry, and vice-president of the International Union of Chemistry [N. N., 1929]. Fichter and Arnal worked on the chemical oxidation of metals, but they also prepared fluorine via the electrolysis of molten potassium bifluoride [Junta, 1933]. As a result of their collaboration Fichter and Arnal published an article together in the prestigious Swiss journal *Helvetica chimica acta* [Fichter and Arnal, 1931]. Other (male) Spanish chemists, like León Le Boucher Villén (1904–1937) and Enrique Moles Ormella (1883–1953) also travelled abroad and published papers on fluorine analyses [Moles, 1931; Ruff and Le Boucher, 1934]. A group of scientific women at the Casares' lab was also very active in fluorine analyses.

Women in the Lab: A Research Strand on Fluorine in Organic Products

During the 1920s and 1930s a group of young female pharmacy graduates specialised in chemical analyses, and some of them focussed on fluorine chemistry. This was the case for María del Carmen Brugger Romaní (b. 1899), and Trinidad Salinas Ferrer (d. 1965). Both studied pharmacy, trained in analytical chemistry with José Casares, specialised in fluorine,

---

[1]The first female director of a department of physics and chemistry in a Spanish secondary school was the chemist Ángela García de la Puerta (1903–1992, from 1928) [Núñez *et al.*, 2012].

and were among the few women completing a PhD in the Faculty of Pharmacy at Madrid in the first third of the twentieth century, which was, until the 1950s, the only university in Spain offering doctoral degrees.

María del Carmen Brugger was born into a bourgeois family of Valencia. Her father, Arturo Brugger Gascón was a businessman and member of the city council. She and her sister first studied at the University of Barcelona, obtaining top marks. The local press of Barcelona published their names highlighting they were two of the seven students finishing with honours, and the only women to whom degrees were awarded that year [N. N., 1925]. In 1924 the Brugger sisters moved to Madrid to finish their pharmacy degrees at the University of Madrid in 1925. Then, María del Carmen Brugger studied techniques in physics (a course taught by José Casares), as well as mineralogy [N. N., 1924]. Both sisters continued their studies further and prepared for a PhD in pharmacy (Figure 1).[2] María del Carmen Brugger's PhD dissertation (published in 1927) was supervised by Casares, and investigated the analytical chemistry of fluorine [Brugger, 1927].

Trinidad Salinas initially trained as a primary school teacher and obtained excellent marks [N. N., 1918]. Later she moved to Madrid, where she graduated with honours in pharmacy in 1926 [N. N., 1926a]. She was also a student of Casares, who (as president of the Professional Association of Pharmacists) helped her to obtain the prestigious Carracido award [N. N., 1926b, p. 11]. Salinas' PhD dissertation (published in 1934) was also supervised by Casares, and focussed on quantitative analyses of fluorine, and their application to different animal products [Salinas, 1934]. Of the female fellows, Salinas stayed at the Casares' laboratory for the longest period, as they were working together between 1928 and 1932. In fact, Salinas and Casares prepared an article on the quantitative determination of fluorine in organic substances that was published in the journal of the Spanish Academy of Science [Casares and Salinas, 1935]. That article was quoted in different international journals because of the uncommon limits of fluorine detected in a tooth sample [Roholm, 1937]. This was also relevant because it was the only one of the nineteen articles prepared by Casares which he co-authored with a woman.

---

[2] Her sister, María de los Desamparados, also obtained a PhD in pharmacy in 1927.

MEMORIA

DE LOS

TRABAJOS EFECTUADOS PARA ASPIRAR AL GRADO DE DOCTOR

POR

MARÍA DEL CARMEN BRUGGER Y ROMANÍ

MADRID, 17 NOVIEMBRE DE 1926

VALENCIA: 1927
ARTES GRÁFICAS - J. GAMÓN - TROYA, NÚM. 7
TELÉFONO 342

Figure 1. Title of Brugger's PhD dissertation (Valencia, 1927; reproduced by kind permission of the Universidad Complutense de Madrid).

Opitz *et al.* [2016] have studied the scientific collaboration between men and women, considering domestic sites as the place of knowledge production or focussing on scientific families. The scientific collaboration in families and between couples (like husband and wife or parents and sons) has also been shown to have been a productive historiographical issue [Lykknes *et al.*, 2012]. In this case, Casares was open to work with some talented students, and created a school of research on water and food analyses with a special focus on fluoride [Suay-Matallana, 2014]. The collaboration between Casares and his students (most of them women) was mutually fruitful. The annual reports published by the Spanish Board for Advanced Studies in that period showed the progress made by Brugger and Salinas in their fluorine research, and highlighted their analytical work on fluorine in animals and plants (Figure 2). Brugger's research compared the different techniques and instruments employed in fluorine chemistry as well as their results. She showed that calcium chloride and

Figure 2. A group of women scientists at the lab of chemical analyses of the Faculty of Pharmacy at Madrid (Madrid, 1934; reproduced by kind permission of the Agencia EFE).

lead dichloride were better reagents than barium chloride for qualitative analyses of fluorine [Brugger, 1927]. Salinas's work stressed the problem of using quartz and silica instead of special glass, and developed a more precise experimental technique to determine fluoride. According to Salinas, she applied her technique to different samples of fossils, and bones of birds, fishes, mammals and humans, obtaining "more accurate results than previous studies" [Salinas, 1934, p. 40]. The original research done by both women scientists was the beginning of research that allowed Casares to continue publishing articles on such issues, either alone or in collaboration with other students.

War, Gender, and the Impact on Fluorine Research in Spain

The Spanish War (1936–1939) and Franco's dictatorship had an immense impact on Spanish science, like the death or the imprisonment of some of the chemists mentioned in this paper, including Le Boucher and Moles. The effect of the dictatorship on most Spanish women working on fluorine chemistry was less dramatic, but nevertheless impactful. As other historians have pointed out, the main role for women during the early decades of the dictatorship was as mothers performing household duties. In countries like Nazi Germany laws even forbade women to work once they were married [Van Tiggelen and Lykknes, 2012]. Similarly, during the early decades of Franco's regime, only very few unmarried women, with strong social connections and family support, were allowed to take part in scientific research.[3]

Casares' attitude towards the role of scientific women dramatically changed after the war. He was reinstated as the dean and director of the customs laboratory, became the president of the scientific sections of the new Spanish Research Council (CSIC), a member of the Francoists Courts, and the president of the Royal Academies of Pharmacy and Sciences. But he did not use his relevant political influence to facilitate the scientific careers of women after the war, and avoided any conflict

---

[3]This was the case for the physicists María Teresa Vigón, María Egües and Olga García Riquelme, who were supported by José María Otero de Navascués, a nobleman, physicist, army officer, and president of the Nuclear Energy Board [Romero, 2016].

with the Francoist government to secure his institutional positions. Recent scientific biographies have pointed out the serious discussions between Casares and some women interested in working at the CSIC laboratories. Only very few women (always unmarried) were allowed, like Piedad de la Cierva Viudes (1913–2007), who specialised in radio-chemistry, because "she was already working before the war, and he [Casares] was friend of her family" [Romero, 2017, p. 331].[4] Another woman affected by Casares' repressive attitude was the chemist María Josefa Molera Mayo (1921–2011), who specialised in chemical kinetics, and had to postpone her doctoral research until the arrival of new directors at the CSIC [Ulises, 2013].

María del Carmen Brugger never married, neither did Trinidad Salinas or Vicenta Arnal. Despite their marital status, which was allegedly an important prerequisite for scientific women after the Spanish war, they did not continue their scientific careers during the dictatorship. The three women became fully integrated into the Spanish and international fluorine chemistry during the 1920s and 1930s, dedicating a lot of time to studying and building their careers, and working hard to obtain first rank research results. This included scientific articles in co-authorship with Nobel Laureates like the article by Fichter and Arnal, the article published by Salinas and Casares, and the PhD dissertations by Salinas and Brugger quoted in different international scientific publications, like treatises on toxicology and biology, and scientific journals. But later, they abandoned their academic activities, and even disappeared from the public sphere. In fact, only very little is known about Brugger and Salinas after 1939. Perhaps, Brugger's and Salinas' families were directly affected by the war, or the two women may not have had sufficient social and political contacts to justify a continuation of their research. Trinidad Salinas continued to live in Madrid until her death in 1965, and María del Carmen Brugger moved back to Valencia, where she co-owned, with her sister, a pharmacy in one of its wealthiest streets [N. N., 1965]. The case of Vicenta Arnal is

---

[4]Another exception was the pharmacist Sara Borrell Ruiz (1917–1999), who specialised in biochemistry. She was born into a wealthy family of pharmacists, never married, and was allowed to continue her scientific career because of her friendship with the family of Román Casares López [Santesmases, 2017].

different, because she left research behind and did not publish on chemistry after the war, but continued her professional life as secondary school teacher.

The war and the dictatorship also affected the scientific outcomes from Casares' group. In the following years he only published one more article on the detection of fluorine in ashes with one of his new students [Casares and Moreno, 1944]. In the 1940s José Casares' scientific authority and most of his institutional positions were transferred to Román Casares López (1908–1990), his grand-nephew, former student, and successor as chair of chemical analysis in the Faculty of Pharmacy. In 1954 Román Casares founded and directed the Bromatology School of the University of Madrid. This institution, which focussed on food science, had a strong programme in water and food analysis. This research programme had been started by Arnal, Brugger and Salinas, but because of the political events, their work was disrupted, their names rendered invisible, and their research appropriated by other (male) students of Casares. In any case, this case study shows the relevance of scientific collaboration, which was essential to the preparation of original work with international impact, to the teaching of new generations of students, and to the foundation of new directions in research, in this case, on the analysis of fluorine in Spain. Arnal's, Brugger's and Salinas' work opened new avenues of research on applied analyses — not just of water, but also of many different organic products —, which was of benefit to other Spanish scientists in the following decades.

References

Brugger Romaní, M. C. (1927). *Contribución al estudio de la química analítica del flúor* (Artes Gráf, J. Gamón, Valencia).

Casares Gil, J. and Salinas Ferrer, T. (1935). Sobre la determinación cuantitativa del flúor y su aplicación a algunos productos naturales, *Revista de la Academia de Ciencias de Madrid*, 32, pp. 88–119.

Casares Gil, J. and Moreno Martín, F. (1944). Investigación de flúor en cenizas vegetales, *Anales de Física y Química*, 11, pp. 685–692.

Delgado Martínez, Mª. Á. and López Martínez, J. D. (2004). De analfabetas científicas a catedráticas de física y química de instituto en España: El esfuerzo de un grupo de mujeres para alcanzar un reconocimiento profesional y científico, *Revista de educación*, 333, pp. 255–268.

Fichter, F. and Arnal, V. E. (1931). Oxydationen mit Fluor XVIII: Einwirkung von Fluor auf Cer(III)sulfat und auf Jodate, *Helvetica chimica acta*, 14:4, pp. 862–867.

Junta para Ampliación de Estudios e Investigaciones Científicas. (1933). *Memoria correspondiente a los cursos 1931 y 1932* (Junta para Ampliación de Estudios e Investigaciones Científicas, Madrid).

Lykknes, A, Opitz, D. L. and Van Tiggelen, B, eds. (2012). *For Better or For Worse? Collaborative Couples in the Sciences* (Springer Birkhäuser, Heidelberg).

McDowell, L. R. (2017). *Mineral Nutrition History: The Early Years* (First Edition Design Pub., Sarasota, FL).

Magallón, C. (1997). Mujeres en las ciencias físico–químicas en España: El Instituto Nacional de Ciencias y el Instituto Nacional de Física y Química (1910–1936), *Llull*, 20, pp. 529–574.

Moles, E. (1931). The Atomic Weight of Fluorine, *Nature*, 128, pp. 966–967.

N. N. (1918). Escuela Normal Central de Maestras: Premios de honor, *La Correspondencia de España,* 10 June 1918, p. 2.

N. N. (1924). Registro de Identidad Escolar, Archivo General de la Universidad Complutense de Madrid, ref: ES, AGUCM, EA–119.

N. N. (1925). De enseñanza nacional, *La Vanguardia*, 10 June 1925, p. 4.

N. N. (1926a). Apertura de curso, *La Libertad*, 19 February 1926, p. 3.

N. N. (1926b). Real Colegio de Farmacéuticos de Madrid, *La Farmacia española*, 01 June 1926, p.11.

N. N. (1929). Expediente de Jenara Vicenta Arnal Yarza, Archivo de la Junta para la Ampliación de Estudios, ref: JAE/11–531.

N. N. (1965). Esquelas. *ABC*, 8 December 1965, p. 64.

Núñez Valdés, J, Rodríguez Antón, B. and Rodríguez Remesal, E. (2012). *Primeras mujeres doctoras en Química en España, Investigación y género, inseparables en el presente y en el futuro* (Universidad de Sevilla, Sevilla).

Opitz, D. L, Bergwik, S. and Van Tiggelen, B., eds. (2016). *Domesticity in the Making of Modern Science* (Palgrave Macmillan, London).

Partington, J. R. (1989). *A Short History of Chemistry* (Dover Publications, New York, NY).

Roholm, K. (1937). Fluorine Intoxication: A Clinical-Hygienic Study. With a Review of the Literature and Some Experimental Investigations (H. K. Lewis & Co. Ltd, London; Nyt Nordisk Forlag, Copenhagen).

Romero De Pablos, A. (2016). Pioneras pero invisibles: Las calculistas del laboratorio y taller de investigación del Estado Mayor de la Armada, *Revista Clepsydra*, 15, pp. 49–61.

Romero De Pablos, A. (2017). Mujeres científicas en la dictadura de Franco: Trayectorias investigadoras de Piedad de la Cierva y María Aránzazu Vigón, *Arenal*, 24:2, pp. 319–348.

Ruff, O. and Le Boucher, L. (1934). Der Dampfdruck von ZnF_2, CdF_2, MgF_2, CaF_2, SrF_2, BaF_2, AlF_3, *Zeitschrift für anorganische und allgemeine Chemie*, 219, pp. 376–381.

Salinas Ferrer, T. (1934). *Estudios sobre la determinación cuantitativa del flúor y su aplicación a varios productos del reino animal* (Universidad Central, Madrid).

Santesmases, M. J. (2017). Género, afición y mérito: Una biografía de Sara Borrell Ruiz, *Arenal*, 24:2, pp. 287–318.

Suay-Matallana, I. (2014). La colaboración científica y los espacios de la química: Un estudio de caso español en la primera mitad del siglo XX, *Revista Española de documentación científica*, 37:4, pp. 1–11.

Suay-Matallana, I. (2017). Scientific Families, and the Shaping of an Expertise in Analytical Chemistry in Modern Spain, *HOST — Journal of History of Science and Technology,* 11, pp. 97–116.

Ulises Acuña, A. and Santiuste, J. Mª. (2013). María Josefa Molera: Cinética, fotoquímica y cromatografía en la España de 1940 a 1980, *Anales de química*, 109:1, pp. 31–33.

Van Tiggelen, B. and Lykknes, A. (2012). Ida and Walter Noddack through Better and Worse: An *Arbeitsgemeinschaft* in Chemistry, in *For Better or For Worse? Collaborative Couples in the Sciences*, eds. Lykknes, A., Opitz, D. L. and Van Tiggelen, B. (Springer Birkhäuser, Heidelberg), pp. 103–147.

Select Bibliography

Banks, R. E. (1986). *Fluorine: The First Hundred Years (1886–1986)* (Elsevier Sequoia, Lausanne and New York, NY).

Magallón-Portolés, C. (2017). Science from Women's Lives: Better Science? How Gendered Studies Improve Science and Lives, *Mètode Science Studies Journal: Annual Review*, 7, pp. 104–111.

Magallón-Portolés, C. (2011). Químicas españolas en la Edad de Plata, *Anales de la Real Sociedad Española de Química*, 1, pp. 94–101.

Suay-Matallana, I. (2014). *Análisis químico y expertos en la España contemporánea: Antonio Casares Rodríguez (1812–1888) y José Casares Gil (1866–1961)* (ProQuest UMI Dissertations Publishing, Ann Arbor, MI).

Wisniak, J. (2002). The History of Fluorine–From Discovery to Commodity, *Indian Journal of Chemical Technology*, 9, pp. 363–372.

Four Women Chemists Review the Elements

Ian D. Rae

School of Chemistry, University of Melbourne,
Victoria 3010, Australia

In 1912 the English chemist John Albert Newton Friend (1881–1966) conceived a multi-volume *Textbook of Inorganic Chemistry*, and to assist in its production he recruited writers — thirteen men and four women — from colleges, universities and industry, mainly in Birmingham and Leeds. Running to twenty-one volumes, the *Textbook* was published over the period 1914 to 1937. The physicist Dr William Garnett (1850–1932) wrote in the publisher's centenary volume that "it must have required some courage to have embarked upon the venture of publishing" the *Textbook* but he felt that it was of high quality, in particular because of the association of some writers with industry [Garnet, 1920, p. 92]. Some of the writers who prepared volumes for Friend were, indeed, employed in industry, while others came from technical colleges or universities in the industrial north of England, where an association with industry was more common than in universities like Oxford and Cambridge. In their coverage of the elements and their compounds, there is frequent reference to industrial production and uses, in addition to the generally more scholarly coverage of physical and chemical properties.

Other compilations such as Ure's *Dictionary of Arts, Manufactures and Mines* [Ure, 1839] and Thorpe's *Dictionary of Applied Chemistry* [Thorpe, 1890–1893] were organised around applications rather than the elements themselves. They were published in successive editions for almost a century, as applications grew and new authors were engaged. Closer to the style of Friend's *Textbook* were a number of German language compilations, notably Gmelin's *Handbuch der Chemie* that evolved into the *Handbuch der anorganischen Chemie*, which ran to eight printed editions and was considered to be "unrivalled" [Gmelin, 1872–1881; Sutton, 2017, p. 4]. Gmelin's *Handbuch* did not appear in the English language until the 1980s.

Friend was a graduate of the University of Birmingham (Bachelor of Science 1902, Master of Science 1903) who received his PhD from the University of Würzburg in 1908 [Kauffman, 1972]. After several short-term appointments he made his career (1920–1946) at the Birmingham College of Advanced Technology. Always referred to as Newton Friend, his research interests ranged from valence theory through metallic corrosion to the history of science and witchcraft. His scrupulous editorship ensured that the *Textbook* covered a very broad range of chemistry, with information drawn from British, American, French, German and Italian journals. It was comprehensive, and the references were up to date.

Structure and Authors

Friend noted in his general introduction to the series in the first volume that while the need for "text-books dealing more or less comprehensively with each branch of chemistry had largely been met for the fields of physical and analytical chemistry," there was a need for coverage of inorganic chemistry that the present series of volumes was designed to meet [Friend, 1916, p. vii]. By dividing the subject matter along the lines of the periodic arrangement of the elements, he envisaged revision or rewriting of particular volumes as new knowledge accumulated, but apart from volume I, of which

A TEXT-BOOK OF
INORGANIC CHEMISTRY.

EDITED BY

J. NEWTON FRIEND, D.Sc., Ph.D., F.I.C.,
CARNEGIE GOLD MEDALLIST.

VOLUME III. PART I.
THE ALKALINE EARTH METALS.

BY

MAY SYBIL BURR (née Leslie),
D.Sc.(Leeds).

With Frontispiece and Illustrations.

LONDON:

CHARLES GRIFFIN & COMPANY, LIMITED,

EXETER STREET, STRAND, W.C. 2.

1925.

Figure 1. Title page from an early volume in J. Newton Friend's *Textbook of Inorganic Chemistry* (public domain).

three editions were published, this did not occur. In thanking his authors, Newton Friend also seems to have taken the opportunity to respond to some resistance to his editorial authority, writing:

> In order that the series shall attain the maximum utility, it is necessary to arrange for a certain amount of uniformity throughout, and this involves the suppression of the personality of the individual author to a corresponding extent for the sake of the common welfare [Friend, 1916, p. ix].[1]

In a comparative study of early responses to the periodic system in different countries several contributors commented that chemists first used Dmitri Mendeleev's (1834–1907) periodic table as a way of indicating where new elements could be searched for and identified, but by the early years of the twentieth century they had begun to use it, especially in textbooks, as an organising principle for the presentation of inorganic chemistry [Kaji *et al.*, 2015]. Newton Friend adopted this principle in his *Textbook*, with each volume based on the elements of a particular group that was identified in a subtitle by a Roman numeral or a name such as "alkaline earth metals" (Figure 1) [Burr, 1925]. Within each volume, the author(s) noted trends in the physical and chemical properties of the elements within the group and their compounds.

Women Authors, Their Co-Authors and Elements

Four women contributed to Friend's *Textbook*. One of them was Annie Roberts Russell (b. 1893), who completed a Bachelor of Science degree at the University of Glasgow in 1915 and studied chemistry and biochemistry at the University of Birmingham (1924–1926). She was the co-author of volume VII part II, on sulphur, selenium and tellurium [Vallance, Twiss and Russell, 1931], but little is known about her. Her co-authors were Douglas F. Twiss (1883–1951), chief scientist of the Dunlop Rubber company [Anon. b, 1951] and Reece Henry Vallance (1888–1942) [Friend, 1943], who was Newton Friend's colleague in the chemistry department of the Municipal Technical School, Birmingham. Because of the use of sulphur in the vulcanisation of rubber, Twiss was in a strong position to

---

[1]The general introduction was included in all volumes of the *Textbook*.

write about this element, but there is no evidence that his co-authors had any specific experience with any of the three elements.

Another female author was Dorothy Goddard née Webster (1894–?), who was a graduate of the University of Birmingham (Bachelor of Science Honours 1920, Master of Science 1921). In 1921, she married another Birmingham chemistry graduate, Archibald Edwin Goddard (1893–1970) (Bachelor of Science 1914, Master of Science 1920). Following his graduation, Archibald Goddard was junior or assistant lecturer at Birmingham from 1920 to 1924. Dorothy did not hold a university appointment but they published a number of research papers together [Goddard and Goddard, 1922a; 1922b]. Their research involved the chemistry of substances with thallium-carbon bonds, and that probably influenced Friend to invite them to write about organometallic compounds in volume XI part I [Goddard and Goddard, 1928]. Archibald went on to prepare three more volumes [Goddard 1930; 1936; 1937] on organometallic compounds for Newton Friend, thanking Dorothy for "encouragement" [Goddard, 1937, p. xii] and help with the index [Goddard, 1936, p. xiii], activities not inconsistent with the euphemistic "home duties" that replaced the professional careers of many married professionally qualified women like Dorothy Goddard [Abir-Am and Outram, 1987; Lykknes *et al.*, 2012].

May Sybil Burr

Better known, May Sybil Burr née Leslie (1887–1937) graduated from the University of Leeds in 1908 with First Class Honours [Dawson, 1938]. Supported by a scholarship, she worked with Marie Curie (1867–1934) in Paris, and then with Ernest Rutherford (1871–1937) in Manchester, studying the radioactive decay of thorium.[2] After wartime work as a chemist in the munitions industry, she returned to Leeds as a junior staff member in 1918, but she was sufficiently prominent for Newton Friend to invite her to prepare volume III part I in his monumental series [Burr, 1925], and to share the authorship of part II [Gregory and Burr, 1926]. The remaining alkaline earth elements, beryllium and magnesium, together with some

---

[2] See Rae on Leslie in this volume.

from group 2B (mercury, cadmium and zinc) were covered in volume III, part II of the *Textbook*, for which Burr was joined as her co-author by Joshua Craven Gregory (d. 1964), a lecturer in chemistry and her colleague at the University of Leeds [Gregory and Burr, 1926].

Burr's maiden name, Leslie, by which she was perhaps better known in scientific circles in the early 1920s, was included on the title pages of both volumes (Figure 1). Her volume covered the alkaline earth metals — calcium, strontium, barium and radium —, and in the preface she acknowledged "her husband, Mr A. H. Burr, for much valuable assistance in proof-correcting and indexing" [Burr, 1925, p. xii].

"The history of radium may, to a large extent, be regarded as the history of radioactivity," Burr wrote [Burr, 1925, p. 267]. The attention shown by chemists and physicists to the radioactivity of radium has sometimes caused chemists to overlook the fact that, on account of the electronic structure of its atom and its chemical properties — in other words, the periodic classification — radium is placed in group 2 along with the other elements that Burr wrote about: calcium, strontium and barium. Consequently, while others might well have been able to write about the more common-place alkaline earth elements, Burr's experience in Paris made her especially suited to being the author of this volume.

Margaret (Maggie) Millen Jeffs Sutherland

The fourth female contributor, Margaret (Maggie) Sutherland (1881–1972) (Figure 2) was an organic chemist who worked with Forsyth James Wilson (1880–1944) and Isidor (Ian) Morris Heilbron (1886–1959) at the Royal Technical College Glasgow, and continued to collaborate with Wilson after Heilbron moved to Liverpool in 1920 [Rayner-Canham and Rayner-Canham, 2008, pp. 292–293; Sutherland, 1945]. Sutherland took her Bachelor of Science degree at the University of Glasgow in 1908, and in 1914 was awarded a doctorate by that university. She was lecturer in chemistry at the Royal Technical College from 1913 to 1947, and the sole author of volume X of Newton Friend's *Textbook* [Sutherland, 1928].

The volume deals with the "metal ammines," the compounds in which ammonia molecules are bonded to metal ions in what came to be known as coordination compounds. This field of chemistry had only recently

Figure 2. Staff of Royal Technical College Glasgow. Rear: Sergeant, Anderson, Jno Agnew, Ian Heilbron. Front: Professor Wilson, Professor Henderson, Dr. Sutherland (Archives and Special Collections, University of Strathclyde Library, Glasgow).

been organised, following the work of Alfred Werner (1866–1919). Sutherland wrote that "inorganic chemistry has widened considerably in recent years, and it becomes more and more evident that methods of representation of the type used in organic chemistry must be employed" so as "to show graphically the union of the constituents of the complex" [Sutherland, 1928, p. 3]. This, of course, was the view of a practicing organic chemist, but the tide of structural depiction was already turning as she wrote. Sutherland showed how the ideas about the metal ammines had developed, culminating in the acceptance of Alfred Werner's (1866–1919) proposals from the 1890s of replacing August Kekulé's (1829–1896) one-directional valence bonds in organic compounds with a more flexible system attracting forces working in all directions. In her text, Sutherland depicted, for example, the trivalency of cobalt in the species $[Co(NH_3)_6]$, using dots where we would now write positive charges [Sutherland, 1928, p. 134]. She also provided examples of isomerism, a phenomenon that until then had mainly been encountered only in the field of organic chemistry. It was exemplified in a number of ammines with the same chemical formula but different structures and consequently different properties, such as the dichlorocobalt tetrammines.

The book was organised along the lines of the periodic classification, with coverage of ammines of groups 1 to 7 metals being preceded by a general and historical introduction. The final chapters covered the ammines of metals in the (then) three transition series on which most of work of Werner and other chemists had centred.

Textbook or Handbook?

A review of volume III part II and volume VII part III appeared in *Nature* in early 1927: "They conform so closely to the pattern already established by earlier volumes of the series," the anonymous reviewer observed, "that very little special comment is called for," but the review went on to criticise the title given to the publication.

> It may, however, be said that the type to which the successive volumes conform with increasing definiteness is that of a work of reference, which contains an array of facts which makes it impossible for the casual reader to discover any sections which can be read consecutively with any degree of enjoyment. For this reason, indeed, it appears likely that many of those who appreciate at their full value the merits of the 'Text-book' will put the new volumes in their appropriate places on the shelf without attempting to "read" them, but with the definite object of referring to them when a suitable opportunity arises [N. N., 1927, p. 16].

To be fair to Newton Friend, the series is supposed to be a repository of information that can be consulted when need arises, rather than offering the reader a succinct account of advances in the various fields of knowledge, in which case it would probably be better called a Handbook. However, the reviewer then launched a direct attack on something that the general editor had specifically made a point of principle.

> It may perhaps be regretted that the general editor of the series has discouraged his contributors from 'letting themselves go', since the whole work might have been transformed by the introduction of a number of chapters in which the individuality of the authors was allowed to stand

out, especially when dealing with the more exhilarating options [N. N., 1927, p. 16].

A less critical reviewer wrote that "the history of the determination of the atomic weight of beryllium is vividly and very interestingly portrayed" and that "many portions of the book show, what is often absent in a compilation such as this, real literary ability" [Lucas, 1927, p. 535] but it is impossible to know whether Gregory or Burr should take credit for this. Coming from a professional group in industry whom Newton Friend's authors had made some effort to cater for, the reviewer also wrote that "the industrial chemist would appreciate the book more if it contained more manufacturing details, more extensive notes on alloys and uses of the compounds" but noted that there was an excellent reference list to assist readers who wanted such information [Lucas, 1927, p. 535].

***

When Newton Friend began his great project for publishers Charles Griffin & Co. in 1912 there was nothing quite like it in the market. Earlier compilations had become outdated in the face of the enormous growth of chemistry in the late nineteenth and early twentieth century. It is true that a competitor soon appeared on the market, in the form of J. W. Mellor (1869–1938) with his *Comprehensive Treatise on Theoretical and Inorganic Chemistry*. The first volume was published in 1922 by Longmans, Green and Co. [Mellor, 1922], and the sixteenth in 1937, shortly before Mellor's death. Unlike Newton Friend, who recruited a small army of authors and co-authors, Mellor was the sole author. His biographer wrote of "twenty years of courageous effort" [Green, 1939, p. 576] and said that the completion of the work was "a source of complete bewilderment ... [because] ... he wrote every word and every reference" [Green, 1939, p. 576]. This also demonstrates the need for such reference books in chemistry, and to some extent answers the criticisms Friend was subjected to.

Newton Friend's approach to his task gave opportunities to four women chemists with a flair for writing, judgement about their subject and the discipline to complete an arduous task. They had their reward in red-bound volumes appearing on library shelves all around the world. What their competitor, Mellor, wrote in the flowery dedication of his *Treatise* can be applied to them: "To the privates in the great army of workers in chemistry. Their names have been forgotten. Their work remains" [Mellor, 1922, p. v]. Luckily, Newton Friend's *Textbook* provided an opportunity for four women chemists not to be forgotten.

References

Abir-Am, P. G., and Outram, D., eds. (1987). *Uneasy Careers and Intimate Lives: Women in Science, 1787–1979* (Rutgers University Press, New Brunswick, NJ).

Burr, M. S. (1925). *A Text-Book of Inorganic Chemistry*, ed. Friend, J. N., vol. 3, part 1: *The Alkali Metals and their Congeners* (Charles Griffin, London).

Dawson, H. M. (1938). May Sybil Burr (1887–1937), *J. Chem. Soc.*, pp. 151–152.

Friend, J. N. (1916). *A Text-Book of Inorganic Chemistry*, vol. 1 (Charles Griffin, London).

Friend, J. N. (1943). Reece Henry Vallance 1888–1942, *J. Chem. Soc.*, xx, p. 44.

Garnett, W. (1920). Griffin's Publications (1820–1920) in Pure Science, in *The Centenary Volume of Charles Griffin and Company Ltd* (Charles Griffin, London), pp. 85–98.

Gmelin, L. (1872–1881). *Handbuch der anorganischen Chemie* (Carl Winter's Universitätsbuchhandlung, Heidelberg).

Goddard, A. E. (1930). *A Text-Book of Inorganic Chemistry*, ed. Friend, J. N., vol. 11: *Organometallic Compounds,* part 2: *Derivatives of Arsenic* (Charles Griffin, London).

Goddard, A. E. (1936). *A Text-Book of Inorganic Chemistry*, ed. Friend, J. N., vol. 11: *Organometallic Compounds,* part 3: *Derivatives of Phosphorus, Antimony and Bismuth* (Charles Griffin, London).

Goddard, A. E. (1937). *A Text-Book of Inorganic Chemistry*, ed. Friend, J. N., vol. 11: *Organometallic Compounds,* part 4: *Derivatives of Selenium, Tellurium, Chromium and Platinum* (Charles Griffin, London).

Goddard, D. and Goddard, A. E. (1922a). Organo-Derivatives of Thallium. Part IV: Action of Thallium Chlorides on the Grignard Reagent and on Organo-Derivatives of Tin, Lead and Bismuth, *J. Chem. Soc.*, 121, pp. 256–261.

Goddard, A. E. and Goddard, D. (1922b). Organo-Derivatives of Thallium. Part V: The Preparation of Thallium Diaryl Salts, *J. Chem. Soc.*, 121, pp. 482–488.

Goddard, A. E. and Goddard, D. (1928). *A Text-Book of Inorganic Chemistry*, ed. Friend, J. N., vol. 11: *Organo-metallic Compounds*, part 1: *Derivatives of the Elements from Groups I to IV* (Charles Griffin, London).

Green, A. T. (1939). Joseph William Mellor. 1869–1938, *Obituary Notices of Fellows of the Royal Society*, 2:7, pp. 572–576.

Gregory, J. C. and Burr, M. S. (1926). *A Text-Book of Inorganic Chemistry*, ed. Friend, J. N., vol. 3, part 2: *Beryllium and its Congeners* (Charles Griffin, London).

Kaji, N., Kragh, H. and Palló, G., eds. (2015). *Early Responses to the Periodic System* (Oxford University Press, Oxford).

Kauffman, G. B. (1972). Friend, John Albert Newton, in *Dictionary of Scientific Biography*, vol. 5, ed. Gillespie, C. J. (Scribner, New York, NY), pp. 189–190.

Lucas, A. R. (1927). Review of A Text-Book of Inorganic Chemistry, vol III, part II, *J. Soc. Chem. Ind.*, 46: 23, pp. 534–535.

Lykknes, A., Opitz, D. L. and Van Tiggelen, B., eds. (2012). *For Better or For Worse? Collaborative Couples in the Sciences* (Springer Birkhäuser, Heidelberg, New York, Dordrecht, London).

Mellor, J. W. (1922). *A Comprehensive Treatise on Theoretical and Inorganic Chemistry*, vol. 1 (Longmans, Green and Co., London).

N. N. (1927). A Text-Book of Inorganic Chemistry. Edited by Dr J. Newton Friend. (Griffins Scientific Text-Books.) Vol. 3, Part 2: *Beryllium and its Congeners*. By Joshua C. Gregory and Dr May Sybil Burr née Leslie. Vol 7, Part 3: *Chromium and its Congeners*. By Reece H. Vallance and Arthur A. Eldridge (London: Charles Griffin and Co., Ltd, 1926), *Nature*, 119, Suppl., p. 16.

N. N. (1951). The late Dr D. F. Twiss, *The Engineer*, 191, p. 738.

Rayner-Canham, M. and Rayner-Canham, G. (2008). *Chemistry Was Their Life: Pioneer British Women Chemists, 1880–1949* (Imperial College Press, London).

Sutherland, M. M. J. (1928). *A Text-Book of Inorganic Chemistry*, ed. Friend, J. N., vol. 10: *The Metal-Ammines* (Charles Griffin, London).

Sutherland, M. M. J. (1945) Forsyth James Wilson 1880–1944, *J. Chem. Soc.*, pp. 723–725.

Sutton, M. (2017). The First Chemical Database, *ChemistryWorld*, 17 May 2017 (https://www.chemistryworld.com/features/200-years-of-gmelins-hand-book/3007265.article, last accessed 10 January 2019).

Thorpe, T. E. (1890–1893). *Dictionary of Applied Chemistry* (Longmans Green, London).

Ure, A. (1839). *A Dictionary of Arts, Manufactures and Mines* (Longman, London).

Vallance, R. H., Twiss, D. F. and Russell, A. R. (1931). *A Text-Book of Inorganic Chemistry*, ed. Friend, J. N., vol 7, part 2: *Sulphur, Selenium, and Tellurium* (Charles Griffin, London).

Select Bibliography

Andraos, J. (2000–2014). Contributions of Women to Named Things in Chemistry and Physics (http://www.chem.york.ca/NAMED, last accessed 21 September 2018).

Brush, S. G. (1996). The Reception of Mendeleev's Periodic Law in America and Britain, *Isis*, 87, pp. 595–628.

Rayner-Canham, M. and Rayner-Canham, G. (2008). *Chemistry was their Life: Pioneer British Women Chemists, 1880–1949* (Imperial College Press, London).

Part 3

New Fields and
Instrumental Methods

Clara Haber, née Immerwahr: In and Out of her Element

Bretislav Friedrich* and Dieter Hoffmann[†]

*Fritz Haber Institute of the Max Planck Society,
Faradayweg 4-6, D-14195 Berlin
[†]Max Planck Institute for the History of Science,
Boltzmannstraße 22, D-14195 Berlin

When in April 1901 the German Association for Electrochemistry convened in Freiburg for its annual meeting, there was, for the first time, a woman among the participants. This was Clara Immerwahr from Breslau, who had completed her PhD just a few months earlier. Immerwahr thus ranks among a small, yet significant group of women scientists who entered the exclusively male domain of scientific research at the turn of the twentieth century. Clara's research on the electrochemical properties of cadmium, copper, lead, mercury, and zinc in equilibrium with their salts in solution broadened the empirical basis for the notion of "electroaffinity," as coined by her PhD adviser, Richard Abegg (1869–1910). Electroaffinity (electronegativity in modern nomenclature) varies characteristically along the rows and columns of the periodic system, and complements the organizing principle encapsulated in it.

Biographical Context

Clara was born on the estate of Polkendorf near Breslau on 21 June 1870, where her father, a PhD chemist, had retreated to after the failure of his young chemical company [Szöllösi-Janze, 1998, pp. 124–131]. Apart from becoming a highly successful agronomist in Polkendorf and surroundings, he co-owned a prosperous specialty store that carried carpets and fine fabrics in Breslau, where the family also kept an apartment.

The Immerwahr family was part of Breslau's well-to-do Jewish middle class. Breslau's Jewish community was the third largest in Germany, after Berlin and Frankfurt/Main, and it was highly academically oriented, with its members belonging to the city's intellectual aristocracy. The Immerwahrs were a part of this elite. They were assimilated Jews participating in communal cultural life but would only rarely go to the synagogue, if at all. The political attitudes of the Immerwahrs were liberal, but entailed the cultivation and demonstration of a certain degree of Prussian-German national awareness and patriotism, not least after the German unification of 1871 [Clark, 2007]. Prussian was also the simple lifestyle of the family, which was frugal — not because of want but out of principle. So despite the family's wealth, Clara and her siblings were brought up in modesty [Johnson, 1998]. Apart from the virtues of simplicity, frugality and modesty, great value was attached to education — not just for the son and heir, but also for the three daughters. This was typical for the German Jewish middle class, and in particular for Breslau, as 40 per cent of the female students at the higher schools in Breslau were Jewish.

Clara's path to education — and the life that came along with it — was shaped by the constraints imposed on women at the time. She started her studies at a "Höhere Töchterschule" (or Women's College) in Breslau [Szöllösi-Janze, 1998, p. 124], which was supplemented with instruction provided by a private tutor at the Polkendorf estate during the summer months. Clara graduated from the Töchterschule at the age of 22. The school was supposed to provide a basic education for young women that was compatible with their social status and to prepare them for what was considered to be their "natural purpose," that is, as companions of their husbands, housewives, and mothers. Nevertheless, Clara was up for more, and after graduating from the Töchterschule she entered a teachers' seminary, which was the only type of institution available that offered women a higher professional education at the time [Szöllösi-Janze, 1998, p. 124]. However, the graduates of the teachers' seminary only qualified to teach at girls' schools and remained ineligible for a university education. So in order to qualify for the university, Clara had to take intensive private lessons and pass an exam equivalent to the "Abitur." This exam — that certified a certain degree of intellectual maturity — was administered by a special committee set up at a "Realgymnasium" in Breslau, and Clara passed it successfully at Easter 1896 [Szöllösi-Janze, 1998, p. 125], when she was 26 years old.

Subsequently, she began her studies at the University of Breslau, however as a guest auditor only, since in Prussia women would become legally admissible as university students only as late as 1908. Prior to this, starting in 1895, women were only allowed to attend lectures as guests, and even that was contingent upon the support of the professor and faculty and a permission from the ministry of education; the ministry required a certificate of good conduct ("Führungszeugnis"), character references, etc. It is difficult to imagine today what it must have meant for women to break into the male domain of academia, and what kind of discrimination and humiliation was connected with it. Clara was also required to apply to the university curator's office for a permission to attend lectures in experimental physics as a guest auditor. And she had to proceed in a similarly awkward manner with all the other lecture courses that she wished to take.

From early on, Clara (Figure 1) developed a keen interest in the field of physical chemistry. Richard Abegg (Figure 2), one of the pioneers in this new field and a friend of Fritz Haber, Clara's husband-to-be, played a key role in fostering Clara's interest in physical chemistry while paying little heed to Clara's guest-auditor status [Szöllösi-Janze, 1998, pp. 126–127]. It was also Abegg who supervised Clara's PhD thesis — a part of the graduation requirement in chemistry — and who wrote a joint paper with Clara in 1899. This paper [Abegg and Immerwahr, 1900] must have been perceived by the young female chemist as a proof of her success and an accolade. In the following year Clara submitted her dissertation and applied to be admitted to the so-called "Rigorosum" final, which entailed exams in chemistry, physics, mineralogy, and philosophy. She passed the exams in the fall and defended her thesis in the university's main auditorium

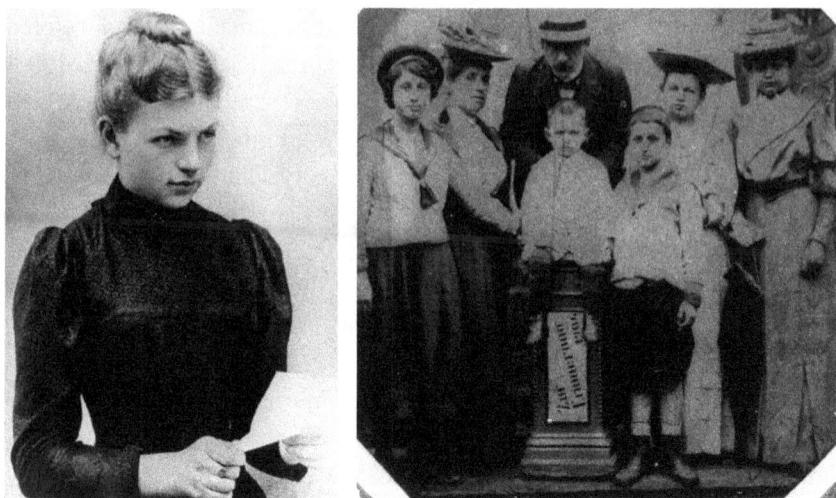

Figure 1.   Left: Clara Immerwahr during her studies at the University of Breslau, in the1890s. Right: Clara Haber, née Immerwahr (second from right); a family photo from 1906 with her son Hermann (seated in the centre), her husband Fritz Haber (standing behind Hermann) and the landlady (second from left) of the Habers' Karlsruhe apartment with her children; on the right is the Habers' maid servant (Archiv zur Geschichte der Max-Planck-Gesellschaft, Berlin-Dahlem).

on December 22, 1900 (Figure 3). Clara graduated *magna cum laude* and her graduation was mentioned in the daily press, as Clara was the first woman ever upon whom the University of Breslau conferred a doctoral degree [Szöllösi-Janze, 1998, p. 128].

Clara Immerwahr's Work on "Electroaffinity": Cadmium, Copper, Lead, Mercury, and Zinc in Equilibrium with their Salts

Clara's research concerned solution chemistry, one of the main preoccupations of physical chemistry at the time, and revolved about the connections among solubility, degree of dissociation, ion concentration, osmotic pressure, and electrochemical potential. Clara's systematic experimental investigations of the elements cadmium, copper, lead, mercury, and zinc, as well as of their salts, were aimed at expanding the empirical base for assessing the elements' "electroaffinity."

Figure 2. Richard Abegg (1869–1910). Abegg graduated in chemistry from the University of Berlin (1891) under August von Hofmann, received his "Habilitation" (1894) under Walther Nernst at the University of Göttingen, was "Extraordinarius" at the University of Breslau (1899–1909), "Ordinarius" at the "Technische Hochschule" Breslau (1909), a Member of the Leopoldina (1900) and the editor of the *Zeitschrift für Elektrochemie* (1901). Photo: Arrhenius, Svante. 1910. Richard Abegg. *Zeitschrift für Elektrochemie* 16: 554–557.

The notion of electroaffinity was proposed in 1899 by Clara's PhD adviser Richard Abegg and his collaborator Guido Bodländer (1855–1904) as a new organizing principle in chemistry [Abegg and Bodländer, 1899]. Gleaned from a body of electrochemical and solution-chemistry data, electroaffinity was intended to express the tendency of an atom or a group of atoms in a molecule to form a negatively charged ion upon the molecule's dissolution in a given solvent, typically water. According to Abegg and Bodländer [1899],

> [electroaffinity] offers to treat the immense body of molecular compounds systematically, and especially to bring into the forefront their connection to the properties of the chemical elements.[1]

---

[1] All translations from the German are the authors', unless otherwise indicated.

Aus dem chemischen Institut der Universität Breslau.

Beiträge zur Löslichkeitsbestimmung schwerlöslicher Salze des Quecksilbers, Kupfers, Bleis, Cadmiums und Zinks.

Inaugural-Dissertation,

welche

nebst den beigefügten Thesen

mit Genehmigung einer hohen

philosophischen Facultät der Königl. Universität Breslau

zur

Erlangung der philosophischen Doctorwürde

Sonnabend, den 22. Dezember 1900, Mittags 12 Uhr

in der Aula Leopoldina

öffentlich verteidigen wird

Clara Immerwahr.

Opponenten:

Herr cand. phil. **Franz Goldschmidt.**

Herr cand. phil. **Otto Sackur.**

Figure 3.   Announcement of Clara Immerwahr's PhD thesis defense at the main auditorium of the University of Breslau on 22 December 1910. Clara's "*Kommilitones*" Franz Goldschmidt and Otto Sackur were appointed as referees (Archiv zur Geschichte der Max-Planck-Gesellschaft, Berlin-Dahlem).

As Abegg and Bodländer noted, the electrochemical theory of Jöns Jacob Berzelius (1779–1848),[2] which made use of a similar notion [Jensen, 1996; 2018], turned out to be inadequate because of a lack of quantitative understanding of the relationship between electrochemical potential and concentration or solubility. This had changed in 1888 with the formulation of Nernst's equation that provided a way of establishing ion concentrations from the measured electrochemical potentials [Nernst, 1888]. Combined with the advent of the theory of chemical affinity and its relationship to other physical properties of molecules, Abegg and Bodländer felt encouraged to make use of electroaffinity in a new systematic approach to elucidating polar compounds that dissociate in solution into ions. Such compounds are mostly to be found in the realm of inorganic chemistry, while most organic compounds considered by Abegg and Bodländer were nonpolar. The periodic system of elements was to provide additional guidance concerning atomic properties.

In rows, from left to right in the periodic table, the tendency to form a cation (affinity towards positive charge) was shown to decrease while the tendency to form anions (affinity towards negative charge) to increase. Oxygen and particularly fluorine are electronegative (their tendency to form an anion is most pronounced). In columns, the electronegativity decreases from top to bottom. So the most electronegative of all elements, fluorine, is at the top of the halogen column (group 17), while caesium, the least electronegative element, is at the bottom of the alkalis' column (group 1). When other data were not available, solubility and the strength of acids were used as a measure of electronegativity. Electroaffinity was then used as an overarching notion, with electronegativity and electropositivity as its extremes.

In the 1930s, the American chemist Linus Pauling (1901–1994) would become Abegg's *Doppelgänger*, without knowing about it — or acknowledging it.[3] It was not just the notion of electronegativity that Pauling put on a firm quantum mechanical footing [Pauling, 1932] and that could be seen

---

[2]The electrochemical theory of Berzelius, also referred to as electrochemical dualism assumed that atoms were charged, and that all combinations of different atoms (molecules, in modern terms) consisted of one positively and one negatively charged part.

[3]In his own words: "I had become interested in the question of the nature of the chemical bond, after having read the 1916 paper on the shared-electron-pair chemical bond by

as a culmination of the effort that started with Berzelius, but also his work on valence [Pauling, 1931]. In 1904, Abegg formulated a rule whereby

> every element possesses a positive as well as a negative maximum valence whose sum is always eight. The maximum positive valence is the same as the number of the periodic group to which the element belongs [Abegg, 1904].

If Pauling was unaware of Abegg's contribution, the American chemist Gilbert Newton Lewis (1875–1946) certainly was and referred in 1916 to the above observation as "Abegg's rule" when formulating his cubical atom model as well as the "rule of eight" [Lewis, 1916] (dubbed the "octet rule" by Irving Langmuir (1881–1957) [Langmuir, 1919]). And so did the American chemist Arthur Amos Noyes (1866–1936) in his 1935 review on valence [Noyes, 1935]. We note that the American theoretical chemist Robert Mulliken (1896–1986) in his take on electronegativity actually used Abegg and Bodländer's term "electroaffinity" [Mulliken, 1934].

Clara Immerwahr wrote three research papers, the first co-authored with Richard Abegg, the other two solo. The paper with Abegg [Abegg and Immerwahr, 1900], which expands on the ideas of the 1899 Abegg and Bodländer paper, largely determined the topic and methodology of Clara's thesis paper. Her thesis paper [Immerwahr, 1901] deals in a more systematic way with the interplay between the solubility of a selection of heavy metal salts, the concentrations of the ions formed, and the potentials of the electrodes made out of these metals when immersed in the heavy-metal solutions (the potentials were measured as the electromotive force between the immersed heavy-metal electrode and a standard electrode). Not only did she find the ion concentrations ordered according to the ions' electroaffinities, but, in addition, the electroaffinities turned out to be additive. That is, the affinities are a property of each ion (atom, group) and do not depend on the way the ions are bound together in compounds. Clara's empirical work had thus anticipated an approximate result that would later be derived from molecular quantum mechanics by Pauling, Mulliken, and others.

G.N. Lewis and the several 1919 and 1920 papers by Irving Langmuir on this subject" [Pauling, 1992, p. 99].

Clara's second paper [Immerwahr, 1900] aimed to expand the solubility data base to include copper salts, using the ideas and methods developed by the founders of physical chemistry, Wilhelm Ostwald (1853–1932) and Walther Nernst (1864–1941), together with Friedrich Wilhelm Küster (1861–1917) [Küster, 1898]. The last was Clara's professor at the University of Breslau, and also deserves credit for arousing her interest in physical chemistry. He moved to the "Bergakademie" in Clausthal in 1899, and it was in Küster's Clausthal laboratory that Clara undertook the measurements reported in her second paper. As she noted, her data could be regarded as a corroboration of the Nernst-Ostwald-Küster theory of the electrochemical potential, and represented a methodological basis of Clara's work on electroaffinity.

As mentioned above, Clara's PhD adviser Richard Abegg was already well known during his lifetime for his work on valence that led to the octet rule. Clara's work on electroaffinity was related to this line of Abegg's research, but her contribution was not deemed significant enough to warrant Clara's inclusion in the Swedish physical chemist Svante Arrhenius's (1859–1927) list of half a dozen or so of Abegg's former affiliates who had contributed to Abegg's research in physical chemistry the most [Arrhenius, 1910]. Admittedly, another pioneer in this field, Clara's friend Otto Sackur (1880–1914), was not on that list either. However, Sackur made a name for himself in a research area that lay outside of Abegg's range of interests and did not publish his key work until after Abegg's death.

Apart from her work as a researcher, as a young graduate Clara also delivered public lectures on the broad topic of science in the household. Inspired by Dr Lassar-Cohn's popular book on chemistry in daily life [Lassar-Cohn, 1897], Clara's lectures attracted audiences of up to one hundred women [Szöllösi-Janze, 1998, p. 194]. None of this was related to the periodic system — only to a social system exclusive of women in academia and other walks of life dominated by men that was in place during the period spanned by Clara's life.

Clara's Marriage to Fritz Haber

At her debut scientific appearance at the annual meeting of the German Association for Electrochemistry in April 1901 Clara met Fritz Haber (Figure 4), another Breslau native, whom she knew from the dancing

Figure 4. Fritz Haber (1868–1934). Haber graduated in chemistry from the University of Berlin (1891) under August von Hofmann, received his "Habilitation" (1894) under Hans Bunte at the "Technische Hochschule" Karlsruhe, was "Extraordinarius" (1898) and "Ordinarius" (1906) at the "Technische Hochschule" Karlsruhe, founding Director of the "KWI für Physikalische Chemie und Elektrochemie" (Kaiser Wilhelm Institute for Physical Chemistry and Electrochemistry) (1911–1933), "Honorarprofessor" (1912–1920) and "Ordinarius" (1920–1933) at the University of Berlin, and a Member of the Prussian Academy (1914). He was awarded the Nobel Prize in Chemistry for 1918. Haber willed to be buried alongside Clara, a wish that was heeded by Hermann, the son of Clara and Fritz and the will's executioner. (Archiv zur Geschichte der Max-Planck-Gesellschaft, Berlin-Dahlem.)

classes of their teenage years. Their subsequent love affair quickly resulted in marriage. As Margit Szöllösi-Janze, both Fritz and Clara Haber's biographer, pointed out, their wedding on 3 August 1901 marked the end of "the chapter 'chemical science' in Clara's book of life" [Szöllösi-Janze, 1998, p. 129]. Nevertheless, during the initial years of her married life Clara continued her public lectures on science in the household and appeared at the lectures as well as in the laboratories of the "Technische Hochschule" in Karlsruhe, where her husband would soon become the founding director of its Institute for Physical Chemistry and Electrochemistry.

Moreover, it seems that at the time Fritz Haber would involve his wife in his research and share with her his scientific ideas, as suggested by the dedication of his 1905 classic textbook *Thermodynamics of Technical Gas-Reactions* [Haber, 1908]: "To my dear wife Clara Haber, Ph.D., in gratitude for her silent co-operation."

That Clara's involvement in Haber's research entailed more than a silent co-operation transpires in her correspondence with Abegg, in which she reports on Haber's progress in writing the textbook, discusses academic appointments, and solicits advice about her own public talks. However, the dream of an equitable and reciprocal scientific marriage — like that of Pierre and Marie Curie in Paris — did not materialize.

The turning point likely occurred with the birth of their son Hermann in 1902[4] and/or later when Haber was appointed "Ordinarius" at Karlsruhe in 1906 and/or became the founding director of the Kaiser Wilhelm Institute for Physical Chemistry and Electrochemistry in Berlin-Dahlem (Figure 5) in 1911. Clara was able to partake in it all not as a scientist, but

Figure 5. Kaiser Wilhelm Institute for Physical Chemistry and Electrochemistry, founded in 1911 in Berlin-Dahlem. The building on the right is the directorial mansion. (Archiv zur Geschichte der Max-Planck-Gesellschaft, Berlin-Dahlem.)

---

[4]This appears similar to the crisis in the marriage between Albert Einstein (1879–1955) and Mileva Maric (1875–1948), cf. [Renn and Schulmann, 2000].

as a scientist's spouse — a difference that the sensitive and earnest Clara surely must have reflected upon. This was aggravated by Haber's obsession with his work and career, which left little room for Clara's professional development and reduced her more and more to a mother and housewife.

The strains and conflicts between Clara and Fritz further aggravated after the outbreak of World War I. Haber applied himself in extraordinary ways to aid the German war effort and, in the process, earned the epithet "father of chemical warfare" for his role in orchestrating the first chlorine cloud attack on 22 April 1915 at Ypres, Belgium. Haber celebrated the "success" at Ypres — and his promotion to the rank of captain — at a gathering in his directorial mansion in Dahlem. The celebration took place on the evening of 1 May 1915. In the early hours of 2 May, Clara Haber committed suicide. She shot herself with Haber's army pistol in the garden of their mansion.

Clara Haber's Suicide

Since the 1990s and in particular after the publication of Gerit von Leitner's book *Der Fall Clara Immerwahr: Leben für eine humane Wissenschaft* ["The Case of Clara Immerwahr: Life for a Humane Science," 1993], Clara's suicide was thought to have resulted from her opposition to chemical warfare and to the role that Fritz Haber had played in it. The image of Clara as a beacon against weapons of mass destruction and for a humane science [Leitner, 1993] struck a chord with the *Zeitgeist*, and had served at the time as a vehicle for furthering the causes of the peace movement, feminism, and antimilitarism. However, this idealized image is not supported by the available historical sources, which do not admit a mono-causal explanation. Furthermore, in light of the available evidence, including recently discovered correspondence from May 1915 between those close to Clara [Henning, 2016], Clara Haber's suicide appears to have resulted from a "catastrophic failure" (to borrow an engineering term as a metaphor) brought about by a most unfortunate confluence of a host of circumstances. These included, apart from her unfulfilling life, Haber's philandering, the premature deaths of her PhD adviser and confidant Richard Abegg and of her friend and Kommilitone Otto Sackur, as well as the death and destruction of the war itself, amplified by the perversions of chemical warfare.

This recent perspective questions the "myth of Clara Immerwahr" [Hoffmann and Friedrich, 2019], but without belittling Clara's actual achievements and courage. Her admirable feat of graduating *magna cum laude* in chemistry in 1900 Prussia was not only unusual but also difficult, or unusual *because* it was difficult. It took the first thirty years of her life to achieve it, for she was denied a straight educational path, free from hurdles and harrassment, and had to take lengthy detours. However, we should refrain from projecting our contemporary ideas of women's rights activists or peace activists on Clara Haber in an ahistorical way [Friedrich and Hoffmann, 2016; 2017; 2019]. What she achieved in her time does not need to be embellished with exaggerations or even wishful thinking fashioned by present-day aspirations. Her achievements speak for themselves, and should not be degraded or even compromised by "upgrading" them by wishful thinking.

References

Abegg, R. and Bodländer, G. (1899). Die Elektroaffinität, ein neues Prinzip der chemischen Systematik, *Zeitschrift für anorganische Chemie*, 20, pp. 453–499.

Abegg, R. and Immerwahr, C. (1900). Notiz über das Elektrochemische Verhalten des Fluorsilbers und des Fluors, *Zeitschrift für physikalische Chemie, Stöchiometrie und Verwandtschaftslehre*, 32, pp. 142–144.

Abegg, R. (1904). Die Valenz und das periodische System: Versuch einer Theorie der Molekularverbindungen, *Zeitschrift für anorganische Chemie*, 39, pp. 330–380.

Arrhenius, S (1910). Richard Abegg, *Zeitschrift für Elektrochemie*, 16, pp. 554–557.

Clark, C. (2007). *Iron Kingdom: The Rise and Downfall of Prussia, 1600–1947* (Penguin, London).

Friedrich, B. and Hoffmann, D. (2016). Clara Haber, nee Immerwahr (1870–1915): Life, Work, and Legacy, *Zeitschrift für anorganische und allgemeine Chemie*, 642, pp. 437–448.

Friedrich, B. and Hoffmann, D. (2017). Clara Immerwahr: A Life in the Shadow of Fritz Haber, in *One Hundred Years of Chemical Warfare: Research, Deployment, Consequences*, eds. Friedrich, B., Hoffmann, D., Renn, J., Schmaltz, F. and Wolf, M. (Springer Open).

Haber, F. (1908). *Thermodynamics of Technical Gas-Reactions* (Longmans, Green, and Co., London).

Haber, F. (1924). *Fünf Vorträge aus den Jahren 1920–1923* (Springer, Berlin).

Henning, E. (2016). Freitod in Dahlem (1915): Unveröffentlichte Briefe von Edith Hahn und Lise Meitner über Dr. Clara Haber geb. Immerwahr, *Zeitschrift für anorganische und allgemeine Chemie*, 642, pp. 432–436.

Hoffmann, D. and Friedrich, B. (2019). Der "Mythos Clara Immerwahr" (in preparation).

Immerwahr, C. (1900). Potentiale von Kupferelektroden in Lösungen analytisch wichtiger Kupferniederschläge, *Zeitschrift für anorganische Chemie*, 24, pp. 269–278.

Immerwahr, C. (1901). Beiträge zur Kenntnis der Löslichkeit von Schwermetallniederschlägen auf elektrochemischem Wege, *Zeitschrift für Elektrochemie*, 7, pp. 477–483.

Jensen, W. B. (1996). Electronegativity from Avogadro to Pauling. Part 1: Origins of the Electronegativity Concept, *Journal of Chemical Education*, 73, pp. 11–20.

Jensen, W. B. (2018). Richard Abegg and the Periodic Table, in *Mendeleev to Oganesson: A Multidisciplinary Perspective on the Periodic Table*, eds. Scerri, E. and Restrepo, G. (Oxford University Press, Oxford), pp. 245–265.

Johnson, J. (1998). German Women in Chemistry, 1895–1925 (part I), *NTM Zeitschrift für Geschichte der Wissenschaften, Technik und Medizin*, 6, pp. 1–21.

Küster, F. W. (1898). Vorlesungsversuche, *Zeitschrift für Elektrochemie*, 5, pp. 503–505.

Langmuir, I. (1919). The Arrangement of Electrons in Atoms and Molecules, *Journal of the American Chemical Society*, 41, pp. 868–934.

Lassar-Cohn, D. (1897). *Die Chemie im täglichen Leben: gemeinverständliche Vorträge* (L. Voss, Hamburg).

Leitner, G. von. (1993). *Der Fall Clara Immerwahr: Leben für eine humane Wissenschaft* (C.H. Beck, München).

Lewis, G. N. (1916). The Atom and the Molecule, *Journal of the American Chemical Society*, 38, pp. 762–785.

Mulliken, R. S. (1934). A New Electroaffinity Scale: Together with Data on Valence States and on Valence Ionization Potentials and Electron Affinities, *Journal of Chemical Physics*, 2, pp. 782–793.

Nernst, W. (1888). Zur Kinetik der in Lösung befindlichen Körper: Theorie der Diffusion, *Zeitschrift für physikalische Chemie*, 2, pp. 613–637.

Noyes, W. A. (1935). Electronic Theories, *Chemical Reviews*, 17, 1–26.

Pauling, L. (1931). The Nature of the Chemical Bond: Application of Results Obtained from the Quantum Mechanics and from a Theory of Paramagnetic

Susceptibility to the Structure of Molecules, *Journal of the American Chemical Society*, 53, pp. 1367–1400.

Pauling, L. (1932). The Nature of the Chemical Bond. IV. The Energy of Single Bonds and the Relative Electronegativity of Atoms, *Journal of the American Chemical Society*, 54, pp. 3570–3582.

Pauling, L. (1992). How I Became Interested in the Chemical Bond: A Reminiscence, in *The Chemical Bond: Structure and Dynamics*, ed. Zewail, A. (Academic Press, San Diego, CA), p. 99.

Renn, J. and Schulmann, R. (2000). *Albert Einstein, Mileva Maric: The Love Letters* (Princeton University Press, Princeton, NJ).

Stern, F. (2011). Fritz Haber: Flawed Greatness of Person and Country, *Angewandte Chemie International Edition*, 51, pp. 50–56.

Szöllösi-Janze, M. (1998). *Fritz Haber (1868–1934): Eine Biographie* (C.H. Beck, München).

Select Bibliography

Clark, C. (2007). *Iron Kingdom: The Rise and Downfall of Prussia, 1600–1947* (Penguin, London).

Dunikowska, M. and Turko, L. (2011). Fritz Haber: The Damned Scientist, *Angewandte Chemie International Edition*, 50, pp. 10050–10062.

Friedrich, B. and Hoffmann, D. (2016). Clara Haber, nee Immerwahr (1870–1915): Life, Work, and Legacy, *Zeitschrift für anorganische und allgemeine Chemie*, 642, pp. 437–448.

Friedrich, B. and Hoffmann, D. (2017). Clara Immerwahr: A Life in the Shadow of Fritz Haber, in *One Hundred Years of Chemical Warfare: Research, Deployment, Consequences*, eds. Friedrich, B., Hoffmann, D., Renn, J., Schmaltz, F. and Wolf, M. (Springer Open).

Henning, E. (2016). Freitod in Dahlem (1915): Unveröffentlichte Briefe von Edith Hahn und Lise Meitner über Dr. Clara Haber geb. Immerwahr, *Zeitschrift für anorganische und allgemeine Chemie*, 642, pp. 432–436.

Johnson, J. (1998). German Women in Chemistry, 1895–1925 (part I), *NTM Zeitschrift für Geschichte der Wissenschaften, Technik und Medizin*, 6, pp. 1–21.

Stern, F. (2011). Fritz Haber: Flawed Greatness of Person and Country, *Angewandte Chemie International Edition*, 51, pp. 50–56.

Stern, F. (2016). *Einstein's German World* (Princeton University Press, Princeton, NJ).

Stoltzenberg, D. (2004). Fritz Haber: Chemist, Nobel Laureate, German, Jew (Chemical Heritage Foundation, Philadelphia, PA).

Szöllösi-Janze, M. (1998). *Fritz Haber (1868–1934). Eine Biographie* (C.H. Beck, München).

Szöllösi-Janze, M. (2009). The Natural Sciences and Democratic Practices: Albert Einstein, Fritz Haber, and Max Planck, in: *Bulletin. German Historical Institute Washington*, 44, pp. 9–22 (http://www.ghi-dc.org/files/publications/bulletin/bu044/bu44_009.pdf, last accessed 18 December 2018).

Cecilia Payne-Gaposchkin and the Elemental Composition of Stars

Matthew Shindell

Space History Department,
Smithsonian National Air and Space Museum,
Washington, D.C., USA

Today we take for granted that we know that stars like our Sun are made up primarily of hydrogen (73%) and helium (25%), and are powered by nuclear fusion. Fewer than 100 years ago, however, this was such an unlikely truth that even the discoverer doubted her own findings. Atomic fusion was not discovered until the 1930s, and scientists believed that the Sun had a similar elemental composition to that of the Earth. So when the British-born astrophysicist, Cecilia Payne-Gaposchkin (1900–1979) reported the results of her spectral analysis of the solar atmosphere in her PhD dissertation, suggesting that hydrogen and helium atoms far outnumbered any other elements, even she was reluctant to believe her own findings. Today, we recognise Payne-Gaposhkin as the first astronomer to propose this correct composition. Although her own career reflected the limitations placed on women in science in the early twentieth century, her persistence in pursuing her scientific work helped to open up opportunities for women at universities and observatories. She became, as she

Figure 1.   Cecilia Payne-Gaposchkin (The Smithsonian Institution Archives).

described it, "a thin wedge" in making space for women in astronomy [Kidwell 1996, p. 28].

***

Cecilia Payne (Figure 1) was born 10 May 1900 in the town of Wendover, England. Like many well-known scientific heroes, her love of science began at a young age. A bookish girl with great enthusiasm for learning, she omnivorously sought out whatever knowledge she could find about the natural world. When she arrived at Cambridge University's Newnham College, a women's college, in 1919, she initially set out to study the natural sciences, including botany, chemistry, and physics. It was not until the end of her freshman year that she decided to limit herself to one primary field of study. She found herself drawn into a heroic age of astronomy, during which great scientific expeditions were being sent around the world to observe planetary transits and solar eclipses. It was also an era in which the physicist Albert Einstein's (1879–1959) theories of relativity transformed the nature of the universe and what it meant to be an astronomer. Payne attended a lecture by Arthur Eddington (1882–1944), a Cambridge astronomer recently returned from an expedition to the island of Principe to observe the solar eclipse of 29 May 1919. Eddington's observations of the eclipse, and of the bending of starlight around the mass of the Sun revealed during the eclipse, confirmed Einstein's theory of general relativity and earned Eddington worldwide fame [Stanley, 2007, pp. 78–123].[1] Eddington's tales of his eclipse expedition and the implications of his observations for the theory of relativity inspired Payne to change her focus from botany to physics.

According to Payne herself, she faced resistance from the physics faculty, and in particular from the eminent experimental physicist Sir Ernest Rutherford (1871–1937), who singled her out as the only woman in his lectures and made her feel unwelcome. Although other testimonies[2] suggest that Rutherford was generally supportive towards women, this is not how Payne felt about it. Rutherford's apparent condescension led her to leave the study of physics. She instead decided to approach Eddington to see if she could study with him. Payne remembered in an oral history interview with historian Owen Gingrich,

> I went to Eddington and asked him if he would give me some research to do. And first he handed over to me some plates which had been taken of a galactic cluster, Messier 36, and said that I could use the observatory's measuring machines to determine the proper motions [Gingerich, 1968].

Payne worked on Eddington's project in her spare time. The work required that she learn advanced mathematics, which she successfully taught herself from volumes held at the British Museum. When she had finished this first set of work for Eddington, she brought her results back to the famous astronomer. Eddington, pleased with her work, sent it to the Royal Astronomical Society under her own name. "Proper Motion of the Stars in the Neighborhood of M36" became Payne's first publication [Payne, 1923]. Wanting to do more, Payne asked Eddington for another assignment: "I was looking for more worlds to conquer and Eddington

---

[1] Eddington's interpretations have, however, been discussed [Kragh, 1999, pp. 96–98; Ball, 2007].

[2] See e.g. Rayner-Canham and Rayner-Canham on Brooks in this volume. "Rutherford a thug to women" is treated as one of the "Rutherford myths" on www.Rutherford.org.nz (last accessed 10 January 2019).

said he would give me some stellar interiors to compute and he gave me a problem of integrating a stellar structure" [Gingerich, 1968]. The new work required lengthy calculations done by hand, and Payne approached it with tireless enthusiasm.

Payne's work with Eddington turned her physics training toward astronomy. However, although she had now worked with astronomical plates and proved herself capable of mastering the mathematics required for astronomy, she still had little personal experience making her own telescopic observations. Moreover, Eddington and others warned Payne that England was not a friendly environment for women astronomers. Newnham College could not grant degrees to the women studying at Cambridge. Her colleagues suggested that she might find more professional opportunities in the United States. She soon found an American mentor to take her on. While on a visit to England, Harlow Shapley (1885–1972), the newly-appointed Director of the Harvard College Observatory, gave a lecture at the Royal Astronomical Society in London. Payne had never heard of Shapley, but attended the lecture at the suggestion of a male astronomy student she knew from Cambridge. She found the lecture, and Shapley's animated lecture style, inspiring. He made the work of the astronomer in the observatory come to life before her eyes. As Payne later remembered, "I was completely fascinated because for the first time I had a vision of the way in which astronomy conducts observations, [I] had seen the plates at Cambridge but … somehow they didn't seem alive" [Gingerich, 1968]. After his lecture, Payne asked the American if she could come and work under his guidance at Harvard.

An exchange of letters confirmed the arrangement, and Payne left Cambridge to study at the Harvard College Observatory in 1923. By this point, the results of her study of the motion of stars in the cluster M36 had appeared in the *Monthly Notices of the Royal Astronomical Society* [Payne, 1923]. In America, Payne worked under Shapley's direction and enrolled in Radcliffe College, Harvard University's associated women's college. Here, for the first time, she benefited from support directly aimed at women in the sciences. Shapley awarded her an Edward C. Pickering Fellowship (a fund established in 1916 to support women working at the Observatory). She also received a Rose Sidgwick Fellowship from the

American Association of University Women. As she had under Eddington, Payne threw herself into her work, and within two years she was awarded a PhD in astronomy from Radcliffe College — making her not only the first woman, but the first student to receive a PhD in astronomy from Radcliffe or the Harvard College Observatory. Payne's research at the observatory centered on the study of stellar spectra. She set out to test the theory, proposed by the Indian astrophysicist Meghnad N. Saha (1893–1956), that the spectrum of a star was a function of the degree of ionisation of the atoms in its atmosphere. Saha's work inspired astronomers to look more closely at stellar spectra [DeVorkin, 1995]. Payne attempted to use ionisation theory to determine the relative abundances of elements in stellar atmospheres. By the time she was awarded her PhD, she had published six papers on stellar spectra and ionisation theory, as well as a book that pulled together the results of this work.

Payne used optical spectroscopy to detect and measure the elemental "signatures" contained in the starlight collected by the telescope. Payne's study of the relative abundance of the elements in stellar atmospheres yielded results that surprised many of her contemporaries. Most astronomers of the time followed the American astronomer Henry Norris Russell's (1877–1957) view that the most common metallic elements in the Sun's atmosphere existed in roughly the same ratios as they did in the Earth's crust [DeVorkin, 2000, pp. 177–198]. Russell was Payne's external thesis advisor, and so she also deferred to his authority on this matter. Payne's work confirmed that silicon, carbon, and common metals did in fact appear in about the same ratios in stellar atmospheres as in the Earth's crust. However, hydrogen and helium existed in much greater ratios. Her work suggested that stars were made primarily of these two elements. Although she had made a major discovery — one that we now believe to be true — she was not confident of her results. When she published them in the *Proceedings of the National Academy of Sciences*, she described the high levels of hydrogen and helium as "almost certainly not real" [Payne, 1925]. Looking back almost four decades later, the astronomers Otto Struve (1897–1963) and Velta Zebergs called her dissertation "undoubtedly the most brilliant PhD thesis ever written in astronomy" [Struve and Zebergs, 1962, p. 220]. For the time being, however, Payne's publications from this period established her as a talented astronomer. Even though he didn't believe Payne's hydrogen measurements, Russell wrote to Shapley that it was possibly the best thesis he had ever read [Kidwell, 1996, p. 20]. Her work signified to Russell and his contemporaries that astronomers were on the verge of understanding the composition of stars. By 1929, the accumulation of spectroscopic data convinced Russell that Payne's findings were correct, and he credited her with the discovery [Russell, 1929].

***

Despite the positive reception of her work and her growing reputation as an astronomer, Payne struggled to advance in her career. Although women did work in astronomy, they would typically only be hired in positions supporting their male colleagues, in which they made contributions that went largely unrecognised in the final published findings [Sobel, 2016]. In the 1920s and '30s, the schools that housed the best observatories did not hire women to faculty appointments. Ironically, even Radcliffe College, the women's college where she had received her PhD, did not hire women. Observatories were also male-dominated institutions, often located in remote areas. As Kidwell notes: "Those in charge often thought it was too dangerous to have women observe alone and improper for them to spend the night in the company of men" [Kidwell, 1996, p. 25].

Payne remained at Harvard for the rest of her career. There she taught introductory graduate courses without a formal appointment and at a salary considerably lower than that of her male peers. Because A. Lawrence Lowell (1856–1943), the president of Harvard, felt it was inappropriate to appoint women as faculty members, Payne was paid for her work indirectly, as Shapley's assistant. Her lack of recognition by Harvard, and her inability to define her own research agenda apart from Shapley's, discouraged Payne to the point that she almost left Harvard to find better opportunities elsewhere. Upon learning this, Shapley did what little he could to try and rectify Payne's marginalisation. He was able to secure a raise for her, and nominated her to the Harvard Faculty Club. Shapley and Russell also lobbied for her outside of Harvard, and Payne was elected to membership in the American Philosophical Society and the National Academy of Sciences. In 1938, after James B. Conant (1893–1978) took over

Harvard's presidency, Shapley managed finally to appoint Payne-Gaposchkin (she married Sergei Gaposchkin (1898–1984) in 1934) to the faculty. However, it was not until 1956 that Payne-Gaposchkin was finally promoted to full professor, becoming the first woman to hold this rank at Harvard.

References

Ball, P. (2007). Arthur Eddington was Innocent!, *Nature*, 7 September 2007 (https://www.nature.com/news/2007/070910/full/070903-20.html, last accessed 10 January 2019).

DeVorkin, D. H. (1995). Meghnad Saha's Influence in Astrophysics, *Journal of Astrophysics and Astronomy Supplement*, 16, pp. 35–36.

DeVorkin, D. H. (2000). *Henry Norris Russell: Dean of American Astronomers* (Princeton University Press, Princeton, NJ).

Gingerich, O. (1968). Interview with Cecilia Payne-Gaposchkin, 5 March 1968, Niels Bohr Library & Archives, American Institute of Physics, College Park, MD (www.aip.org/history-programs/niels-bohr-library/oral-histories/4620, last accessed 5 February 2019).

Kidwell, P. A. (1996). An Historical Introduction to "The Dyer's Hand," in *Cecilia Payne-Gaposchkin: An Autobiography and Other Recollections*, ed. Haramundanis, K., pp. 11–37.

Kragh, H. (1999). *Quantum Generations: A History of Physics in the Twentieth Century* (Princeton University Press, Princeton, NJ).

Payne, C. H. (1925). Astrophysical Data Bearing on the Relative Abundance of the Elements, *Proceedings of the National Academy of Science*, 11, pp. 192–198.

Payne, C. H. (1923). Proper Motions of the Stars in the Neighbourhood of M 36 (N. G. C. 1960), *Monthly Notices of the Royal Astronomical Society*, 83, p. 334.

Russell, H. N. (1929). On the Composition of the Sun's Atmosphere, *The Astrophysical Journal*, 70, pp. 11–82.

Sobel, D. (2016) *The Glass Universe: How the Ladies of the Harvard Observatory Took the Measure of the Stars* (Viking, New York, NY).

Stanley, M. (2007). *Practical Mystic: Religion, Science, and A.S. Eddington* (University of Chicago Press, Chicago, IL).

Struve, O. and Zebergs, V. (1962). *Astronomy of the 20th Century* (The Macmillan Company, New York, NY).

Select Bibliography

Boyd, S. L. (2014). *Portrait of a Binary: The Lives of Cecilia Payne and Sergei Gaposchkin* (Penobscot Press, Rockland, ME).

DeVorkin, D. H. (2010). Extraordinary Claims Require Extraordinary Evidence: C. H. Payne, H. N. Russell and Standards of Evidence in Early Quantitative Stellar Spectroscopy, *Journal of Astronomical History and Heritage*, 13, pp. 123–144.

DeVorkin, D. H. (2000). *Henry Norris Russell: Dean of American Astronomers* (Princeton University Press, Princeton, NJ).

Haramundanis, K., ed. (1996). *Cecilia Payne-Gaposchkin: An Autobiography and Other Recollections* (Cambridge University Press, Cambridge).

Kidwell, P. A. (1987). Cecilia Payne-Gaposchkin. Astronomy in the Family, in *Uneasy Careers and Intimate Lives: Women in Science, 1787–1979*, eds. Abir-Am, P. G., and Outram, D. (Rutgers University Press, New Brunswick, NJ), pp. 216–238.

Sobel, D. (2016). *The Glass Universe: How the Ladies of the Harvard Observatory Took the Measure of the Stars* (Viking, New York, NY).

Struve, O. and Zebergs, V. (1962). *Astronomy of the 20th Century* (The Macmillan Company, New York, NY).

Ida Noddack, the Eka-Manganeses and Nuclear Fission

Brigitte Van Tiggelen

Mémosciences asbl, Voie du Vieux Quartier 18,
1348 Louvain-La-Neuve, Belgium
Science History Institute, 315 Chestnut Street,
Philadelphia, PA 19106, USA

On 5 September 1925, a young woman finished her presentation on the newly discovered eka-manganese elements at the annual general assembly of the "Verein Deutscher Chemiker" ("Association of German Chemists") in Nuremberg. After she had finished, the chair of the chemical society, Professor Friedrich Quincke (1865–1934), referred to this talk as a historic moment — not because of the discovery of the new elements, which had been named "masurium" and "rhenium," but rather because, for the first time, a "Kollegin" (a female colleague) had addressed the Association. He finished by expressing the hope that this example would be followed by many other "Chemikerinnen" (female chemists) [Tacke, 1925a]. These hopes, like the acceptance of the new elements, would meet a somewhat different future. While rhenium was eventually accepted into the periodic system and figures in it to this day, element 43 is now known as "technetium," and its discovery credited to Carlo Perrier (1886–1948) and Emilio

209

Gino Segrè (1905–1989). A year after the meeting the young woman, Ida Tacke (1896–1978), married Walter Noddack (1893–1960), with whom she had been searching for the eka-manganese elements for several years, which made it difficult for her to pursue a career in her own right. Most of her female contemporaries experienced the same fate, especially as Nazism increasingly branded women as wives and mothers, not as professionals, including scientists, except if in support of their husbands' careers. Noddack continued her life as a wife and scientist, mostly because she was the wife of a scientist, and this granted her access to a laboratory and the necessary literature. Her expertise in the chemical properties of missing elements lead to her 1934 proposal of the possibility of what would later be called "nuclear fission," at a time when the "radioactivists" were unable to imagine this process.

Ida and Walter

Born in 1896, Ida was the daughter of Adelberg Tacke, the owner of a varnish and lacquer factory in Lackhausen. She enjoyed a good education in the nearby town of Wesel at first, and then attended the Gymnasium in Aachen. Probably intending to help her father with the family business, but also due to her penchant for scientific subjects, she enrolled at the "Technische Hochschule," in Charlottenburg, in Berlin, in 1916, and graduated as "Diplom-Ingenieur" in 1919, winning first prize in chemistry and metallurgy. In 1921, she received a PhD as "Doktor Ingenieur" ("doctor of engineering"); her thesis had been supervised by Professor David Holde (1864–1938) and was on organic chemistry, specifically on the anhydrides of high molecular weight fatty acids. Ida's curriculum was quite exceptional. The "Technische Hochschule" had only been accepting female students for six years when Ida began her studies there, and when she was awarded her doctorate, she was one out of twenty women across Germany who received that degree in the same year [Van Tiggelen and Lykknes, 2012]. In some ways her doctoral research directly related to the manufacture of varnish, since she had researched the drying properties of linseed oil as a part of her work. But Ida never returned to work in the family factory: she had met Walter Noddack, and became engaged with his personal scientific project — the quest for missing elements.

Walter was only a few years older than Ida, but had had to interrupt his university education during World War I. Born in Berlin in 1893, he had studied at the University of Berlin and obtained his doctorate in 1920, supervised by Walther Nernst (1864–1941; Nobel Laureate in Chemistry in 1920). In the same year, Nernst offered him a position as assistant at the University, and in 1922, when Nernst was offered the presidency of the "Physikalisch-Technische Reichsanstalt" (the Imperial Physical Technical Institute, abbreviated as PTR), Noddack followed his mentor and continued his career at the PTR. Most of the work was dedicated to standardisation and technical processes, but Walter was toying with the idea of filling in a few blanks in the periodic system by finding the yet unknown elements. Since Henry Moseley (1887–1915) had discovered the link between the element's atomic number and its X-ray spectrum in 1913, and proposed the shift from atomic weight to atomic number, the predictive power of the periodic system became fully evident [Kragh, 2012, pp. 104–107]. When Walter started thinking about the yet unknown elements, only a handful of them were still missing in the gaps before the heaviest element known, uranium (atomic number 92), and those were elements 43, 61, 72, 75, 85 and 87. By focusing on the manganese homologues — the elements close to each other in properties and occupying the spaces beneath manganese — he hoped to find at least two elements, numbers 43 and 75, also named eka-manganese and dvi-manganese, and hopefully also element 93, the first of the transuranium elements (i.e. elements heavier than uranium) [Van Tiggelen, 2001]. One has to remember that at that time, there was no actinides series, and element 93 was placed under Mn, after elements 43 and 75 (Figure 1).

As early as in 1914, Walter looked into the platinum ores for the eka-manganeses, but he did not succeed and could not devote more time to this, as his PhD research in the field of photochemistry required all his energy. Ida was fascinated both by the endeavour and the man undertaking it. Although she was employed at the "Allgemeine Elektrizität Gesellschaft" ("General Electrical Company") in Berlin, a producer of electrical equipment, immediately after the completion of her PhD, she spent much time at the PTR and in libraries, and in early 1923 she resigned from her position in order to focus fully on the search for the eka-manganese elements. She started with a comprehensive literature review of the properties of

1	2	3	4	5	6	7	8	9	10	11	12	13	14	15	16	17	18
																1 H	2 He
3 Li	4 Be											5 B	6 C	7 N	8 O	9 F	10 Ne
11 Na	12 Mg											13 Al	14 Si	15 P	16 S	17 Cl	18 A
19 K	20 Ca	21 Sc	22 Ti	23 V	24 Cr	25 Mn	26 Fe	27 Co	28 Ni	29 Cu	30 Zn	31 Ga	32 Ge	33 As	34 Se	35 Br	36 Kr
37 Rb	38 Sr	39 Y	40 Zr	41 Nb	42 Mo	43 Ma	44 Ru	45 Rh	46 Pd	47 Ag	48 Cd	49 In	50 Sn	51 Sb	52 Te	53 J	54 X
55 Cs	56 Ba	57 * La	72 Hf	73 Ta	74 W	75 Re	76 Os	77 Ir	78 Pt	79 Au	80 Hg	81 Tl	82 Pb	83 Bi	84 Po	85 —	86 Rd
87 —	88 Ra	89 Ac	90 Th	91 Pa	92 U	93 —	94 —	95 —	96 —								

*	58 Ce	59 Pr	60 Nd	61 —	62 Sm	63 Eu	64 Gd	65 Tb	66 Dy	67 Ho	68 Er	69 Tu	70 Yb	71 Cp

Figure 1. The periodic system as it was presented in Ida Noddack (1934), Das Periodische System der Elemente und seine Lücke, *Angewandte Chemie*, 47, p. 301.

eka-manganeses and the neighbouring elements, and the analytical methods used to identify them. During this period Ida was accepted as (an unpaid) guest at the PTR,[1] and Walter had become the head of the chemistry department [Van Tiggelen and Lykknes, 2012].

Methodic Search for New Elements: Different Uses of the Periodic System

The periodic system was important for the Ida-Walter team — they were not yet a married couple — beyond their search for the eka-manganese elements. In fact, the periodic system really governed their entire systematic search. Following their broad literature review, they focused on the description of the missing elements by the deduction of the physical and chemical properties of elements 43 and 75 in comparison with their neighbours [Van Tiggelen, 2001]. The aim was twofold: to establish precise means of identification for the missing elements, and to develop analytical procedures to isolate them from others that bear a close resemblance to them in their chemical and physical behaviour (Figure 2).

---

[1] Many women had the status unpaid guest researchers, see Roqué on Meitner, and Johnson on Cremer in this volume. Cremer was at the PTR as well.

V	Cr	Mn	Fe
Atomgew. 51	Atomgew. 52	Atomgew. 55	Atomgew. 56
Dichte 5,6	Dichte 6,7	Dichte 7,3	Dichte 7,9
Smp. (^{0}abs.) 2100	Smp. (^{0}abs.) 1800	Smp. (^{0}abs.) 1500	Smp. (^{0}abs.) 1800
V_2O_5	CrO_3	Mn_2O_7	FeO_4
Smp. (^{0}C) 660	Smp. (^{0}C) 196	Smp. (^{0}C) —25	unbekannt
flüchtig > 1200^0	flüchtig > 200^0	flüchtig > 4^0	
Farbe braungelb	Farbe dunkelrot	Farbe rotviolett	

Nb	Mo	43	Ru
Atomgew. 94	Atomgew. 96	Atomgew. 98—99	Atomgew. 102
Dichte 8,4	Dichte 10,2	Dichte 11,5	Dichte 12,3
Smp. (^{0}abs.) 2000	Smp. (^{0}abs.) 2650	Smp. (^{0}abs.) 2300	Smp. (^{0}abs.) 2400
Nb_2O_5	MoO_3	X_2O_7	RuO_4
Smp. —	Smp. (^{0}C) 791	Smp. (^{0}C) 350	Smp. (^{0}C) 22,5
flüchtig —	flüchtig > 500^0	flüchtig > 250^0	siedet > 100^0
Farbe	Farbe	Farbe hell	Farbe orange
kalt weiß, heiß gelb	kalt weiß, heiß gelb	gelb bis rosa	

Ta	W	75	Os
Atomgew. 182	Atomgew. 184	Atomgew. 187-188	Atomgew. 191
Dichte 16,6	Dichte 19,1	Dichte 21	Dichte 22,5
Smp. (^{0}abs.) 3100	Smp. (^{0}abs.) 3600	Smp. (^{0}abs). 3300	Smp. (^{0}abs.) 2800
Ta_2O_5	WO_3	X_2O_7	OsO_4
Smp. —	Smp. (^{0}C) 1800?	Smp. (^{0}C) 600	Smp. (^{0}C) 55
flüchtig —	flüchtig > 1300^0	flüchtig > 400^0	siedet > 100^0
Farbe weiß	Farbe gelb	Farbe weiß	Farbe weiß

Figure 2. The expected properties of element 43 and 75, from Ida Tacke (1925), Zur Auffindung der Ekamangane, *Angewandte Chemie*, 38, p. 1157.

In that respect, the team was following a tradition initiated by Dmitri Mendeleev (1834–1907) himself, who predicted the properties of gallium, germanium and scandium from their position in his periodic system. Their geochemical interpretation of the periodic system was more innovative. These elements had not yet been identified, and Noddack and Tacke inferred that this was due to their extreme rarity, i.e. that they were even rarer than the known elements [Van Tiggelen, 2001] (Figure 3). Because of the expected scarcity, the team knew for certain that all samples would need to be enriched considerably, and one of their early accomplishments was a 100,000-fold concentration of element 75 out of gadolinite [Noddack and Noddack, 1933, p. 22].

Noddack and Tacke also rejected the idea that the missing elements would form specific ores, and proposed that they were hidden in ores

Period	1	2	3	4	5	6	7	8	9	10	11	12	13	14	15	16	17	18
I	H (1) 10^{-2}																	He (2) $2\cdot10^{-9}$
II	Li (3) $2\cdot10^{-1}$	Be (4) 10^{-4}											B (5) $2\cdot10^{-4}$	C (6) $2\cdot10^{-3}$	N (7) $5\cdot10^{-4}$	O (8) $5\cdot10^{-1}$	F (9) $4\cdot10^{-4}$	Ne (10) $2\cdot10^{-8}$
III	Na (11) $1{,}8\cdot10^{-2}$	Mg (12) $1{,}7\cdot10^{-2}$											Al (13) $5\cdot10^{-2}$	Si (14) $2{,}5\cdot10^{-1}$	P (15) $6\cdot10^{-4}$	S (16) $6\cdot10^{-4}$	Cl (17) 10^{-3}	A (18) $2\cdot10^{-8}$
IV	K (19) 10^{-2}	Ca (20) $1{,}4\cdot10^{-2}$	Sc (21) 10^{-9}	Ti (22) $2\cdot10^{-3}$	V (23) $3\cdot10^{-5}$	Cr (24) $3\cdot10^{-5}$	Mn (25) $7\cdot10^{-4}$	Fe (26) $5\cdot10^{-2}$	Co (27) $3\cdot10^{-6}$	Ni (28) $3\cdot10^{-5}$	Cu (29) 10^{-7}	Zn (30) 10^{-6}	Ga (31) 10^{-10}	Ge (32) 10^{-9}	As (33) 10^{-7}	Se (34) 10^{-8}	Br (35) $2\cdot10^{-5}$	Kr (36) $3\cdot10^{-11}$
V	Rb (37) $2\cdot10^{-7}$	Sr (38) $2\cdot10^{-5}$	Y (39) $2\cdot10^{-6}$	Zr (40) $6\cdot10^{-5}$	Nb (41) 10^{-7}	Mo (42) 10^{-7}	Ma (43) $\sim10^{-13}$	Ru (44) $2\cdot10^{-12}$	Rh (45) 10^{-11}	Pd (46) 10^{-11}	Ag (47) 10^{-8}	Cd (48) 10^{-8}	In (49) 10^{-9}	Sn (50) $7\cdot10^{-6}$	Sb (51) $7\cdot10^{-8}$	Te (52) $6\cdot10^{-9}$	J (53) $2\cdot10^{-7}$	X (54) $4\cdot10^{-12}$
VI	Cs (55) $5\cdot10^{-9}$	Ba (56) 10^{-4}	La*) (57) $6\cdot10^{-7}$	Hf (72) $6\cdot10^{-6}$	Ta (73) $5\cdot10^{-7}$	W (74) $5\cdot10^{-5}$	Re (75) $\sim10^{-12}$	Os (76) $2\cdot10^{-11}$	Ir (77) $2\cdot10^{-11}$	Pt (78) 10^{-9}	Au (79) 10^{-9}	Hg (80) 10^{-8}	Tl (81) $4\cdot10^{-9}$	Pb (82) $4\cdot10^{-7}$	Bi (83) 10^{-9}	Po (84) 10^{-17}	— (85)	Em (86) $3\cdot10^{-19}$
VII	— (87)	Ra (88) $7\cdot10^{-14}$	Ac (89)	Th (90) $7\cdot10^{-8}$	Pa (91)	U (92) $7\cdot10^{-8}$												

Die Ordnungszahlen stehen eingeklammert unter den Symbolen, darunter die Häufigkeiten als Anteil der Erdrinde. Die Elemente gerader Ordnungszahl sind fettgedruckt.

*) Die auf Lanthan folgenden Erden (58—71) sind fortgelassen.

Figure 3. Abundance of elements in the earth's crust, given as a fraction according to Ida Tacke (1925). *Zur Auffindung der Ekamangane, Angewandte Chemie*, 38, p. 1158. For instance, according to this table, rhenium is expected to be 10^9 times less abundant than manganese in the earth's crust.

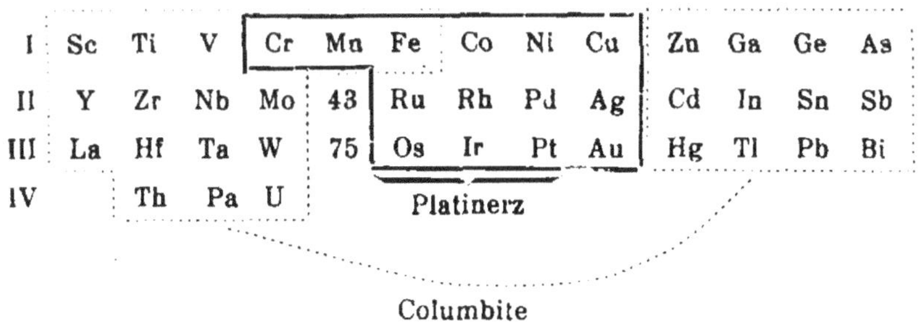

I	Sc	Ti	V	Cr	Mn	Fe	Co	Ni	Cu	Zn	Ga	Ge	As
II	Y	Zr	Nb	Mo	43	Ru	Rh	Pd	Ag	Cd	In	Sn	Sb
III	La	Hf	Ta	W	75	Os	Ir	Pt	Au	Hg	Tl	Pb	Bi
IV		Th	Pa	U			Platinerz						

Columbite

Figure 4. The expected joint occurrence of the elements surrounding element 43 and 75 in mineral ores, i.e. a visualisation of the other elements together with which they were expected to appear in nature. From W. Noddack and I. Tacke (1925), Die Ekamangane: Chemischer Teil, *Die Naturwissenschaften*, 13, p. 568.

containing neighbouring elements: in the case of specific ores, the ores containing primarily 43 and 75 would have long been discovered, they argued. This can be seen in the figure accompanying the article announcing the discovery of the missing elements 43 and 75 (Figure 4).

Moreover, the expectation to find both eka-manganese elements at once (i.e. in the same ores) was linked to a similar deduction: in their mind, the search for one element would also lead to the discovery of the second manganese homologue. The geochemical approach also indicated the kinds of ores to be inspected. They first selected manganese ores, but soon realised that they would have to broaden the search to platinum ores, and even beyond.

Finally, the use of the atomic number, the ordering feature of the periodic system as of 1921,[2] was also key to the identification process. The use of chemical procedures for enriching the selected ores and for testing for the presence of one or both eka-manganeses — procedures that had been popular at the beginning of the search for the eka-manganese elements — soon gave way to optical spectroscopy, and then X-ray spectroscopy. Optical spectroscopy is both difficult to execute and uncertain in its results, as it requires the identification of the faint lines of an unknown element amidst thousands of lines, some defined and some blurry, of

---

[2]It was not until 1921 that the German Atomic Weight Commission decided to base the new version of the periodic system on atomic numbers [Kragh, 2012, p. 34].

well-known elements. X-ray spectroscopy, by contrast, seemed to offer both precision and certainty: the spectral signature of yet unknown elements could unequivocally be detected, since the X-ray spectrum of each element is unambiguously linked to its atomic number. For access to such an instrument and expertise the team consulted Siemens und Halske, a company with a strong industrial interest in manganese homologues. The predicted properties of element 75 were, indeed, most attractive: this element would be a major competitor for tungsten (wolfram), especially for use in light bulbs, "Glühlampen."[3] A contract was signed, according to which Siemens und Halske would provide the team with the X-ray spectroscopic analysis, and an employee, Dr Otto Berg (1873–1939), would take care of this part of the investigation [Van Tiggelen, 2001]. Henceforth the work was divided between two locations reflecting the chemical and spectroscopical parts of the research respectively. In the summer of 1924 Berg was joined by Ida, who learned how to use the X-ray spectrometer and performed some measurements on her own, while the chemical work (the enrichment and analytical separation of other know neighbouring elements) was performed at the PTR laboratory by both Walter and Ida.

In short, the team knew exactly what to look for (chemico-physical properties and behaviour), where to look for it (ores and enrichment procedures), and how to look for it (characteristic chemical reactions and spectroscopical lines), and all this was based on different ways of making use of the periodic system.

Success and Failure

In May 1925, at last, the X-ray measures seemed to indicate the presence of both elements 43 and 75. The discovery was first mentioned to a very select audience, the "Preussische Akademie der Wissenschaften" ("Prussian Academy of Sciences") [Noddack and Tacke, 1925a; Berg and Tacke, 1925a], announced during an academic session by Walther Nernst, who was a member of the Academy. Soon afterwards, detailed results were published in *Die Naturwissenschaften* [Noddack and Tacke, 1925b; Berg and Tacke, 1925b]. Both times the discovery was presented in two

---

[3]The Noddacks secured several patents [Santos, 2014, p. 375].

steps: a first section, entitled the "chemical part," presented the analytical and geochemical facts of the investigation, with both Walter Noddack and Ida Tacke as authors, whereas the second section dealt exclusively with the X-ray spectroscopy, and was written by Ida Tacke and Otto Berg.

For Walter and Ida, however, the quest was not yet finished. Walter declared that the crucial point to proving the existence of these two new elements was the production of samples in sufficient amounts to do further research on their chemistry, and allow for their colleagues' confirmation (through handling samples) of the existence of the elements that were now named "masurium" (Ma) and "rhenium" (Re) [Van Tiggelen, 2001]. Walter and Ida (who had just married) decided to explore new ways of enrichment and new ores, and between 1926 and 1930 they travelled to Scandinavia. Because masurium was expected to be even rarer and more difficult to enrich than rhenium, they decided to focus on the latter. With its expected industrial applications rhenium would also gather more financial support for the team.

Controversies arose, but the methodical preparation of their search for eka- and dvi-manganese had also prepared the Noddacks for this: their geochemical and analytical approaches covered roughly all aspects of the controversies. The first gram of rhenium was eventually obtained in 1929, and very soon afterwards, rhenium was found in large amounts in the molybdenum glance, a mineral ore found in Mansfeld, Germany. By the mid-1930s, about 120 kg of one of the scarcest elements on the earth's crust was produced every year, and available for use in thermocouples, or as a catalyst. However, it was the acknowledgement of rhenium as an element by the German Atomic Weight Commission in 1929 that concluded the controversies with the first establishment of an atomic weight of 188.71 suggested by Ida and Walter Noddack and based on rhenium disulfide [N. N., 1929, pp. 19–20]. Otto Hönigschmid (1878–1945) and Rudolf Sachtleben (1897–1966) provided a corrected value in 1930 [Hönigschmid and Sachtleben, 1930]. A year later, in 1931, Walter provided the British chemist Francis William Aston (1877–1945) with a rhenium sample, and via mass spectrometry Aston was able to identify the two main isotopes of Re (185 and 187), and gave a value of 186.22 for the atomic weight, which was consistent with Hönigschmid's latest value of 186.31 [Aston, 1931].

In 1931, the existence of rhenium was generally acknowledged and the discovery attributed to the Noddacks. After ten years of continuous labour they jointly received the Liebig Medal of the German Chemical Society — a recognition from the national chemical community. Both Ida and Walter were elected members of the "Deutsche Akademie der Naturforscher Leopoldina" in 1934. Ida received the Scheele Medal of the Swedish Chemical Society in 1934, and other evidence of international recognition came forward in the form of nominations for the Nobel prize: Ida was nominated jointly with Walter in 1933, 1935 and 1937, and Walter was further nominated in 1932 and 1934.

The story of masurium is quite different from rhenium's success. Once the Noddacks had established rhenium, it would have been element 43's turn to be manufactured. Though laboratory notebooks show that the couple was working on this into the 1930s and 1940s, the failure to document the discovery of element 43 undermined the Noddacks' credibility. Interestingly though, there were no contenders: during all these years, no other claims for having found 43 appeared. We now know that element 43 is a radioelement, and that it does not occur in nature other than in elusive traces.[4] The production of element 43 was eventually achieved by two types of nuclear means. In 1937, Carlo Perrier and Emilio Segrè obtained several isotopes of element 43 by bombarding a molybdenum plate with a strong deuteron beam. In 1939, Otto Hahn (1879–1968) and Friedrich Wilhelm (Fritz) Strassmann (1902–1980) identified element 43 as a product of the nuclear fission of uranium. By 1940, Segrè and Chien-Shiung Wu (1912–1997)[5] were able to confirm this second way of producing element 43. The discovery of this element was thus achieved following a very different approach compared to the manufacturing of rhenium through analytical and enrichment procedures. In 1947, element 43 was given its current name, "technetium,"

---

[4] A controversy arose in 1988 when Pieter Van Assche from Leuven, Belgium, aimed to demonstrate that the Noddacks could, indeed, have "seen" 43 (without later being able to reproduce the spectrum). While there are minute traces of 43 in nature, it is now accepted that the instrument the Noddacks used was insufficiently sensitive to detect it if it was present in their samples [Fontani *et al.*, 2014 pp. 310–319, esp. pp. 314–317].

[5] See Robinson on Wu in this volume.

because it was the first artificially produced element, a decision left unchallenged by the Noddacks who, however, continued to hope that the "natural" element 43 would be found, and mentioned it in a presentation given in 1955 [Van Tiggelen, 2001, p. 123].

Walter's career advanced alongside these events: he was appointed as "ordentlicher Professor der physikalischen Chemie" ("ordinary full professor of physical chemistry") at Freiburg in 1935, and from 1942 he was head of the "Institut für Photochemie" (Institute for Photochemistry) at Strasbourg as well as a lecturer in physical chemistry (Figure 5).

Figure 5. Walter and Ida Noddack in Freiburg, ca. 1937 (KU Leuven Universiteitsarchief, PP 209 Noddack, 52).

Figure 6. Ida Noddack in the laboratory at the Reichsuniversität in Strasburg, press photograph, dated 1944 (KU Leuven Universiteitsarchief, PP 209 Noddack, 1277).

Ida worked as an unpaid research associate throughout her career, except when in Strasbourg (Figure 6) where, somewhat ironically, the Reichsuniversität[6] employed a married woman, against the Nazi principles which placed particular emphasis on the home duties of the "Hausfrau" (housewife). The promotion to Strasbourg ruined Walter's chances of an academic career in a university after the war. Even though "denazified," due to his former affiliation with the Strasbourg university he had fallen into disrepute. Walter found an appointment in Bamberg, a "Hochschule" at that time. Once he was settled there, in 1956, he founded a Geochemical

---

[6]The Reichsuniversität in Strasburg was founded in 1941 under German occupation and was a Nazi stronghold — 80% membership of the faculty were members of the National Socialist Party (NSDAP, Nationalsozialistische Deutsche Arbeiterpartei).

Institute, which soon flourished. Ida was once more accepted only as a guest researcher there [Van Tiggelen and Lykknes, 2012].

The Limits of Understanding Matter

The contrasting stories of masurium and rhenium shed light on a change in the search for missing elements. The successful discovery of rhenium has many parallels with the Curie's career: they, too, were a married couple who worked as a team in their discoveries, they shared the emphasis on the analytical enrichment procedure, as well as the determination to produce weighable samples. However, the Noddacks' inability to employ new methods, such as nuclear reaction, tied them to a tradition that had served them well but was doomed to fail in the case of element 43. It seems as if Ida and Walter could simply not deal with the emergence of a new type of element which was man-produced, elusive, and mostly known through specific manufactured isotopes in samples of much less than a gram. They were not alone in this dilemma: other element hunters, such as Horia Hulubei (1896–1972) in France, were also caught in this paradigm of the natural element, that is the idea that, even for man-made elements, there is always a "natural" version to be found in nature.[7] For the Noddacks this is rather surprising, since they were among the first who witnessed the exponential development of radiochemistry in Berlin, and Walter had had the opportunity to discuss such matters with Otto Hahn. The most ironic element in this story, however, is Ida's proposal of nuclear fission.

As discoverers of at least one new element, the Noddacks were well placed to evaluate claims of discoveries of other elements. For instance, they wrote a commentary on an alleged finding of element 61[8] [Noddack and Tacke, 1926], and Ida, in particular, acquired substantial authority in this area. In 1934, she analysed a sample of a substance provisionally

---

[7] See Van Tiggelen on Cauchois in this volume.

[8] Element 61 had been allegedly discovered by the Italian scientists Luigi Rolla (1882–1960) and Lorenzo Fernandes (1902–1977) at the Institute of Chemistry of the University of Florence, and independently by the American scientists B. Smith Hopkins (1873–1952), Leonard Francis Yntema (1892–1976), and J. Allen Harris (1901–1972) at the University of Illinois; both claims were made in 1926.

named "bohemium" by the Czech engineer Odolen Koblic (1897–ca.1959), who claimed that he had isolated element 93. Ida convinced him to withdraw his claim [Van Tiggelen and Lykknes, 2012, p. 117]. Her expertise here was also based on the initial goal pursued by the team Noddack-Tacke: to find all eka-manganeses, including, hopefully, 93. It was actually this chemical expertise on the expected properties of the first transuranium element that brought Ida to comment — negatively — Enrico Fermi's (1901–1954) claim that he had produced elements 93 and 94 through bombardment of uranium nuclei with neutrons.

In her 1934 paper "Über das Element 93" (On Element 93) [Noddack, 1934b], Ida criticised Fermi for not having "compared his new radioelement with all known elements" [Noddack, 1934b, p. 654; Noddack, 1971, p. 18]. Indeed, Fermi had stopped his chemical identifications at lead, which was an established end-result for a radioactive series. And this is when Ida boldly wrote: "When heavy nuclei are bombarded by neutrons, it is conceivable that the nucleus breaks up into several large fragments, which would of course be isotopes of known elements but would not be neighbors of the irradiated element" [Noddack, 1934b, p. 654; Noddack, 1971, p. 18], thereby proposing what would later be known as nuclear fission. Her comments were ignored, and more frustratingly, continued to be ignored even at the very end of 1938, when Otto Hahn, Lise Meitner[9] (1878–1968) and co-workers eventually came up with the idea of nuclear fission. Ida protested in vain that no mention had been made of her proposal four years earlier [Noddack, 1939]. This is not the place to go into details about how and why her proposal was ignored, but in the context of a volume dedicated to women scientists and elements, it is essential to stress how each period, each community, and each individual found their way into the understanding of matter, and sometimes met their limitations. In the same way in which the Noddacks had been unable to fathom and integrate the new conception of elements, and the corresponding nuclear means of discovering them, the radiochemical community proved unable to imagine nuclear fission, a process that was not substantiated by the nuclear theory of the time.

---

[9] See Roqué on Meitner in this volume.

The author wishes to express her gratitude for Annette Lykknes' and Anke Timmermann's careful editorial work on this contribution.

References

Aston, F. W. (1931). Constitution of Rhenium, in *Nature*, 127, p. 591.

Berg, O. and Tacke, I. (1925a). Zwei neue Elemente der Mangangruppe: Röntgenspektroskopischer Teil, *Sitzungsberichte der physikalisch-mathematischen Klasse vom 11. Juni 1925 (Königlich Preussische Akademie der Wissenschaften zu Berlin)*, pp. 405–409.

Berg, O. and Tacke, I. (1925b). Die Ekamangane: Röntgenspektroskopischer Teil, *Die Naturwissenschaften*, 13, pp. 571–574.

Fontani, M., Costa, M. and Orna, M. V. (2014). *The Lost Elements: The Periodic Table's Shadow Side* (Oxford University Press, Oxford).

Hönigschmid, O. and Sachtleben, R. (1930). Revision des Atomgewichtes des Rheniums: Analyse des Silberperrhenats, *Zeitschrift für anorganische und allgemeine Chemie*, 191, p. 309.

Kragh, H. (2012). *Niels Bohr and the Quantum Atom: The Bohr Model of Atomic Structure 1913–1925* (Oxford University Press, Oxford).

N. N. (1929). Bericht der Deutschen Atomgewichts-Kommission, *Berichte Der Deutschen Chemischen Gesellschaft (A and B Series)*, 62:1, pp. 19–20.

Noddack, I. (1934a). Das periodische System der Elemente und seine Lücken, *Angewandte Chemie*, 47, pp. 301–305.

Noddack, I. (1934b). Über das Element 93, *Angewandte Chemie*, 47, pp. 653–656.

Noddack, I. (1939). Bemerkung zu den Untersuchungen von O. Hahn, L. Meitner und F. Straßmann über die Produkte, die bei der Bestrahlung von Uran mit Neutronen entstehen, *Die Naturwissenschaften*, 27, pp. 212–213.

Noddack, I. (1971). On Element 93, transl. by H. G. Graetzer, in Graetzer, H. G. and Anderson, D. L. (1971). *The Discovery of Nuclear Fission: A Documentary History* (Van Nostrand Reinhold, New York, NY), pp. 16–20.

Noddack, W. and Tacke, I. (1925a). Zwei neue Elemente der Mangangruppe: Chemischer Teil, *Sitzungsberichte der physikalisch-mathematischen Klasse vom 11. Juni 1925 (Königlich Preussische Akademie der Wissenschaften zu Berlin)*, pp. 405–409.

Noddack, W. and Tacke, I. (1925b). Die Ekamangane: Chemischer Teil, *Die Naturwissenschaften*, 13, pp. 567–571.

Noddack, W. and Tacke, I. (1926). Über das Element 61, *Die Metallbörse*, 16, pp. 985–986.

Noddack, W. and Noddack, I. (1933). *Das Rhenium* (Leopold Voss, Leipzig).

Santos, G. (2014). A Tale of Oblivion: Ida Noddack and the "Universal Abundance" of Matter, *Notes Rec. R. Soc. Lond.*, 68:4, pp. 373–389.

Tacke, I. (1925a). Zur Auffindung der Ekamangane, *Angewandte Chemie*, 38, p. 794.

Tacke, I. (1925b). Zur Auffindung der Ekamangane, *Angewandte Chemie*, 38, pp. 1157–1160.

Van Tiggelen, B. (2001). The Discovery of New Elements and the Boundary Between Physics and Chemistry in the 1920s and 1930s: The Case of Elements 43 and 75, in *Chemical Sciences in the 20th Century: Bridging Boundaries*, ed. Reinhardt, C. (Wiley-VCH, Weinheim), pp. 131–144.

Van Tiggelen, B. and Lykknes, A. (2012). Ida and Walter Noddack Through Better and Worse: An Arbeitsgemeinschaft in Chemistry, in *For Better or For Worse? Collaborative Couples in the Sciences,* eds. Lykknes, A., Opitz, D. and Van Tiggelen, B. (Science Networks, Historical Studies no. 44) (Springer Birkhäuser, Heidelberg), pp. 103–147.

Select Bibliography

Fontani, M., Costa, M. and Orna, M. V. (2014). *The Lost Elements: The Periodic Table's Shadow Side* (Oxford University Press, Oxford).

Noddack, I. (1934). Das periodische System der Elemente und seine Lücken, *Angewandte Chemie*, 47, pp. 301–305.

Noddack, I. (1934). Über das Element 93, *Angewandte Chemie*, 47, pp. 653–656.

Noddack, W. and I. (1933). *Das Rhenium* (Leopold Voss, Leipzig).

Santos, G. (2014). A Tale of Oblivion: Ida Noddack and the "Universal Abundance" of Matter, *Notes Rec. R. Soc. Lond.*, 68:4, pp. 373–389.

Tacke, I. (1925b). Zur Auffindung der Ekamangane, *Angewandte Chemie*, 38, pp. 1157–1160.

Van Tiggelen, B. (2001). The Discovery of New Elements and the Boundary Between Physics and Chemistry in the 1920s and 1930s: The Case of Elements 43 and 75, in *Chemical Sciences in the 20th Century: Bridging Boundaries*, ed. Reinhardt, C. (Wiley-VCH, Weinheim), pp. 131–144.

Erika Cremer and Ortho-/Parahydrogen

Jeffrey Allan Johnson

Villanova University, Villanova, PA 19085, USA

Erika Cremer (1900–1996) was a German (later Austrian) physical chemist who is today best known for her pioneering but belatedly recognised work with her students in developing the technique of gas-solid chromatography, beginning in 1944 [Ettre, 2008; Kolomnikov *et al.*, 2018, pp. 112–113; Johnson, 2019]. But before taking up chromatography, Cremer had worked for more than a decade (1932–1943), under extremely difficult conditions, on the quantum chemistry of molecular hydrogen, which she first encountered while working at the Kaiser Wilhelm Institute for Physical Chemistry in Berlin. To understand the course of this investigation and the difficulties that Cremer had to deal with in conducting it, it is necessary to discuss Cremer's work and the obstacles to her academic career in the political and social context of Weimar and National Socialist Germany from the 1920s to the 1940s.

***

Erika Cremer was born in 1900 into an academic family in Munich [Beneke, 1999; Oberkofler, 1998]. Her father Max Cremer (1865–1935), a professor of physiology, encouraged her to get an academic education at a time when German women were only just beginning to gain access to higher education, and only a very few could pursue academic careers [Cremer, 1994]. Like her father, who applied physics to the study of physiological problems, Erika chose very early to apply physics to the study of chemical problems [Beneke, 1999, pp. 311–312]. She first studied physical chemistry at the University of Berlin under Walther Nernst (1864–1941), who had become the leading German in the field after the retirement of its co-founder Wilhelm Ostwald (1853–1932) in 1906. Cremer completed her doctorate in 1927 under Nernst's successor Max Bodenstein (1871–1942), who specialized particularly in reaction kinetics, the factors affecting the speed and outcome of chemical reactions, which would become Cremer's main area of interest for the first half of her career. Her dissertation explored the nature of chemical chain reactions in chlorine-hydrogen-oxygen explosions in light [Cremer, 1927]. When later followed up by the Soviet Russian Nikolai N. Semenov (1896–1986), who in 1932 invited her to work with him for several weeks in Leningrad (today St. Petersburg), this line of research ultimately earned Semenov a share in the 1956 Nobel Prize for Chemistry with the British chemist Sir Cyril Hinshelwood (1897–1967). In his Nobel lecture, Semenov cited no work before his own first publication in 1928, thus failing to acknowledge Cremer's priority [Semenov, 1964]. This was not the only occasion in which Cremer's pathbreaking research failed to receive appropriate or timely recognition [Oberkofler, 1998].

Cremer's dissertation did not lead to a conventional academic position; instead, she would see more than a decade of "journeyman years" ("Wanderjahre"), always working — and with some of the best scientists of the day — but only rarely being paid, and then only for temporary postdoctoral assistantships, stints as a guest researcher (as with Semenov), or routine summer jobs [Cremer, 1994; Oberkofler, 1998, pp. 15–30]. Initially she worked on quantum theoretical problems of photochemistry, a topic related to her dissertation, with Karl Friedrich Bonhoeffer (1899–1957) at the Kaiser Wilhelm Institute (KWI) for Physical Chemistry under Fritz Haber (1868–1934). She and Bonhoeffer became good friends, and he probably introduced her to the problem of ortho- and parahydrogen, which became a central topic of research at the KWI when Cremer was first there in 1927–1928.

In 1927 Werner Heisenberg (1901–1976) and Friedrich Hund (1896–1997) had predicted the existence of two forms of molecular hydrogen, ortho- and para-, based on the alignment of the quantum spins of the two protons that made up the hydrogen molecule: either parallel with odd quantum numbers (ortho-) or anti-parallel with even quantum numbers (para-). Figure 1 provides a non-mathematical visualization of these structures.

The expectation was that the two forms existed in a ratio of 1:3 (para- to ortho-) at room temperature, but that lower temperatures would favour the para form, "if one could somehow rescind the ban on transition between even and odd rotational states imposed by quantum mechanics" [Farkas, 1989, p. 93]. In 1928–1929 Bonhoeffer and Paul Harteck (1902–1985) demonstrated various conditions that might favour conversion at different rates in either direction (temperature, pressure, electric discharge, catalysts) [Farkas, 1989, pp. 93–94], and the Hungarian student Adalbert Farkas (1906–1995) then completed his dissertation under Bonhoeffer on the ortho-parahydrogen conversion in 1930 [Nye, 2011, p. 119]. The ortho-/parahydrogen problem became an important test case for the new theories of quantum mechanics, and it thus attracted considerable attention from physicists and physical chemists.

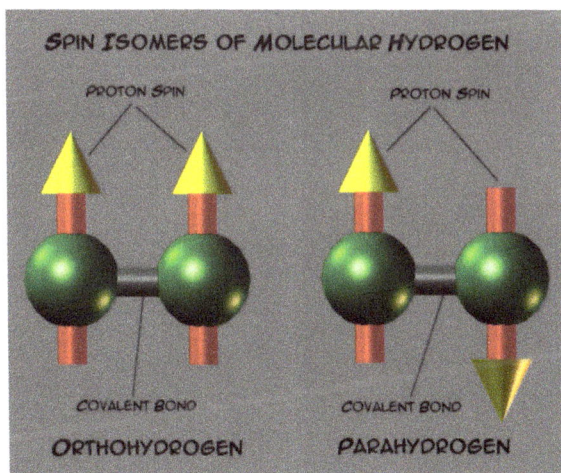

Figure 1. A schematic representation of ortho- and parahydrogen as spin isomers, showing quantum spins in parallel (ortho-) or antiparallel (para-) orientations. The covalent bonds indicate the two shared electrons in the diatomic hydrogen molecule (Xaa (Jim Farris), public domain).

Cremer did not participate in this initial round of work on ortho- and parahydrogen at the KWI, because she had been able to obtain a postdoctoral fellowship to work with the future Nobel laureate George de Hevesy (1885–1966) in the southwestern German town of Freiburg, studying problems of catalysis related to rare-earth oxides, as well as zircon and hafnium sulphates [Oberkofler, 1998, pp. 15–16]. During this period she also collaborated briefly on a study of activation energy in chemical reactions with Georg-Maria Schwab (1899–1984), whom she had met when he was Bodenstein's assistant [Cremer and Schwab, 1929; Oesper, 1952].

By the time Cremer returned to the KWI in Berlin in October 1930, Bonhoeffer had taken a professorship in Frankfurt am Main. Cremer now became an unpaid assistant to the brilliant and versatile scientist (later philosopher) Michael Polanyi (1891–1976) [Nye, 2011]. Figure 2 shows Cremer with the Polanyi research group in 1933.

Figure 2. The Polanyi research group at the Kaiser Wilhelm Institute for Physical Chemistry in Berlin-Dahlem, early 1933. Erika Cremer is second from left in the back row, behind Michael Polanyi (centre) (Archive of the Max Planck Society, Berlin-Dahlem; reproduced by kind permission of the Archive of the Max Planck Society).

Cremer now demonstrated her command of quantum theory and her experimental dexterity while studying various aspects of the quantum chemistry of hydrogen. One of her projects was to apply the so-called quantum tunnelling theory to heterogeneous catalysis in the case of the hydrogenation of the organic compound styrene [Cremer and Polanyi, 1932]. Heterogeneous catalysis involves a catalytic substance in a different phase than the substances whose reaction is being accelerated (e.g., a solid catalyst used with reactants in a gaseous phase); quantum theory predicted that the presence of a catalyst would allow the reaction to "tunnel through" an otherwise insurmountable energy barrier. Cremer also investigated the problem of the catalytic transformation of ortho- to parahydrogen, particularly in the solid state (below 259.2 degrees Celsius), and in the process she was the first to observe that solid oxygen could function as a catalyst in this transformation; this was quickly followed up by others at the KWI, particularly Adalbert's brother Ladislaus Farkas (1904–1948) and Hans Sachsse (1906–1992), whose work has tended to be cited without reference to her priority [Cremer and Polanyi, 1933; Günther, 1938; Farkas and Sachsse, 1933; Chayut, 1994, p. 245]. Cremer also calculated the reaction mechanism using the quantum-mechanical formulation for the magnetic spin effect developed by another Hungarian physicist at the KWI, Eugene Wigner (1902–1995, Nobel Prize for Physics 1963); she also made her first use of thermoconductivity cells as detectors, which she would later apply to her gas chromatography technique [Bobleter, 1997, pp. 398–399; Beneke, 1999, pp. 324–325]. The Polanyi-Cremer collaboration led to six publications but was terminated in 1933, when the Nazis forced the dismissal of almost the entire Institute staff and left Cremer again without an affiliation.

Feeling that her work with Polanyi had not yet produced sufficiently "positive" results due to the limits of the KWI's laboratory facilities and then the Nazis' interruption of the project [Bobleter, 1990a, minute 11], Cremer returned to the problem of clarifying factors affecting the reaction speed in the catalytic transformation of ortho- and parahydrogen between 1935 and 1939, including a follow-up study of the catalytic role of solid oxygen [Cremer, 1935]. She produced several publications, even though she did not have a paid position for most of this time [Bobleter, 1997, p. 398; Beneke, 1999, pp. 314–315]. After an interim period (perhaps with a small Rockefeller grant) at the Institute for Physical Chemistry in Munich (until its director, the Polish Jew Kasimir Fajans (1887–1975), was forced out in 1935 [Hurwic, 2000, pp. 104–106]), Cremer returned to Berlin. There she continued her research on hydrogen for two years as an unpaid and unofficial guest at the low-temperature laboratory at the Imperial Physical Technical Institute ("Physikalisch-Technische Reichsanstalt," PTR),[1] then presided over by the Nazi physicist Johannes Stark (1874–1957). Because Stark did not want any women at the PTR, Cremer recalled that her friends on the staff had to hide her when Stark inspected the laboratory; however, this may have been somewhat romanticized in her memory [Cremer, 1994; Oberkofler, 1998, pp. 26–27; Beneke, 1999, p. 316].

It was probably during this period that one of Cremer's colleagues remarked, "if I were in your position, with so little chance of becoming a professor, I certainly wouldn't work so hard." To which she replied, "I'm not doing it to become a professor. I'm doing it because it's fun" [Cremer, 1994; Beneke, 1999, p. 315]. Unfortunately not all the work was fun. She was seriously injured (suffering a concussion) doing a low-paid routine summer job at the Bio-Climatic Research Station on the North Sea coast [Bobleter, 1997, p. 398]. After this Bodenstein took pity on her and allowed her to work for a semester as his private assistant [Oberkofler, 1998, p. 27]. He then persuaded the German chemist Otto Hahn (1879–1968) to arrange a research fellowship for Cremer in the laboratory for catalysis and surface chemistry at the KWI for Chemistry in 1937 [Beneke, 1999, p. 318]. By the end of 1938 Cremer was finally able to bring her hydrogen research to a substantial conclusion. She completed a conclusive analysis of the ortho-para conversion, and along with that a much more precise (four decimal places better than before) determination of the constant of self-diffusion of hydrogen in the solid state (essentially the probability that a molecule of hydrogen would change position if not acted on by other chemical forces) [Cremer, 1939]. She submitted this and her other work to the Berlin University Faculty for Natural Sciences to qualify for lecturing. The faculty referees considered this to be her finest work to date, and thus, in February 1939, she received her formal "Habilitation,"

---

[1] This was just after Ida Noddack (see Van Tiggelen on Noddack in this volume) had been an unpaid guest researcher at the PTR (1925–1935).

or qualification for university lecturing [Günther, 1938; Oberkofler, 1998, pp. 30, 91]. Unfortunately there was a catch: all university lecturers first had to pass through a political indoctrination camp, and between 1933 and 1939 no women were allowed into these camps [Cremer, 1994; Bobleter, 1997, p. 398; Vogt, 1998, p. 34].

Just before receiving the mixed news about her qualification for lecturing, Cremer became an eyewitness to an exciting but ominous development in Hahn's KWI section regarding an element at the exact opposite end of the periodic table: the fission of uranium. As she recalled, the talk in the KWI immediately touched on the possibility of an explosive device based on a chain reaction, the nuclear equivalent of what she had studied in her dissertation on chemical chain reactions. Thus in 1939–1940 she was recruited (informally and unofficially, of course) into the "Uranium Club" based in the neighbouring KWI for Physics to study the possibility of producing a nuclear weapon for the German military, and her hydrogen expertise may also have proved useful in discussing techniques for producing concentrated heavy water as a moderator in a controlled reactor [Bobleter, 1990b, p. 472, and 1997, pp. 398–399; Stöger, 1990; Beneke, 1999, pp. 321–323].

Ironically, the war the Nazis started in 1939 finally forced them to open the gates of academe to German women, including Erika Cremer. As young male scientists joined the military and left the universities short-staffed, in 1940 the government finally decided to permit women to go through the political indoctrination camps that would make it possible for them to become university instructors. Cremer did so and found herself enjoying the unexpected pleasure of being able to choose between several offers. Wanting to be near her home in Munich and close to her beloved Alps, she chose the Institute for Physical Chemistry at the once and future Austrian University of Innsbruck, which was part of Germany from 1938 to 1945 [Oberkofler, 1998, pp. 33–34; Beneke, 1999, p. 323].

Cremer was initially given lecturing duties in the new Institute, then being equipped (using remodelled rooms in the building of the Institute for Chemistry) under its new director Carl Angelo Knorr (1894–1960, but she did not receive the formal title of "Dozentin" (instructor) until 1942 [Oberkofler, 1998, pp. 33–34]. By this time, she had largely completed her final round of work on the ortho-parahydrogen problem. Her last two major publications on this topic appeared as entries in G.-M. Schwab's widely respected, multi-volume *Handbuch der Katalyse* ("Handbook of Catalysis"), summarizing her previous work and that of other scientists on the homogeneous and the heterogeneous forms of catalytic transformation between ortho- and parahydrogen [Cremer, 1941; 1943]. Of these probably the second is the more significant contribution, because heterogenous catalysis (in which the catalyst is in a different phase than the reactants, e.g. solid vs. liquids or gases) is of more general interest than homogeneous catalysis (which involves catalysts in the same phase as the reactants, e.g. as a mixture of gases). Unfortunately, because Schwab's "handbook" appeared in German during the war, scientists on the other side did not gain access to it until several years after its appearance, and it is difficult to find citations to Cremer's contributions in later English-language publications. This may be in part because catalysis has long been something of a "black art" [Smith, 2011, p. 311]. Although it plays a crucial role in the chemical industry, much of the research is done in corporate laboratories that discourage publications. Meanwhile Cremer had begun to move in new scientific directions. Her 1941 collaboration with Knorr on the catalytic hydrogenation of acetylene, which was peripherally related to the ortho-and parahydrogen problem, raised questions that would lead her to develop gas chromatography beginning in 1944 [Bobleter, 1997, p. 399].

Following the war, Cremer remained in Innsbruck. Her Institute had been bombed and was operating in temporary quarters; all the male members of staff had been dismissed for political activity or returned to Germany, so Cremer became acting director. For several years she and her students worked under the difficult circumstances of occupation and gradual reconstruction (Austria was divided into four zones for a full decade from 1945 to 1955, but unlike Germany the country was then reunified with the withdrawal of all occupying forces). Despite the difficulties during this period, up to 1951 they made significant contributions to the new analytical technique of gas-solid chromatography. For various reasons this work did not begin to gain wider recognition until 1959, when Cremer provided a German translation for and a major supplement to the second edition of the standard work on gas chromatography [Keulemans *et al.*, 1959]. This finally resulted in her formal appointment to the chair of physical chemistry in Innsbruck, which she held until her retirement in

1970. She was much-honoured before her death in 1996, more for chromatography than for her work on hydrogen. Nevertheless, the latter must command special respect, given her ability to make major contributions to our understanding of hydrogen in the absence of a salaried academic position and with a minimum of formal institutional support, and despite the fact that, as mentioned above, her contributions were often followed up by others who received greater recognition than she. The interconversion of the two forms of hydrogen became particularly important in the post-war era because of the need for large quantities of hydrogen in military applications (bombs and rocketry) as well as in fuel cells for generating clean power. The exothermic nature of the spontaneous conversion of ortho- to parahydrogen could produce problems including the substantial loss of gas during low-temperature storage, unless the hydrogen was first catalytically converted to the para- form. This has continued to be the subject of ongoing research, which has, in effect, continued Cremer's pioneering studies, albeit generally crediting others as the pioneers [typical is Weitzel *et al.*, 1958; an exception is Petitpas *et al.*, 2014, p. 6536].

References

Beneke, K. (1999). Erika Cremer (20.05.1900 München – 21.09.1996 Innsbruck): Pionierin der Gaschromatographie, in *Biographien und wissenschaftliche Lebensläufe von Kolloidwissenschaftlern, deren Lebensdaten mit 1996 in Verbindung stehen*, Beiträge zur Geschichte der Kolloidwissenschaften, VIII, Mitteilungen der Kolloid-Gesellschaft, ed. Beneke, K. (Verlag Reinhard Knof, Nehmten), pp. 310–334.

Bobleter, O. (1990a). Interview in German with Erika Cremer (1990), in Video History of Catalysis (uploaded Jan 15, 2015), ed. Davis, B. (https://www.youtube.com/watch?v=YT0VVaomuho, last accessed 1 June 2018).

Bobleter, O. (1990b). Professor Erika Cremer — A Pioneer in Gas Chromatography, *Chromatographia*, 30, pp. 471–476.

Bobleter, O. (1997). In memoriam em. Univ.-Prof. Dr. phil. Dr. rer. nat. H.c. Erika Cremer (1900–1996): 96 Jahre eines Forscherlebens, *Ber. nat.-med. Verein Innsbruck*, 84, pp. 397–406.

Chayut, M. (1994). From Berlin to Jerusalem: Ladislaus Farkas and the Founding of Physical Chemistry in Israel, *Historical Studies in the Physical and Biological Sciences*, 24, pp. 237–263.

Cremer, E. (1927). Die Reaktion zwischen Chlor, Wasserstoff und Sauerstoff im Licht, *Z. Physik. Chem.*, 128, pp. 285–317; published form of Cremer, E. (1927). Über die Reaktion zwischen Chlor, Wasserstoff und Sauerstoff im Licht, unpublished PhD dissertation (University of Berlin).

Cremer, E. (1935). Kinetik der heterogenen o-p-Wasserstoffumwandlung an festem Sauerstoff, *Z. Physik. Chem.*, B 28, pp. 383–392.

Cremer, E. (1939). Bestimmung der Selbstdiffusion in festem Wasserstoff aus dem Reaktionsverlauf der Ortho-Para-Umwandlung, Habilitationsschrift (University of Berlin).

Cremer, E. (1941). Homogene o- und p-Wasserstoffkatalyse, in *Handbuch der Katalyse*, ed. Schwab, G.-M., vol. I: *Allgemeines und Gaskatalyse* (Springer Verlag, Vienna), pp. 325–384.

Cremer, E. (1943). Heterogene o- und p-Wasserstoffkatalyse, in *Handbuch der Katalyse*, ed. Schwab, G.-M., vol. VI: *Heterogene Katalyse III* (Springer Verlag, Vienna), pp. 1–35.

Cremer, E. (1994). Retired Professor of Chemistry, University of Innsbruck, Austria, telephone interview with Jeffrey A. Johnson, 29 July 1994.

Cremer, E. and Polanyi, M. (1932). Eine Prüfung der "Tunneltheorie" der heterogenen Katalyse am Beispiel der Hydrierung von Styrol, *Z. Physik. Chem.*, B 19, pp. 443–450.

Cremer, E. and Polanyi, M. (1933). Die o-p-Wasserstoffumwandlung in festem Zustand, *Z. Physik. Chem.*, B 21, pp. 459–468.

Cremer, E. and Schwab, G.-M. (1929). Zusammenhang zwischen Aktivierungswärme und Aktivität. *Z. Physik. Chem.*, 144, p. 243.

Ettre, L. S. (2008). The Beginnings of Gas Adsorption Chromatography 60 years Ago, *LCGC North America*, 26, pp. 48–60.

Farkas, A. (1989). Paul Harteck the Triumphant Decade 1925–1934, *Ambix*, 36, pp. 91–102.

Farkas, L. and Sachsse, H. (1933). Über die homogene Katalyse der Para-Orthowasserstoffumwaldung unter Einwirkung paramagnetischer Moleküle I, *Z. Physik. Chem.*, B 21, pp. 1–18.

Günther, P. (1938). Evaluation of the Work of Erika Cremer, Transcribed in Oberkofler, G. (1998). *Erika Cremer (1900–1996): Ein Leben für die Chemie* (Studien Verlag, Innsbruck), pp. 90–92.

Hurwic, J. (2000). *Kasimir Fajans (1887–1978), Lebensbild eines Wissenschaftlers*, Berliner Beiträge zur Geschichte der Naturwissenschaften und der Technik 25 (ERS-Verl., Berlin).

Johnson, J. A. (2019). Erika Cremer and the Origins of Gas–Solid Adsorption Chromatography, 1944–1947, in *The Posthumous Nobel Prize in Chemistry, vol. 2: Ladies in Waiting for the Nobel Prize*, eds. Mainz, V. V. and Strom, E. T.

(ACS Books, Washington, DC), pp. 183–198; online (2018) at https://pubs. acs.org/isbn/9780841233911.

Keulemans, A. I. M., Cremer, E. and Verver, C. G. (1959). *Gaschromatographie* (Verlag Chemie, Weinheim).

Kolomnikov, I. G., Efremova, A. M., Tikhomirovaa, T. I., Sorokinaa, N. M. and Zolotova, Y. A. (2018). Early Stages in the History of Gas Chromatography, *J. Chromatogr.*, A, 1537, pp. 109–117.

Nye, M. J. (2011). *Michael Polanyi and His Generation: Origins of the Social Construction of Science* (University of Chicago Press, Chicago, IL).

Oberkofler, G. (1998). *Erika Cremer (1900–1996): Ein Leben für die Chemie* (Studien Verlag, Innsbruck).

Oesper, R. E. (1952). Georg-Maria Schwab, *J. chem. Educ.*, 29, p. 247.

Petitpas, G., Aceves, S. M., Matthews, M. J. and Smith, J. R. (2014). Para-H2 to Ortho-H2 Conversion in a Full-Scale Automotive Cryogenic Pressurized Hydrogen Storage up to 345 Bar, *Int. J. Hydrogen Energy*, 39, pp. 6533–6547.

Semenov, N. N. (1964). Some Problems Relating to Chain Reactions and to the Theory of Combustion. Nobel Lecture, December 11, 1956, in *Nobel Lectures: Chemistry* 1942–1962, Nobelstiftelsen (Elsevier, Amsterdam), pp. 487–514.

Smith, J. K. (2011). The Catalyst Club: Contentious Chemistry and Confounding Innovation, *Technology and Culture*, 52, pp. 310–334.

Stöger, M. (1990). *Ein Leben für die Wissenschaft: Erinnerungen der Forscherin Erika Cremer*, videofilm in 2 parts (Aquamarin-Filmproduktion, Vienna).

Vogt, A. (1998). Aufbruch und Verdrängung: Wissenschaftlerinnen an der Berliner Universität zwischen 1918 und 1945/46, in *Frauen an der Humboldt-Universität 1908–1998: Vorträge anläßlich der Festveranstaltung 90 Jahre Frauen an der Berliner Universität*, ed. Kramer, G. (Humboldt University, Berlin), pp. 21–48.

Weitzel, D. H., Loebenstein, W. V., Draper, J. W. and Park, O . E. (1958). Ortho-Para Catalysis in Liquid-Hydrogen Production, *J. Res. Natl. Bur. Stand.*, 60, pp. 221–227.

Select Bibliography

Bobleter, O. (1990). Professor Erika Cremer — A Pioneer in Gas Chromatography, *Chromatographia*, 30, pp. 471–476.

Cammack, R., Frey, M. and Robson, R., eds. (2001). *Hydrogen as a Fuel: Learning from Nature* (Taylor & Francis, London).

Ettre, L. S. (2008). The Beginnings of Gas Adsorption Chromatography 60 Years Ago, *LCGC North America*, 26, pp. 48–60.

Farkas, A. (1935). *Orthohydrogen, Parahydrogen and Heavy hydrogen*, The Cambridge Series of Physical Chemistry (Cambridge University Press, Cambridge).

Johnson, J. A. (2019). Erika Cremer and the Origins of Gas–Solid Adsorption Chromatography, 1944–1947, in *The Posthumous Nobel Prize in Chemistry, vol. 2: Ladies in Waiting for the Nobel Prize*, eds. Mainz, V. V. and Strom, E. T. (ACS Books, Washington, DC), pp. 183–198; online (2018) at https://pubs.acs.org/isbn/9780841233911.

Nye, M. J. (2011). *Michael Polanyi and His Generation: Origins of the Social Construction of Science* (University of Chicago Press, Chicago, IL).

Oberkofler, G. (1998). *Erika Cremer (1900–1996): Ein Leben für die Chemie* (Studien Verlag, Innsbruck).

Rigden, J. S. (2002). *Hydrogen: The Essential Element* (Harvard University Press, Cambridge, MA).

Smith, J. K. (2011). The Catalyst Club: Contentious Chemistry and Confounding Innovation, *Technology and Culture*, 52, pp. 310–334.

The Importance of Carbon: The Work of Dame Kathleen Lonsdale FRS (1903–1971)

Jennifer M. Wilson

Honorary Research Associate,
Department of Science and Technology Studies,
University College London, UK

Carbon is a chemical element with the symbol C and an atomic number of six. It is non-metallic and tetravalent since it has four electrons available to form covalent chemical bonds. The atoms of carbon can bond together in different ways, forming what are known as allotropes, such as diamond and graphite. The physical properties of carbon vary according to the allotropic form. For example, graphite is opaque, one of the softest materials known and is a good electrical conductor. Diamond is highly transparent, the hardest naturally occurring material known, and has a low electrical conductivity. Carbon forms long chains of interconnecting carbon–carbon bonds, a property referred to as catenation. The element occurs in all known organic life and is the basis of organic chemistry. Organic molecules can contain straight chains of carbon atoms, or the carbon atoms can be arranged in a ring as in benzene.

Figure 1. Photograph of Kathleen Lonsdale on her election to the Fellowship of the Royal Society, May 1945 (©Godfrey Argent Studio, reproduced by kind permission of the Royal Society).

This essay is to explain and discuss the work undertaken by Dame Kathleen Lonsdale FRS (1903–1971) on the structures of benzene and diamond. Lonsdale was one of the prominent female X-ray crystallographers of the twentieth century, and in 1945 one of the first two women to be elected a Fellow of the Royal Society of London (Figure 1).[1] Lonsdale's association with carbon started in her early career when she confirmed the structure of benzene. It was her first major contribution to crystallography,

---

[1] The other woman was Marjorie Stephenson (1885–1948) [Mason and King-Hele, 1992]. Biographies of Lonsdale include Hodgkin's contribution to the *Biographical Memoirs of Fellows of the Royal Society* [Hodgkin, 1975] and an account of Lonsdale's married life to a fellow student and Quaker [Julian, 1996]. Baldwin has focused on Lonsdale's scientific career being shaped by her identity as a married woman and mother [Baldwin, 2009]. A tribute by a close colleague has detailed Lonsdale's career at University College London [Milledge, 1975].

and it was an important step in the development of organic chemistry. Later in her career Lonsdale had further associations with carbon, when she devoted considerable time to the study of diamonds, both natural and artificial. Her research contributed to the knowledge of diamonds, as it showed new experimental facts which were supported by X-ray photographs. Lonsdale's work was acknowledged in 1967, when an allotrope of carbon with a hexagonal lattice was first identified from the Canyon Diablo meteorite. It was named Lonsdaleite in her honour.

The Structure of Benzene

Benzene is a very important carbon compound found in crude oil and the major part of gasoline. It is the starting material for making important products such as plastics, resins, synthetic fibres, dyes, drugs and pesticides.

In 1865, the structure of benzene had been proposed by the professor of chemistry at Ghent University, August Kekulé (1829–1896), to be a six-membered ring of carbon atoms with alternating single and double bonds, subsequently drawn as a hexagon. However, by 1925 there was still no confirmation of the ring structure of benzene which had been conjectured by Kekulé sixty years earlier, and which remained a source of argument amongst chemists. Whilst working in the physics department at the University of Leeds, Lonsdale was given crystals of hexamethybenzene $C_6(CH_3)_6$ for X-ray analysis work. Since benzene was a liquid it was not suitable for study, but the substitution of six methyl (CH_3) groups into the benzene ring had provided a solid derivative suitable for analysis. Crystallographers were particularly interested in the benzene structure and hoped that the recently developed technique of X-ray analysis would provide the solution. X-ray crystallography is a technique for determining the structure of a crystalline substance by analysing how it scatters a beam of X-rays.

In 1928, when Lonsdale started her work on hexamethylbenzene, the evidence was confused as scientists debated whether the benzene ring was planar or flat like that in graphite, or had a puckered or zig-zag arrangement found in diamond. From her experimental work Lonsdale deduced that the molecule existed in the crystal as a separate identity, the benzene carbon atoms were arranged in a ring formation, and the ring was hexagonal and planar. She concluded that this supplied a definite proof, from an X-ray point of view, that the chemist's conception of the benzene ring is a true representation of the facts [Lonsdale, 1929a, p. 503]. Lonsdale's investigation then went on to show that there were several similarities with graphite. The size of the ring carbon atom was the same as that in graphite, and the carbon atom of the methyl group was found to lie in the plane of the benzene ring so that, as in graphite, three of the valencies of the ring carbon atom were coplanar, i.e., lay in the same plane [Lonsdale, 1929a, pp. 504–514; 1929b, pp. 360–363]. Lonsdale herself considered this work to be "my most fundamental and satisfying piece of research" [Lonsdale, 1962, p. 599].

Lonsdale continued her research by undertaking an X-ray analysis of hexachlorobenzene $C_6H_6Cl_6$ which is also a solid. In addition to the procedure she used for hexamethybenzene, she also used Fourier analysis — a mathematical method in which general functions are represented in terms of sums of trigonometric functions. Unfortunately, her results did not match those obtained for hexamethylbenzene and she was unable to make the same conclusion. However, Lonsdale was the first to use the Fourier method in the structural analysis of an organic compound [Lonsdale, 1931].

Chemists welcomed the confirmation of the structure of the benzene ring. Lonsdale was particularly pleased that her work on benzene had brought both credibility and prestige to crystallography. Her analysis of hexamethylbenzene was fully confirmed later by Brockway and Robinson using Fourier analysis. Their paper also contained details of other aspects of the structure, for which Lonsdale gave an explanation but which she had been unable to confirm [Brockway and Robinson, 1939].

Natural and Artificial Diamonds

Lonsdale was fascinated by diamonds and spent many years researching their properties. She remarked about diamond:

> everyone must recognise it to be the aristocrat of the crystal world and must be interested to know how it attains such a high degree of perfection in so many of its properties [Lonsdale, 1944, p. 669].

From her research Lonsdale concluded that each diamond is an individual; diamonds may look alike yet behave very differently. Some properties differ widely from one specimen to another, such as the appearance of diamonds when illuminated by ultra-violet light and the colour of diamonds.

Lonsdale became interested in the thermal vibrations of atoms in crystals which accounted for the appearance of diffuse scattering, shown as extra spots and streaks, in addition to the expected sharp reflections on an X-ray photograph. Diffuse reflections are shown by diamonds which led Lonsdale to her work on both natural and artificial diamonds. Robinson *et al.* [1934] had shown that the infra-red absorption and ultra-violet transparency of diamonds had placed them in two types, I and II. Of the two types the first and commoner was opaque to both infra-red and ultra-violet radiation, while type II was rare and transparent to these radiations. It was also found that type II diamonds were much better reflectors of X-rays than type I.

For her work on diamonds, Lonsdale used divergent beam X-ray diffraction in which a strongly divergent source was used instead of the usual highly collimated or parallel beam. Lonsdale found that type II diamonds always gave excellent divergent beam photographs, whereas the best type I diamonds gave poor photographs. She concluded, that this proved that type I diamonds are "perfect" in structure since they contain no point, linear or planar imperfections, but that type II diamonds are "mosaic," being composed of small crystallites (small perfect crystals) which are not regularly arranged. Lonsdale was not only able to determine the quality of the crystals of diamonds, but also to calculate their lattice dimensions and hence the carbon-carbon bond length, which she found was the same for both types [Lonsdale, 1947].

In the nineteenth century a great many attempts were made to find conditions under which carbon could be made to crystallise as diamond rather than as the more stable form, graphite. One attempt was made by James Ballantyre Hannay (1855–1931), who reported that he had succeeded in producing a very small amount of diamond [Hannay, 1880]. Later workers, particularly Sir Charles Parsons (1854–1931), could not reproduce Hannay's research, and his claim to have made diamond by a laboratory method seems never to have been taken seriously. This is confirmed in an article written in 1928 which, having reviewed all the attempts, stated that "the conclusion seems inevitable that diamonds have not yet been produced in the laboratory" [Desch, 1928, p. 800].

The Mineral Department of the British Museum had some twelve, minute specimens, labelled as being diamonds, artificially prepared and presented by Hannay in 1880. Lonsdale in collaboration with staff from the Museum examined these diamonds using various X-ray methods. The research showed that some were pure diamond, others diamond plus a small impurity and only one was not diamond at all. One of the diamonds was found to be of the rare type II variety. The conclusion was that Hannay had in fact succeeded in synthesising diamond although in very small quantities and that his method was capable of producing the rare type of diamond which only occurs to the extent of perhaps one percent among natural diamond single crystals. The fact that Hannay's research could not be reproduced by later workers, and the fact that he made only a minute amount of diamond, suggest that it would be difficult to make artificial diamonds on a large scale [Bannister and Lonsdale, 1943].

Artificial diamonds are those made by man, unlike natural diamonds which formed naturally inside the earth millions of years ago. They were first produced in the 1950s by using high pressures and temperatures to convert graphite into diamond. Synthetic diamonds are a type of artificial diamond, made specifically by chemical synthesis as they are made in the laboratory. They were first produced in the USA by the General Electric Company of New York and in Sweden by the Allmänna Svenska Electriska Aktiebolaget, Västeras.

Lonsdale undertook X-ray studies of the diamonds produced by both companies. She found that the General Electric synthetic diamonds always contained single crystal inclusions of nickel, whereas the Swedish synthetic diamonds did not contain nickel inclusions but were generally less well crystallised than the General Electric specimens. Lonsdale also found that both diamonds gave good diffraction spots, indicating no trace of graphite present and there were no extra spots or streaks typical of type I diamonds. The research also involved a further study of a Hannay diamond to see whether any crystalline inclusions were present that would indicate it must have been made in the laboratory. No traces of nickel lines were seen, concluding that the Hannay diamond is probably a natural and not a synthetic diamond [Lonsdale *et al.*, 1959].

The hardness of diamond and its high dispersion of light are useful properties for both industrial applications and in jewellery. Artificial diamonds found almost immediate industrial applications in cutting, drilling, grinding and polishing. Eighty percent of mined diamonds are unsuitable for gemstones and are relegated for industrial use.

Diamonds are found not only in the earth but in meteorites. When meteors containing graphite strike the earth the impact transforms the graphite into a hexagonal diamond which is an allotrope with a hexagonal lattice. Hexagonal diamond has been found in the meteorites that have fallen in Arizona, USA particularly in the Canyon Diablo meteorite. Hexagonal diamond has been synthesised in the laboratory by the General Electric Company [Bundy and Kasper, 1967].

It was suggested that hexagonal diamond be named Lonsdaleite:

> after the distinguished crystallographer, Professor Kathleen Lonsdale, who has made numerous contributions to our knowledge of diamond. The name was subsequently approved by the Commission on New Mineral Names of the International Mineralogical Association [Frondel and Marvin, 1967, p. 589].

Lonsdale continued with her work on diamonds after retirement and her last paper was an explanation on how Lonsdaleite is formed from graphite [Lonsdale, 1971].

References

Baldwin, M. (2009). "Where Are Your Intelligent Mothers to Come From?": Marriage and Family in the Scientific Life of Dame Kathleen Lonsdale FRS (1903–1971), *Notes and Records of the Royal Society*, 63, pp. 81–94.

Bannister, F. A. and Lonsdale, K. (1943). An X-Ray Study of Diamonds Artificially Prepared by J. B. Hannay in 1880, *Min. Mag.*, 26, pp. 315–324.

Brockway, L. O. and Robertson, J. M. (1939). The Crystal Structure of Hexamethylbenzene and the Length of the Methyl Group Bond to Aromatic Carbon Atoms, *J. Chem. Soc.*, pp. 1324–1332.

Bundy, F. P. and Kasper, J. S. (1967). Hexagonal Diamond — A New Form of Carbon, *J. Chem. Phys.*, 46, pp. 3437–3446.

Desch, C. H. (1928). The Problem of Artificial Production of Diamonds, *Nature*, 121, pp. 799–800.

Frondel, C. and Marvin, U. B. (1967). Lonsdaleite, a Hexagonal Polymorph of Diamond, *Nature*, 214, pp. 587–589.

Hannay, J. B. (1880). On the Artificial Formation of the Diamond, *Proc. R. Soc. Lond.*, 30, pp. 188–189 (preliminary notice) and pp. 450–461.

Hodgkin, D. M. C. (1975). Kathleen Lonsdale. 28 January 1903–1 April 1971, *Biographical Memoirs of Fellows of the Royal Society*, 21, pp. 447–484.

Julian, M. (1996). Kathleen and Thomas Lonsdale, in *Creative Couples in the Sciences*, eds. Abir-Am, P., Pycior, H. and Slack, N. (Rutgers University Press, New Brunswick, NJ), pp. 170–181.

Lonsdale, K. (1929a). The Structure of the Benzene Ring in $C_6(CH_3)_6$, *Proc. R. Soc. Lond. A.*, 223, pp. 494–515.

Lonsdale, K. (1929b). X-Ray Evidence of the Structure of the Benzene Nucleus, *Trans. Faraday. Soc.*, 25, pp. 352–366.

Lonsdale, K. (1931). An X-Ray Analysis of the Structure of Hexachlorobenzene, Using the Fourier Method, *Proc. R. Soc. Lond. A.*, 133, pp. 536–552.

Lonsdale, K. (1944). Diamonds, Natural and Artificial, *Nature*, 53, pp. 669–672.

Lonsdale, K. (1947). Divergent Beam X-Ray Photography of Crystals, *Phil. Trans. R. Soc. Lond. A.*, 240, pp. 219–250.

Lonsdale, K., Milledge (née Grenville-Wells), H. J. and Nave, E. (1959). X-Ray Studies of Synthetic Diamonds, *Min. Mag.*, 32, pp.185–201.

Lonsdale, K. (1962). Reminiscences, in *Fifty Years of X-Ray Diffraction*, ed. Ewald, P. P. (N.V.A. Oosthoek, Utrecht), pp. 595–602.

Lonsdale, K. (1971). Formation of Lonsdaleite from Single-Crystal Graphite, *Am. Min.*, 56, pp. 333–336.

Mason, J. and King-Hele, D. G. (1992). The Admission of the First Women to the Royal Society of London, *Notes and Records of the Royal Society*, 46:2, pp. 279–300.

Milledge, H. J. (1975). Kathleen Lonsdale, 28 January 1903–1 April 1971, *Acta Cryst.*, 31A, pp. 705–708.

Robinson, R., Fox, J. J. and Martin, A. E. (1934). Two Types of Diamond, *Phil. Trans. R. Soc. Lond. A.*, 232, pp. 463–535.

Select Bibliography

Authier, A. (2013). *Early Days of X-Ray Crystallography* (Oxford University Press, Oxford).

Bridgmann, P. W. (1955). Synthetic Diamonds, *Sci. Am.*, 193, pp. 42–46.

Hannemann, R. E., Strong, H. M. and Bundy, F. P. (1967). Hexagonal Diamonds in Meteorites: Implications, *Science*, 155, pp. 995–997.

Hodgkin, D. M. C. (1975). Kathleen Lonsdale. 28 January 1903–1 April 1971, *Biographical Memoirs of Fellows of the Royal Society*, 21, pp. 447–484.

Julian, M. M. (1993). Women in Crystallography, in *Women in Science*: *Righting the Record*, eds. Kass-Simon, G. and Farnes, P. (Indiana University Press, Indianapolis, IN), pp. 335–383.

Lonsdale, K. (1948). *Crystals and X-Rays* (G.Bell & Sons Ltd, London).

Milledge, H. J. (1963). Synthetic Diamonds, *Sci. Prog.*, 51:204, pp. 540–550.

Spying on Shy Elements: Yvette Cauchois and her Spectrograph

Brigitte Van Tiggelen

Mémosciences asbl, Voie du Vieux Quartier 18,
1348 Louvain-La-Neuve, Belgium
Science History Institute, 315 Chestnut Street,
Philadelphia, PA 19106, USA

On 2 August 1938, the French journal *Le Figaro* reported that professor Jean Perrin (1870–1942) of the Sorbonne in Paris had announced the discovery of element 93 by a M. Hulube [*sic*] during a meeting of the *Académie des Sciences* (the French Academy of Sciences). The short note of just three sentences ends with a reference to the newly devised "spectrograph" that had delivered proof of the element's existence — a spectrograph invented by "Melle Cauchois" [N. N., 1938]. This new spectrograph significantly extended the number of applications for the (by then well-established) method of X-ray spectrography, which had been used to distinguish and characterise elements since 1914. Thirteen months later, a publication by Horia Hulubei (1896–1972) and Yvette Cauchois (1908–1999) (Figure 1) in the *Compte rendus de l'Académie des sciences* claimed that the discovery of the "natural element 93" had been confirmed, and suggested that it should be named "sequanium," from the Latin for the river

Figure 1. Yvette Cauchois, in 1965, director of the laboratory of physical chemistry, at the University La Sorbonne (Musée Curie, coll. ACJC, reproduced by kind permission of the Musée Curie).

Seine [Hulubei and Cauchois, 1939b, p. 479]. The use of the qualifier "natural" is to be understood as a contrast to other contemporary scientists' efforts to produce unknown elements by bombarding nuclei of known elements with particles. That approach was chosen by the Italian physicist Enrico Fermi (1901–1954) in 1934, in his attempt to produce elements 93 and 94; however, he was later unable to provide any chemical evidence of their existence.[1] The same fate was met by the "natural element 93," whose claim of existence was later abandoned, and sequanium is now among the "lost" elements gathered in a recent study of the same

---

[1] See Xavier Roqué on Meitner and Brigitte Van Tiggelen on Noddack in this volume.

title, *The Lost Elements* [Fontani *et al.*, 2014, pp. 321–336]. But the new instrument used in its discovery, also known as the "Cauchois spectrograph," justified its existence and utility quite independently, and continued to be used to improve the observation and the identification of elements, both known and unknown. With this, her invention, Yvette Cauchois contributed to the knowledge of the periodic system, and equipped the scientific community with another means to detect the elements, old and new, and even spot their inner structure.

Taking Spectroscopy to the Next Level

When exposed to energy, atoms of elements emit photons with specific energies. This is because the energy absorbed by an atom causes an electron from the inner shell of an atom to move to a higher energy level, and when it returns to its initial energy level, the energy gained by the excitation is emitted as a photon with a characteristic wavelength, which is unique to each element. Optical spectroscopy was used in the identification of elements from the mid-nineteenth century onwards, using wavelengths of the visible light spectrum emitted when substances are heated. But with the advent of X-ray spectroscopy, the identification process became much simpler: as Henry Moseley (1887–1915) showed in 1913, there was a straightforward link between the atomic number of an element and the corresponding X-ray spectrum. However, for some elements that are found in traces in the samples — like rare earths or heavy elements, and also for some gases — the lines are very faint, so that their spectrum is difficult to observe and read. In her doctoral research at the Sorbonne (her PhD thesis was entitled "An Extension of X-Ray Spectroscopy: A Spectrometer Focused with a Curved Crystal; X-Ray Emission Spectra from Gases" [Cauchois, 1933b])[2] Yvette Cauchois developed a technique that made it possible to measure X-ray emission lines of the low intensity typical for rare earths, rare gases and heavy elements.

---

[2]French original title: "*Extension de la spectrographie des rayons X. Spectrographe à focalisation par cristal courbé; spectre d'émission X des gaz.*" All translations from the French are the author's unless otherwise indicated.

X-ray spectrographs were then still based on William Henry Bragg's (1862–1942) concept, according to which X-rays are collected through successive slits to collimate the beam into a fine pencil line, which is reflected onto flat crystalline slates and then diverges to produce a spectrum. The successive slits reduce luminosity — that is, the amount of radiation collected by the instrument to be concentrated on the detecting device (usually a photographic plate) — to gain resolution, i.e. the capacity of the spectrograph to distinguish adjacent wavelengths. Instead of using flat slates, Cauchois chose a curved crystalline slate, akin to the use of curved mirrors and gratings used to focus UV and soft X-rays (X-rays with lower energy and wavelength). Since the curved crystal was thus also a lens, Cauchois did not need to use two slits but only one. That way, the rays reflected from the convex surface of the crystals converged onto a narrow zone, and allowed for a great gain in luminosity. With this method, Cauchois improved both the resolving power and the luminosity, and caught spectral lines that had never been observed before.

She started her work in 1928, when she started her PhD research and entered Perrin's Laboratory of Physical Chemistry at the Sorbonne. Some of her results were published four years later, even before she presented her thesis in 1933 [Cauchois, 1932; 1933a], and gained her national and international recognition for the new instrument — the Cauchois spectrograph — which read elemental spectra with more accuracy and detail than ever before. The fact that this was published so swiftly demonstrates that Perrin and others realised the incredible potential of this instrument. Indeed, thanks to this apparatus, Cauchois provided information about the complete X-ray spectra emitted by xenon and argon (K-spectra),[3] which had never been seen before.

As soon as Cauchois published her results, the community of X-ray spectroscopers and element hunters embraced the new technique.[4] Among them were Manne Siegbahn (1886–1978) of Uppsala, Pieter Zeeman

---

[3] K-lines appear when an electron transitions back to the innermost electronic shell (the K-shell).

[4] In an endnote to the second part of her original publication Cauchois mentioned that as she had been reading the proofs, an article by Manne Siegbahn had appeared, using the same method and reporting similar results [Cauchois, 1932, p. 72].

(1865–1943) of Amsterdam, and scientists from other European countries, the USSR, Japan, the USA and Australia. Cauchois herself pursued the systematic study of weak lines, the so-called "satellites," that accompany the normal X-ray lines. These lines, which are hard to observe, correspond to electronic transitions in ionised atoms. Since Cauchois was able to observe a larger number of these lines than ever before, she was able to provide the energy levels for highly ionised heavy atoms.

Novel Techniques to Spot New (?) Elements

Since the turn of the twentieth century, some chemists speculated about the existence of elements heavier than uranium, called the transuranium elements. While some tried to find them in mineral ores that also contained uranium, others later took advantage of the newly discovered induced radioactivity and techniques used to bombard nuclei to create new nuclei with more than 92 nucleons. Different spectroscopic methods had been used to demonstrate the existence of yet unknown elements, and the new spectrograph devised by Cauchois in 1932 was immediately put to use.

In the year in which Cauchois defended her thesis, another PhD student supervised by Perrin also finished a PhD on X-ray spectroscopy: the Romanian researcher Horia Hulubei (Figure 2).[5] Observing spectral lines that were as yet unidentified, he soon became convinced that these lines were emitted by unknown elements which were still missing in the periodic system, and enlisted Cauchois and her newly invented instrument in the quest for their discovery.

Together they undertook a systematic recording of the spectra of rarely occurring elements, since the technique allowed the analysis of extremely small amounts of material present in samples. Among other things, they observed the L-spectrum[6] of polonium, and were able to fully confirm the atomic number (84) from the establishment of many hitherto unobserved spectral lines.

---

[5]Hulubei thesis, "Contribution to the Study of the Quantum Diffusion of X-rays," (Contribution à l'étude de la diffusion quantique des rayons X) in 1933.

[6]L-lines result when an electron transitions back to the second innermost electronic shell (the L-shell).

Figure 2. A tea party at the laboratory of physical chemistry at the Sorbonne in honour of four doctoral candidates' thesis defenses, Paris, 1933. Left to right: Gheorghe Manu, Marie Curie, Jean Perrin, Yvette Cauchois, Horia Hulubei and (second row) Raymond Gregoire (Musée Curie, coll. Joliot, reproduced by kind permission of the Musée Curie).

In 1936 Hulubei observed lines which he assumed to be the signature of element 87 [Hulubei, 1936]. Not even a year later, convinced he had provided proof of its existence with the help of Cauchois, he daringly suggested the name "moldavium," after Moldavia (then part of Romania), a region he considered to be the edge of Europe, and wanted to be commemorated in the periodic system [Hulubei, 1937]. When Marguerite Perey (1909–1975), another scientist working in Paris, albeit in Marie Curie's (1867–1934) laboratory, announced in January 1939 that she had discovered element 87 while working with the radioactive actinium series,[7] Hulubei maintained his claim, assuring that the accuracy of the spectrograph had allowed him to observe element 87 first [Hulubei, 1939, p. 678]. His argument was that small amounts of element 87 could be

---

[7] See Rayner-Canham and Rayner-Canham on Perey in this volume.

found in nature, and this is an accepted fact today. Since francium-223 is the decay product of actinium 227, there are indeed (small) traces of it in ores containing uranium and thorium. The minerals used by the team, however, did not contain any francium.

At about the same time, in 1938, while examining minerals from Madagascar, Hulubei and Cauchois identified lines they ascribed to element 93 [Hulubei and Cauchois, 1938]. While element 93 was expected to be highly radioactive, they observed only weak radioactivity, and believed that this was the signature of a stable (or long-lived) isotope. Not even a year later, the name "sequanium" was suggested, this time paying homage to Cauchois' home country [Hulubei and Cauchois, 1939b]. In 1940, however, Edwin McMillan (1909–1991) and Philip Hauge Abelson (1913–2004) used nuclear techniques to produce what would be recognised as the first isotope of element 93 isolated by man. After World War II, it was named "neptunium."

Undeterred, Hulubei and Cauchois continued their spectroscopic investigations. When observing three previously unknown principal lines, Hulubei claimed that this time they had discovered the missing element 85 [Hulubei and Cauchois, 1939a; 1940]. Controversies and doubts followed the claim, which continued throughout World War II [Hulubei, 1947; Fontani *et al.*, 2014, pp. 331–333; Thornton and Burdette, 2010]. Reporting on the complete spectrum in October 1944, Hulubei suggested the name "dor," derived from the Romanian word for "longing," as in longing for peace [Hulubei, 1944],[8] but by the end of the 1940s the discovery of element 85 had been ascribed to Dale R. Corson (2014–2012), Kenneth Ross MacKenzie (1912–2002) and Emilio G. Segrè (1905–1989). In 1940 they produced element 85 by means of bombarding bismuth-209 with alpha particles, and were then allowed to name it "astatine." As demonstrated by Berta Karlik (1904–1990) and Traude Bernert (1915–1998),[9] astatine presents as a decay product from uranium and thorium ores, so that the French team might, indeed, have observed 85, and thus have been the first to spot its existence [Thornton and Burdette, 2010, p. 92–94].

---

[8]The suggested symbol for dor was Do. Hulubei's equipment was destroyed during the American bombardment of Bucharest on 15 April 1944 [Fontani *et al.*, 2014, p. 333].
[9]See Forstner in this volume.

During this period there were many claims of discoveries of new elements, and the scientific community debated the question of what exactly constituted an elemental discovery, how the ultimate evidence could be provided, and how priorities could be attributed. The successive but unsuccessful claims did not undermine any of the protagonists' careers. Hulubei's reputation as a skilled spectroscopist remained intact, and he had an internationally respected career in his home country at the University of Bucharest [Romanian National Committee of Physics, 1973]. As for Cauchois, the identification of elements was not her main area, and her spectrograph was used in many projects that did not involve the finding of new elements. In addition to confirming and expanding the spectra of many elements that were already known, she was also able to investigate the inner structure of atoms [Wuilleumier, 2003].

A Life in Science

Born in Paris in 1908, Yvette Cauchois had always been attracted to science, and after attending secondary school she chose to study the physical sciences at the Sorbonne.[10] Her acceptance into Jean Perrin's Laboratory of Physical Chemistry in 1928 proved crucial to her career. The laboratory was a node for the local and international scientific community, and later in life Cauchois would fondly remember the so-called "Monday teas" at the laboratory of physical chemistry at the Sorbonne of the pre-war period, where she was able to meet with foreign as well as the French top scientists of the day. These meetings also included politicians and artists around the "Arcouest colony," a place in Brittany where the Curies, Perrin and others regularly met when they were not in Paris.

Under the guidance of Francis, the son of Jean, Perrin (1901–1992) Cauchois researched the phenomenon of fluorescence,[11] and turned her attention to X-ray spectroscopy for the thesis she defended in July 1933. As mentioned above, this work immediately established Cauchois' reputaiton in the French scientific community and beyond. Even before she was awarded her doctorate, she received one of the first research

---

[10]The biographical details are taken from [Bonnelle, 2001] and [Bonnelle, 2011].

[11]Emission of light after absorption of electromagnetic radiation.

scholarships awarded by the "Caisse Nationale des Sciences," which had been founded two years earlier by her supervisor Perrin, and would later become the "Centre National de la Recherche Scientifique" (CNRS). In 1937 she was offered a permanent position as CNRS research associate. As a member of the French government when the war broke out, Jean Perrin had to leave France for the United States, and between 1940 and 1945 Cauchois was responsible for the laboratory. Perrin died in New York in 1942, and after the liberation of France, in 1945, Cauchois became an assistant professor at the Sorbonne; she became a full professor in 1951. As the director of the laboratory, she supervised the move to larger premises in 1953 and the creation of the "Centre de Chimie Physique" in Orsay, South of Paris.

From 1975 to 1978, Cauchois was the second female president of the "Société Française de Chimie Physique" (the French Society for Physical Chemistry); the first had been Marie Curie. Notably, the first female president of the "Société Chimique de France" merged with the "Société de Chimie Physique" was not elected until 2015. Cauchois was awarded multiple prizes, mostly from French institutions but also from the Czechoslovak spectroscopic society. Once retired, Cauchois continued to researching and working in the lab. Hulubei was one of her earliest laboratory colleagues, along with other Romanians, and this created a link to Romania. Her ties with that country intensified towards the end of her life, and she converted to the Orthodox Christian faith while visiting the Barsana monastery, Romania, at the age of 90. Unfortunately her health deteriorated during this journey, she died back in Paris and was buried in the monastery.

***

Cauchois's achievements cannot be fixed to one element. In fact, the claims made by her long-time colleague Hulubei to which she was associated were controversial. Yet this was the state of affairs in the scientific community at a time when the concept of an element in itself, and the concept of elemental discovery, were undergoing a major shift that would leave wet chemistry and traditional techniques of investigation behind. The Cauchois spectrograph, however, was a definite step towards a deeper knowledge of the elements through their spectroscopic properties. While developing soft X-ray techniques further, Cauchois also undertook research in the area of UV spectroscopy to study light elements and chemical bonding in solids, and promoted the utilisation of synchrotron radiation[12] for atomic and molecular physics [Bonnelle, 2011]. She suggested the use of the newly built storage ring in Orsay as a synchrotron radiation facility, a project that was first turned down by particle physicists, but would eventually lead to the LURE laboratory and the first synchrotron used for atomic physics. Her technical imagination thus led Cauchois towards a more precise knowledge of elements and their characteristic X-ray spectra [Wuilleumier, 2003]. In fact, she developed spectroscopic tools in a much broader way, that allowed her to lead the way into looking into all aspects of the intimacy of matter, at the chemical and electronic levels.

References

Bonnelle, C. (2001). Yvette Cauchois, *Physics Today,* April 2001, pp. 88–89.

Bonnelle, C. (2011). Yvette Cauchois, in *European Women in Chemistry,* eds. Apotheker, J. and Simon Sarkadi, L. (Wiley-VCH, Weinheim), pp. 175–179.

Cauchois, Y. (1932). Spectrographie des rayons X par transmission d'un faisceau non canalisé à travers un cristal courbé (1), *J. Phys.*, VII, 3, pp. 320–336.

Cauchois, Y. (1933a). Spectrographie des rayons X par transmission d'un faisceau non canalisé à travers un cristal courbé (2), *J. Phys.*, VII, 4, pp. 61–72.

Cauchois, Y. (1933b). Extension de la spectrographie des rayons X (Masson, Paris).

Fontani, M., Costa, M. and Orna, M. V. (2014). *The Lost Elements: The Periodic Table's Shadow Side* (Oxford University Press, Oxford).

Hulubei, H. (1936). Recherches relatives à l'élement 87, *Comptes rendus de l'Acad. Sciences Paris*, 202, pp. 1927–1929.

Hulubei, H. (1937). Recherches relatives à l'élement 87- II, *Comptes rendus de l'Acad. Sciences Paris*, 205, p. 854.

Hulubei, H. (1939). Sur l'élément 87 (Ml), *Comptes rendus de l'Acad. Sciences Paris*, 209, pp. 675–678.

Hulubei, H. (1944). Sur l'élément 85, *Bull. Section Scientifique de l'Académie Roumaine*, 27, pp. 124–134.

---

[12] Synchroton radiation is electromagnetic radiation emitted when charged particles are accelerated in a curved path.

Hulubei, H. (1947). État actuel des informations sur les isotopes de numéro atomique 85, J. Chim. Phys., 44, pp. 225–229.

Hulubei, H. and Cauchois, Y. (1938). L'existence de l'élément 93 à l'état naturel, *Comptes rendus de l'Acad. Sciences Paris*, 207, pp. 333–334.

Hulubei, H. and Cauchois, Y. (1939a). Spectres de l'émission propre ondulatoire du radon et de ses dérivés: Raies attribuables à l'élément 85, *Comptes rendus de l'Acad. Sciences Paris*, 209, pp. 39–42.

Hulubei, H. and Cauchois, Y. (1939b). Nouvelles recherches sur l'élément 93 naturel, in *Comptes rendus de l'Acad. Sciences Paris*, 209, pp. 476–479.

Hulubei, H. and Cauchois, Y. (1940). Sur la présence de l'élément 85 parmi les produits de désintégration du radon, *Comptes rendus de l'Acad. Sciences Paris*, 210, pp. 696–697.

N. N. (1938). Académie des Sciences, *Le Figaro*, 2 August 1938, p. 7.

[Romanian National Committee of Physics] (1973). Horia Hulubei, *Europhysics News*, 1, p. 5.

Thornton, B. F. and Burdette, S. C. (2010). Finding Eka-Iodine: Discovery Priority in Modern Times, *Bull. His. Chem.*, 35, pp. 86–96.

Wuilleumier, F. (2003). Yvette Cauchois and her Contribution to X ray and Inner Shell Ionization Processes, in *X-ray and Inner-Shell Processes: 19th International Conference on X-Ray and Inner-Shell Process*, eds. Bianconi, A. Marcelli, A. and Saini, N. L., AIP Conference Proceedings, vol. 652 (American Institute of Physics, Melville, NY), pp. 30–49.

Select Bibliography

Bonnelle, C. (2011). Yvette Cauchois, in *European Women in Chemistry*, eds. Apotheker, J. and Simon Sarkadi, L. (Wiley-VCH, Weinheim), pp. 175–179.

Cauchois, Y. (1932). Spectrographie des rayons X par transmission d'un faisceau non canalisé à travers un cristal courbé (1), *J. Phys.*, VII, 3, pp. 320–336.

Cauchois, Y. (1933). Spectrographie des rayons X par transmission d'un faisceau non canalisé à travers un cristal courbé (2), *J. Phys.*, série VII, IV, pp. 61–72.

Thornton, B. F. and Burdette, S. C. (2010). Finding Eka-Iodine: Discovery Priority in Modern Times, *Bull. His. Chem.*, 35, pp. 86–96.

Wuilleumier, F. (2003). Yvette Cauchois and her Contribution to X-ray and Inner Shell Ionization Processes, in *X-ray and Inner-Shell Processes: 19th International Conference on X-Ray and Inner-Shell Process*, eds. Bianconi, A. Marcelli, A. and Saini, N. L., AIP Conference Proceedings, vol. 652 (American Institute of Physics, Melville, NY), pp. 30–49.

Part 4

Clusters of Women in Radioactivity

Marie Skłodowska Curie — Polonium and Radium

Xavier Roqué

Centre for the History of Science,
Universitat Autònoma de Barcelona, 08193 Bellaterra, Spain

No element is more closely related to a woman scientist than radium is to Marie Skłodowska Curie (1867–1934). The story of the discovery of polonium and radium has been told many times, not least by the Curie family itself. Precisely for this reason it needs to be retold, as important insights into exceptionality and entrepreneurship have recently challenged the heroic tale. The story of Marie Curie and radium is not just about personal resolution and physical force, but also about family cooperation, domestic decisions, and the public value of commercial science.

Marie Skłodowska arrived in Paris from her native Warsaw in 1891. Women were not admitted to Polish universities at the time, but Marie's parents, both teachers, had educated their daughters, and Marie intended to study mathematics and physics at La Sorbonne, to follow into the footsteps of her elder sister Bronisława, who had studied medicine there. The Skłodowska family fully supported Marie's choice. In 1894 she met the French physicist Pierre Curie (1859–1906), and they were married in the following year. The couple "jointly embraced the 'anti-natural path [of devotion to science],' which permitted [them] time only for science and for their extended family" [Pycior, 1987, p. 198]. Curie described it vividly in her 1923 biography of Pierre, in an intriguing chapter devoted to "Marriage and Organization of Family Life":

> We lived a very single life, interested in common, as we were, in our laboratory experiments and in the preparation of lectures and examinations. ... All formal social obligations were excluded from our life. ... Our relations with our families were very restricted on his side as on mine... [Curie, 1923, p. 82, pp. 86, 88].

In her autobiographical notes, included in the 1923 American edition of the biography, Curie would likewise propose the couple's scientific achievements as a consequence of its domestic arrangements: "It was under this mode of quiet living, organized according to our desires, that we achieved the great work of our lives" [Curie, 1923, p. 180; Opitz *et al.*, 2016].

The couple's first daughter, Irène (1897–1956), was born in September 1897, and three months later Marie Curie began work on her dissertation. While her Polish family had seen her through her scientific education, now Pierre and his father, Eugène Curie (1827–1910), provided essential support in the early stages of her scientific career.

Polonium

In 1896, following the discovery of X-rays, the French scientist Henri Becquerel (1852–1908) observed that uranium emitted a new kind of radiation which affected photographic plates and made air conduct electricity. Both effects were barely quantifiable, and Becquerel's "rayons uraniques" did not attract much attention. Marie Curie chose the rays as her dissertation topic because she had access to state-of-the-art electrometric devices: an improved, highly sensitive electrometer of Pierre's design, and a quartz charge generator based on piezoelectricity, the physical phenomenon discovered by Pierre and his elder brother Jacques Curie (1855–1941) in 1880. The "quartz piézoélectrique" created minute electrical charges by applying pressure to a quartz lamella. Charge increased linearly with weight, allowing the steady production of charges of the order of picocoulombs [Boudia and Moulinié, 2009].

The work proceeded at the "École Supérieure de Physique et de Chimie Industrielles de la ville de Paris" (ESPCI Paris), where Pierre taught. He was soon to join in Marie's radioactive researches. Their notebooks, now publicly available through the Bibliothèque nationale de France's digital portal Gallica, allow us follow them into the laboratory [Curie and Curie, 1897–1898, 1898a, and 1898–1900; transcription included in Curie, 1955]. Marie Curie first examined a range of metals, salts, oxides and minerals that were placed in a chamber to which an electric voltage was applied, and that collected ionisation charges (ionisation chamber). The current was fed into the electrometer and the substances classified as "active," i.e. current-producing, or "not active." Marie Curie reported in April 1898 that "all the uranium compounds studied are active and, in general, more active to the extent that they contain more uranium." Thorium compounds were also found to be active [Curie, 1898, p. 1102].[1]

The Curies next combined their devices into a single system that worked as an electrometric balance (Figure 1). Pierre's electrometer was connected to both the ionisation chamber and the quartz charge generator, so that incoming opposite currents would cancel each other. Upon the placement of a finely powdered active substance onto the ionisation chamber's plate, the circuit was closed, and current began to flow into the electrometer. With a carefully gauged movement, Marie Curie exerted pressure on the quartz lamella in order to strike a perfect balance with the ionisation current. The total charge produced by the piezoelectric device in a given period of time measured the ionization current, and thus "radioactivity," as Pierre and Marie Curie [1898b] referred to the new property of matter.

The observation that pitchblende, a uranium mineral, was more active than uranium, put into question the Curies' earlier findings that the radioactivity of uranium compounds was proportional to their uranium content. Suspecting the presence of a new substance, the Curies, with the help of their co-worker Gustave Bémont (1857–1932), set out to isolate it. They submitted pitchblende to chemical processes that produced distinct residues, which were then electrometrically checked to trace the new substance: "Our chemical researches have been constantly guided by the

---

[1]All translations from the French are the author's unless otherwise indicated.

control of the radiant activity of the products separated with each operation" [Curie and Curie, 1898b, p. 176].

Six months into their researches the Curies were getting samples four hundred times more active than uranium. No known element accounted for this activity, and in July 1898 the Curies concluded that the samples contained a new "simple body" or metal and named it "polonium," after Marie's homeland. But what exactly had they discovered? They neither had a pure sample of the new element, nor had they been able to ascertain its atomic weight, nor indeed found a single trace of its spectrum. Polonium was solely identified by the chemical operations distinguising it from every other element: unlike arsenic and antimony, it was "completely insoluble in the ammonium sulphide"; unlike copper, it was precipitated "completely by ammonia," etc. Polonium behaved very much like bismuth, from which it was difficult to distinguish [Curie and Curie, 1898b, pp. 176 and 177].

Radium

By November 1898 the Curies realised that some residues from polonium extraction remained very active. Applying their electrometric techniques in conjunction with chemical separation methods, they tracked a new, strongly radioactive substance "entirely different in its chemical properties" from polonium: it was precipitated "neither by hydrogen sulphide … nor by ammonia"; and its sulphate and its carbonate, unlike polonium's, were insoluble in water [Curie et al.,1898, p. 1215]. The new substance behaved like barium, which was not radioactive, but the Curies and Bémont soon realised that its chloride was less soluble in a water solution of alcohol than that of barium. "Based on this fact, one ought to be able to effect a series of fractionations … We have thus obtained chloride samples nine hundred times more active than uranium" [Curie et al., 1898, p. 1216].

A new element was emerging along with the techniques to concentrate it, but a shortage of material prevented the Curies from pursuing the work further. In December 1898 they announced the existence of a strongly radioactive new substance, radium [Curie et al., 1898]. Radium was as yet little more than the set of chemical operations that yielded increasingly active samples, several hundred times more active than

Figure 1. Pierre and Marie Curie, with their laboratory assistant, Mr Petit, and the electrometric device used to discover radium and polonium at the *École Supérieure de Physique et de Chimie Industrielles de la ville de Paris*, ca. 1898 (Musée Curie, coll. ACJC; reproduced by kind permission of the Musée Curie).

uranium. An expert spectroscopist, Eugène Demarçay (1852–1903), identified one new line which could not be ascribed to any known element, and whose intensity increased with the activity of the samples, "a very weighty reason for attributing it to the radioactive part of our substance" [Curie *et al.*, 1898, p. 1217].

The Curies' electrometric techniques rendered elements visible through their radiation. Their "chemistry of the invisible" had led them to claim the discovery of two new elements. But to convince the scientific community about their existence they needed to produce the new elements in larger amounts and establish their atomic weight and spectrum.

Manufacturing Evidence

A steep increase of activity in ever more concentrated samples demonstrated radium's enormous radioactivity. The average atomic weight of "active barium" ("baryum actif" [Curie *et al.*, 1898, p. 1217]) was barely distinguishable from barium's, meaning that very little radium made a big difference. Realizing that they would need to process pitchblende by the ton, the Curies sought industrial assistance. As early as in 1899 the "Société Centrale de Produits Chimiques" processed a few tons of pitchblende from the rich uranium mines in Bohemia for them. In 1903 a visitor to the Curies' laboratory reported that "all the radium manufactured in France is made under the supervision of Prof. Curie, and is tested and the samples classified by him" [W. J. Hammer, quoted in Roqué, 1997, p. 270]. The isolation of radium depended on the cooperation of chemical companies willing to provide facilities, workers and reagents in exchange for expert advice on the extraction procedures. Manufacture was therefore part of the research process, not a by-product of it. Curie cooperated with

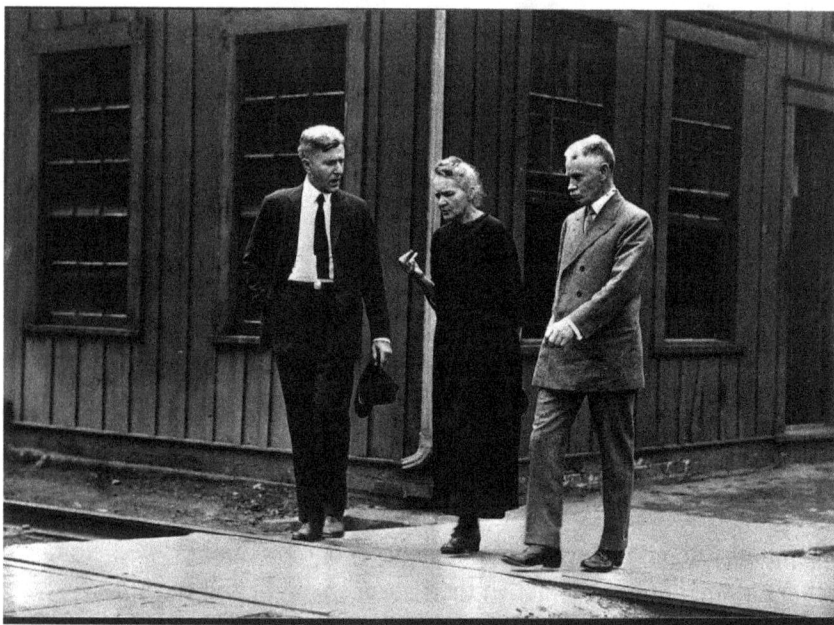

Figure 2. Marie Curie visiting the works of the Standard Chemical Company in Canonsburg (Pittsburgh) during her trip to the United States in 1921, with the company president James C. Gray (to her left) and the works manager Louis Fenn Vogt (to her right) (Musée Curie, coll. ACJC; reproduced by kind permission of the Musée Curie).

the nascent French radium industry in the 1900s, and sought logistic assistance from several radium producers, including the Standard Chemical Company in Canonsburg, Pittsburgh (Figure 2), in the inter-war years [Boudia, 2001; Roqué, 1997].

In 1898 the Curies had only been able to claim the discovery of radium in reference to its concentration procedure. Applications for radium did not emerge until 1901, when the biological effects of radioactivity were first noticed [Boudia and Roqué, 1997]. By the time radium was commercially profitable, it was too late for the Curies to think about patents. However, both Marie Curie's biography of Pierre [Curie, 1923] and Ève Curie's biography of Marie [Curie, 1938], argued that the couple had decided not to patent radium because they supported disinterested research. These highly influential retrospective accounts tell us more about the stakes of radioactivity in the inter-war years — a costly field of research that had developed alongside a profitable world-wide industry — than about the Curies' intentions at the time of discovery.

Recognition

In November 1903 the Swedish Academy of Sciences announced that the Nobel Prize in Physics for that year had been awarded to Becquerel and the Curies. One half of the prize went to Becquerel for "his discovery of spontaneous radioactivity," while the other half was awarded jointly to Pierre and Marie Curie for "their joint researches on the radiation phenomena discovered by Professor Becquerel" [N. N., 2018a]. It was the third time the prize was awarded, and the first time it went to a woman. The decision to acknowledge her contribution to the discovery of radioactivity worked as much to the prize's as to the Curies' advantage, because the couple's award did much to popularize the fledgling prize. Radium, the "conjugal metal" ("métal conjugal", Émile Gauthier in *Le Journal*, 21 December 1903, quoted in [Quinn, 1995, p. 195]), captured the public imagination. Many journalists assumed Marie had played a secondary role, if at all. "Only a few observers," writes Curie's biographer Susan Quinn, "seemed capable of understanding the mutuality of the Curies' relationship" [Quinn, 1995, p. 193; Hemmungs, 2015].

By the time the Curies received the Nobel Prize they had deduced the atomic weight of radium from a sample of radium chloride. Following Pierre Curie's death in an accident in Paris in 1906, Marie pursued the isolation of radium metal, which she achieved in 1910. In the following year she was awarded her second Nobel Prize, this time in Chemistry, for "the discovery of the elements radium and polonium, […] the isolation of radium and the study of the nature and compounds of this remarkable element" [N. N., 2018b]. Before World War II, only one other woman, the Curies' elder daughter Irène Joliot-Curie (1897–1956),[2] would be awarded the Nobel Prize: she received the Nobel Prize in Chemistry together with her husband Frédéric Joliot (1900–1958) in 1935.

***

Marie Curie was an extraordinary woman scientist, and recognised as such in her lifetime. Her career was so remarkable, in fact, that her myth "has both empowered and stigmatized women, liberated and constrained them, often at the same time. … Women scientists have felt as though they cannot measure up to Curie, and of course how could they?" [Des Jardins, 2010, p. 5]. Curie was indeed forceful and determined, but her success entailed domestic and social arrangements that help explain it beyond unique personal traits. Her life and career bear witness to the importance of family support and the careful management of a personal and a professional life. Curie did not hide these arrangements, but rather stressed them in her biographical texts. They not only redress the historical record, but also illuminate current debates on women in science.

References

Boudia, S. (2001). *Marie Curie et son laboratoire: Sciences et industrie de la radioactivité en France* (Éditions des Archives Contemporaines, Paris).

Boudia, S. and Mouliné, P. (2009). Mastering Picocoulombs in the 1890s: The Curies' Quartz-Electrometer Instrumentation, and How it Shaped Early Radioactivity History, *J. Electrostatics*, 67, pp. 524–530.

---

[2] See Jacquemond in this volume.

Boudia, S. and Roqué, X., eds. (1997). Science, Medicine, and Industry: The Curie and Joliot-Curie Laboratories, *Hist. Tech.*, 13:4.

Curie, È. (1938). *Madame Curie* (Doubleday, New York, NY, and Gallimard, Paris).

Curie, M. (1898). Rayons émis par les composés de l'uranium et du thorium, *C. R. Acad. Sci.*, 126, pp. 1101–1103.

Curie, M. (1923). *Pierre Curie*, transl. C. and V. Kellogg (Macmillan, New York, NY).

Curie, M. (1955). *Pierre Curie: Avec une étude des "Carnets de laboratoire" par Irène Joliot-Curie* (Denoël, Paris).

Curie, P. and Curie, M. (1897–1898). "Carnet de la découverte 1," 15 September 1897–18 March 1898 (https://gallica.bnf.fr/ark:/12148/btv1b8451519k. r=curie%20carnets?rk=128756;0, last accessed 1 February 2019).

Curie, P. and Curie, M. (1898a). "Carnet de la découverte 2," 18 March–16 July 1898 (https://gallica.bnf.fr/ark:/12148/btv1b84515207.r=curie%20 carnets?rk=107296;4, last accessed 1 February 2019).

Curie, P. and Curie, M. (1898–1900). "Carnet de la découverte 3," 11 November–13 July 1900 (https://gallica.bnf.fr/ark:/12148/btv1b8451521n. r=curie%20carnets?rk=64378;0, last accessed 1 February 2019).

Curie, P. and Curie, M. (1898b). Sur une nouvelle substance radio-active, contenue dans la pechblende, *C. R. Acad. Sci.*, 127, pp. 1215–1217.

Curie, P., Curie, M. and Bémont, G. (1898). Sur une nouvelle substance fortement radio-active, contenue dans la pechblende, *C. R. Acad. Sci.*, 127, pp. 175–178.

Des Jardins, J. (2010). *The Madame Curie Complex* (The Feminist Press, New York, NY).

Hemmungs Wirtén, E. (2015). *Making Marie Curie: Intellectual Property and Celebrity Culture in an Age of Information* (The University of Chicago Press, Chicago, IL, and London).

N. N. (2018a). The Nobel Prize in Physics 1903 (https://www.nobelprize.org/ prizes/physics/1903/summary, last accessed 8 December 2018).

N. N. (2018b). The Nobel Prize in Chemistry 1911 (https://www.nobelprize.org/ prizes/chemistry/1911/summary, last accessed 8 December 2018)

Opitz, D. L., Bergwik, S. and Van Tiggelen, B., eds. (2016). *Domesticity in the Making of Modern Science* (Basingstoke, Palgrave Macmillan).

Pycior, H. M. (1987). Marie Curie's "Anti-Natural Path": Time Only for Science and Family, in *Uneasy Careers and Intimate Lives: Women in Science, 1789–1979*, eds. Abir-Am, P. and Outram, D. (Rutgers University Press, New Brunswick, NJ), pp. 191–214. Reprinted as: Marie Curie: Time Only for Science and Family, in *A Devotion to their Science: Pioneer Women in Radioactivity*, eds. Rayner-Canham, M. F. and Rayner-Canham, G. F. (McGill's-Queen University Press, Montreal), pp. 31–50.

Quinn, S. (1995). *Marie Curie: A Life* (Simon & Schuster, New York, NY).

Roqué, X. (1997). Marie Curie and the Radium Industry: A Preliminary Sketch. *History and Technology*, 13 (4), pp. 267–291.

Select Bibliography

Boudia, S. (2001). *Marie Curie et son laboratoire: Sciences et industrie de la radioactivité en France* (Éditions des Archives Contemporaines, Paris).

Dry, S. (2003). *Curie* (Haus, London).

Hemmungs Wirtén, E. (2015). *Making Marie Curie: Intellectual Property and Celebrity Culture in an Age of Information* (The University of Chicago Press, Chicago, IL, and London).

Quinn, S. (1995). *Marie Curie: A Life* (Simon & Schuster, New York, NY).

Harriet Brooks: Radon, A "New Gas" from Radium

Marelene Rayner-Canham* and Geoffrey Rayner-Canham[†]

*Department of Physics, Grenfell Campus,
Memorial University, Corner Brook, NL, Canada
[†]Department of Chemistry, Grenfell Campus,
Memorial University, Corner Brook, NL, Canada

The end of the nineteenth and beginning of the twentieth centuries was a particularly exciting time for chemistry: many questions, few clear answers. One of the most perplexing questions was the nature of the substance released from radioactive heavy elements, such as radium. Was it a vapour? Was it a finely divided particulate? Was it a gas? This enigmatic substance was named "emanation" by Ernest Rutherford (1871–1937) to encompass all possibilities. The issue was settled by the bold title of a research article by Rutherford and his research student Harriet Brooks (1876–1933): "The New Gas from Radium" [Rutherford and Brooks, 1901]. We now call "the radioactive gas" the element radon, and this was the first ever mention of it.

The Role of Harriet Brooks in the Radon Discovery Process

The allocation of credit for element "discovery" is rarely a simple proce-dure [Karpenko, 1980]. Here we contend that, for radon, the crucial first step was the realization that a gaseous species could be produced from a solid precursor, and that it had a measurable atomic weight (mass) sig-nificantly different to that of radium from which it came. This was the very first evidence that a different element, a radioactive gaseous element, existed. And this was the significant contribution by Brooks.

To begin, we must go back to the initial step, the discovery in 1899 that something was being released by radium and thorium, apart from rays. The thorough and authoritative account by James L. Marshall and Virginia R. Marshall [2003] makes it clear that it was Ernest Rutherford who had proposed the name "emanation" for this (not Friedrich Ernst Dorn (1848–1916) as is commonly claimed). In Rutherford's earlier paper of January 1900 [Rutherford, 1900, p. 13], he initially proposed "[t]hat the emanation may be due to fine dust particles of the radioactive substance emitted by the thorium compound" or "[t]hat the emanation may be a vapour given off from thorium compounds." On the basis of those experi-ments Rutherford concluded that a vapour was the more likely in that all elements "give off vapour to some degree" [Rutherford, 1900, p. 14].

It was in 1901, in the publication "The New Gas from Radium" by Rutherford and Brooks in the *Transactions of the Royal Society of Canada*, that the possibility of emanation being a radioactive gas was first ever mooted. They used a gas diffusion apparatus to confirm that emana-tion was a previously unknown gas and reported (erroneously) that the gas had to have an atomic weight between 40 and 100. They did not claim at the time that it was a new element, though this seems to be the implication left to the reader.

Later in the same year Rutherford re-published the results under the title of "Emanations from Radio-Active Substances" [Rutherford, 1901] in the more widely read journal *Nature*, but under his name alone. This briefer account contained the comment: "In these experiments, I have been assisted by Miss H. T. Brooks" [Rutherford, 1901, p. 157]. He then repeated the more ambiguous statement from the previous publication that "[w]e must therefore conclude that the emanation is in reality a heavy radioactive vapour or gas" [p. 158]. In his later obituary of Brooks, Rutherford [1933a] gave Brooks more of the credit for the research, and also emphasized the importance of the identification of emanation as a gas:

> With Rutherford, she determined the rate of diffusion of the radium emanation into air and other gases. These experiments were at the time of much significance, for they showed that the radium emanation diffused like a gas of heavy molecular weight — estimated to be at least 100 [Rutherford, 1933a, p. 865].

Their research was at a time when the transmutation of elements was a topic for alchemy, not chemistry. It was generally accepted that one element could not be changed into another element. Thus, Rutherford and Brooks considered the sequence of events to be the conversion of radium into this mysterious radioactive emanation/gas followed by the deposition of a solid radioactive substance, presumably another form of radium, and therefore named radium A:

Radium → Ra emanation → Radium A (and ultimately to Radium G)

We would now depict the process instead as an element transmutation:

Radium-226 → Radon-222 → Polonium-218 (and ultimately to Lead-206)

In 1901 Rutherford and Brooks could not go beyond the claim of a "new gas." If they had claimed the gas to be a previously unknown chemical element — which we now know was the case — it would have opened them to ridicule. It was only subsequently that atomic transmutation became an acceptable concept [Malley, 1979].

A Second Contribution to Element Discovery

In 1904, also while at McGill University, Brooks made a second major discovery of importance to element discovery. She observed that a non-radioactive plate, placed inside a radioactive container, itself became radioactive [Brooks, 1904a]. Published in *Nature* under her name alone, her article ascribed this finding to a volatility of the radioactive material, enabling atoms of it to transfer from one surface to another.

We now explain the observation as the recoil of the radioactive atom: when a particle is released from an atom, the atom itself will be thrust in the opposite direction, often with enough momentum to eject the atom from the radioactive material and deposit it on another surface. By this means the product from the decay can be removed and examined separately from the parent material. Brooks' observation was overlooked, and the effect was rediscovered four years later by the team of Otto Hahn (1879–1968) and Lise Meitner (1878–1968) [Trenn, 1975]. It was by this means that Hahn and Meitner separated the first long-lived isotope of protactinium and were able to definitively study this radioactive element [Sime, 1986].[1] Rutherford [1908] wrote to Hahn, pointing out that Brooks had observed the phenomenon years earlier, and in his later biography, Hahn [1966, p. 268] noted that "Brooks may have been the first researcher to have observed the phenomenon of radioactive recoil."

Who Was Harriet Brooks?

Born on 2 July 1876 in Exeter, Ontario, Harriet Brooks (Figure 1) was the third of the nine children of George Brooks and Elizabeth Worden [Rayner-Canham and Rayner-Canham, 1992]. Although most of the academic women of the time came from fairly affluent middle-class families [Solomon, 1985, p. 63], Brooks grew up in a very different environment, as her father was a commercial traveller for a flour company. On his salary, it involved considerable hardship to feed and clothe them all. In 1894 the family moved to Montreal, where Harriet Brooks enrolled at the women-only Royal Victoria College, an integral part of McGill University, graduating with a Bachelor of Arts (Honours) in Mathematics and Natural Philosophy in 1898.

Rutherford had just arrived in Montreal from England, lured by the promise of a new Physics building at McGill University, and he hired Brooks as his first graduate student researcher. It was working with

---

[1] See also Roqué on Meitner in this volume.

Figure 1. Portrait of Harriet Brooks (circa 1906) (Gorky Museum, Moscow, reproduced by kind permission of the Archivist, Gorky Museum).

Rutherford in this period which led to the discovery of the "new gas" as described above.

In 1901 Brooks obtained leave from Royal Victoria College to take up a position at the women-only Bryn Mawr College as Fellow in Physics to pursue studies towards a PhD. In many of her periodic letters to Rutherford, Brooks indicated her low self-esteem and her reliance upon Rutherford for encouragement and support. For example, in her first letter from Bryn Mawr, she commented: "Please do not give me any more credit than I deserve … You are quite too generous in that respect" [Brooks, 1901].

During her year at Bryn Mawr (1901–1902), she was awarded the prestigious Bryn Mawr President's Fellowship for graduate study in Europe. Brooks informed Rutherford of the good news: "There are of course other difficulties to be met ... money and the objections of my family for instance who will think me wholly out of my mind but I think I can overcome them" [Brooks, 1902]. Rutherford arranged for Brooks to work in the Cavendish Laboratory, Cambridge, with his former supervisor, the famous J. J. Thomson (1856–1940). In one of her letters to Rutherford from Cambridge, she again displayed her lack of confidence: "I am afraid I am a terrible bungler in research work, this is so extremely interesting and I am getting along so slowly and so blunderingly with it. I think I shall have to give it up after this year, there are so many other people who can do so much better and in so much less time than I that I do not think my small efforts will ever be missed" [Brooks, undated].

After her year in England, Brooks did not return to Bryn Mawr to complete a PhD, but instead rejoined Rutherford's research group at McGill University. It was during this period at McGill (1903–1904) that she discovered the phenomenon which we now know to be the recoil of the radioactive atom (see above). After only one year at McGill, Brooks accepted a position as Tutor in Physics at Barnard College, the women's college of Columbia University, New York [Brooks, 1904b]. Although the annual salary was an improvement over that at McGill, the main reason for her move was probably a romantic attachment to a physics professor at Columbia, Bergen Davis (1869–1958) [Webb, 1960], whom she had met while they were both researchers at the Cavendish Laboratory.

At Barnard, most of Brooks' time was occupied by teaching advanced physics, and her life was relatively uneventful until the summer of 1906, when she became engaged to Davis. Brooks wrote Dean Laura Gill at Barnard to inform her of the impending marriage. Gill replied: "I feel very strongly that whenever your marriage does take place it ought to end your official relationship with the college" [Gill, 1906a]. Brooks, not so low in self-esteem anymore, refused to submit to the norms of the time, replying:

> I think also it is a duty I owe to my profession and to my sex to show
> that a woman has a right to the practice of her profession and cannot be
> condemned to abandon it merely because she marries. ... I am sorry to

have thus to appeal from your decision but I cannot acquiesce without violating my deepest convictions of my rights [Brooks, 1906].

Gill [1906b] and the Trustees refused to allow her to stay if her marriage took place. Though the reason is not clear, the engagement was broken off. Brooks initially agreed to continue as Tutor, but in the following month resigned her position at Barnard College.

In 1906 Brooks voyaged across the Atlantic to Paris, where she became an independent researcher in the Curie Laboratory, under the direction of André-Louis Debierne (1874–1949). Although Brooks never published any of her work at the Curie Institute, we know that she commenced a study of the emanation from the element actinium and continued her experiments on the recoil of the radioactive atom for three subsequent articles from the Institute, including one by Curie's research associate, Mlle Lucie Blanquies [1909], which cited her work. Curie invited Brooks to stay for the 1907–1908 year, but Brooks decided instead to apply for a position at the University of Manchester in England.

The position at Manchester was the John Harling Fellowship. Rutherford was about to move to Manchester, and it seems that Brooks was keen to rejoin her former mentor. Brooks was still waiting to hear about the decision on the Harling fellowship when she suddenly decided to terminate her physics research and marry Frank Pitcher (1872–1935), who had been a laboratory demonstrator at McGill University while she was a student. This decision came as a complete shock to Rutherford [1907a]. The numerous courtship letters from Pitcher to Brooks indicate that Brooks initially had a strong reluctance to give up her peripatetic research life, while Pitcher emphasized the need for her to settle down [Rayner-Canham and Rayner-Canham, 1992, pp. 73–76].

After her marriage in London on 13 July 1907, Brooks returned to Montreal with Pitcher. During the following years they raised three children, two of whom died as teenagers (Figure 2). Brooks gave up any active role in science, even though this abandonment ran completely counter to her earlier expressed views at Barnard College. Her acquiescence to a purely domestic role may have been due to the traditionalist values of her husband and those of the upper-class milieu in which she now found herself [Fee, 1986]. Additionally, Rutherford, her mentor, had

Figure 2. Portrait of the Pitcher family. From left to right: Barbara Pitcher, Frank Pitcher, Harriet Brooks Pitcher, Charles Pitcher, with Paul Brooks Pitcher seated in front (reproduced by kind permission of Paul Brooks Pitcher).

moved to Manchester, England, so she did not have the opportunity to continue her research on radioactivity with him. In fact, with Rutherford gone, the radioactivity research programme at McGill had declined considerably from its former glory [Pyenson, 1978].

Harriet Brooks died on 17 April 1933, at the age of 57. From accounts of her lingering illness, it would seem possible that she suffered from a radiation-related disease. Rutherford wrote to Arthur Eve when he heard the news of her death:

> She was a woman of great personal charm as well as of marked intellectual interests. I am afraid her domestic life was not without serious trials which she bore with astonishing fortitude. My wife and I held her in great affection and her premature death is a grievous blow to us [Rutherford, 1933b].

***

Harriet Brooks was repeatedly mentioned by Rutherford in his text and in his research papers, more so than his other students of the period. Brooks was the one at the lab bench performing much of the research. In fact, Rutherford [1907b] wrote a glowing reference for Brooks for the John Harling Fellowship: "Her work on 'Radioactivity' has been of great importance in the analysis of radioactive transformations and next to Mme. Curie she is the most prominent woman physicist in the department of radioactivity" [Rutherford, 1907b]. So why was Brooks overlooked by historians of science? For the discovery of the radioactive gas, the lack of her name on the more widely read publication in *Nature* may have been a factor. In the case of the recoil of the radioactive atom, the loss of her name from scientific history may be due to the fact that she mis-assigned it as volatility. For both discoveries, years passed before the significance became apparent, which is not uncommon in the history of science [Garfield, 1981]. Also, the lack of any previous study of this fascinating scientist may reflect our tendency to focus on the more prestigious figures (mostly male) in the advancement of science.

Another contributing factor for Brooks' absence from the scientific historical record was her decision to leave research. Perceiving little opportunity for a long-term career and doubting her own abilities, and in spite of encouragement by her mentor Rutherford to continue research, she finally reverted to convention, to become a wife and mother.

Acknowledgements

The authors would like to thank C. Rittenhouse, Archivist, Bryn Mawr College, for information on Brooks' time at Bryn Mawr; A. E. B. Owen, Archivist, Cambridge University, for copies of the correspondence between Brooks and Rutherford; P. K. Ballou and L. Manning, Archivists, Barnard College, for copies of the relevant correspondence at Barnard College Archives; M. Bordry, Curie Institute, for copies of the Brooks-Curie correspondence; N. H. Robinson, Librarian, is thanked for supplying the Rutherford-Schuster correspondence at the Royal Society Archives; M. Cohen, McGill University, for a copy of the letter from Rutherford to A. S. Eve; and P. B. Pitcher, Harriet Brooks' son, now deceased, for access to all surviving correspondence of Frank Pitcher to Harriet Brooks.

References

Blanquies, L. (1909). Comparaison entre les rayons α produits par différentes substances radioactives, *Le Radium*, 6, pp. 230–232.

Brooks, H. (1901). Letter to Ernest Rutherford, 8 December 1901, Cambridge University Archives (CUA), GB 12 Add. MS 7653.

Brooks, H. (1902). Letter to Ernest Rutherford, 27 May 1902, CUA, GB 12 Add. MS 7653.

Brooks, H. (1904a). A Volatile Product from Radium, *Nature*, 70, p. 270.

Brooks, H. (1904b). Letter to Laura Gill, undated, Barnard College Archives (BCA), B.C. 5.1 Dean's Office and Department Correspondence, 1904–1960.

Brooks, H. (1906). Letter to Laura Gill, 18 July 1906, BCA, B.C. 5.1 Dean's Office and Department Correspondence, 1904–1960.

Brooks, H. (undated). Letter to Ernest Rutherford, undated, CUA, GB 12 Add. MS 7653.

Fee, E. (1986). Critiques of Modern Science: The Relationships of Feminism to Other Radical Epistemologies, in *Feminist Approaches to Science*, ed. Bleier, R. (Pergamon Press, New York, NY), p. 45.

Garfield, E. (1981). Premature Discovery or Delayed Recognition — Why?, in *Essays of an Information Scientist*, vol. 4 (ISI Press, Philadelphia, PA), pp. 488–493.

Gill, L. (1906a). Letter to Harriet Brooks, 12 July 1906, BCA, B.C. 5.1 Dean's Office and Department Correspondence, 1904–1960.

Gill, L. (1906b). Letter to Margaret Maltby, 30 July 1906, BCA, B.C. 5.1 Dean's Office and Department Correspondence, 1904–1960.

Hahn, O. (1966). *Otto Hahn: A Scientific Biography* (Charles Scribner, New York, NY).

Karpenko, V. (1980). The Discovery of Supposed New Elements: Two Centuries of Errors, *Ambix*, 27, pp. 77–102.

Malley, M. (1979). The Discovery of Atomic Transmutation: Scientific Styles and Philosophies in France and Britain, *Isis*, 70, pp. 213–223.

Marshall, J. L. and Marshall, V. R. (2003). Ernest Rutherford, the 'True Discoverer' of Radon, *Bull. Hist. Chem.*, 28:2, pp. 76–83.

Pyenson, L. (1978). The Incomplete Transmission of a European Image: Physics at Greater Buenos Aires and Montreal, 1890–1920, *Amer. Phil. Soc., Proceedings*, 122, pp. 92–114.

Rayner-Canham, M. F. and Rayner-Canham, G. W. (1992). *Harriet Brooks: Pioneer Nuclear Scientist* (McGill-Queen's University Press, Montreal).

Rutherford, E. (1900). A Radioactive Substance Emitted from Thorium Compounds, *Phil. Mag.*, series 5:49, pp. 1–14.

Rutherford, E. (1901). Emanations from Radio-Active Substances, *Nature*, 64, pp. 157–158.

Rutherford, E. (1907a). Letter to Arthur Schuster, 25 March 1907, Royal Society Archives, (RSA), Schuster collection, MSS 663–65.

Rutherford, E. (1907b). Letter of Reference for the John Harling Fellowship for Harriet Brooks, to Arthur Schuster, 25 March 1907, RSA, Schuster collection, MSS 663–65.

Rutherford, E. (1908). Letter to Otto Hahn, 22 December 1908, CUA, GB 12 Add. MS 7653.

Rutherford, E. (1933a). Obituary: Harriet Brooks (Mrs. Frank Pitcher), *Nature*, 131, p. 865.

Rutherford, E. (1933b) to Arthur S. Eve, 6 May 1933, McGill University Archives, Ernest Rutherford Collection.

Rutherford, E. and Brooks, H. T. (1901). The New Gas from Radium, *Roy. Soc. Canada, Transactions* section 3, pp. 21–25.

Sime, R. L. (1986). The Discovery of Protactinium, *J. Chem. Educ.*, pp. 653–657.

Solomon, B. M. (1985). *In the Company of Educated Women* (Yale University Press, New Haven, CT).

Trenn, T. J. (1975). Rutherford and Recoil Atoms: The Metamorphosis and Success of a Once Stillborn Theory, *Hist. Studies Physical Sci.*, 6, pp. 513–547.

Webb, H. W. (1960). Bergen Davis 1869–1958, *Bio. Mem. Natl. Acad, Sci.*, pp. 63–82.

Select Bibliography

Rayner-Canham, M. F. and Rayner-Canham, G. W. (1992). *Harriet Brooks: Pioneer Nuclear Scientist* (McGill-Queen's University Press, Montreal).

Rayner-Canham, M. F. and Rayner-Canham, G. W., eds. (1997). *A Devotion to their Science: Pioneer Women of Radioactivity* (McGill-Queen's University Press, Montreal and Chemical Heritage Foundation, Philadelphia, PA).

Rayner-Canham, M. F. and Rayner-Canham, G. W. (2002). Harriet Brooks: The Story Behind the Story, *Chem13 News*, pp. 8–11.

Rayner-Canham, M. F. and Rayner-Canham, G. W. (2004). Rutherford, the "True Discoverer of Radon," *Bull. Hist. Chem.*, 29:2, pp. 89–90.

Dr Margaret Todd
and the Introduction of the
Term "Isotope"

John A. Hudson

Independent Researcher, Graythwaite, Loweswater,
Cockermouth, Cumbria CA13 0SU, UK

New scientific concepts need a new name. The term "isotope" was coined to describe atoms of the same element with identical chemical properties but differing in atomic weight. The name was proposed in 1913 by Dr Margaret Georgina Todd (1859–1918) (Figure 1), and it was a brilliant choice, as its derivation from the Greek *isos topos,* meaning "same place," reaffirmed the primacy of the periodic table at a time when the fundamental basis for the ordering of the elements was in question.

A Dinner Party

Todd made her suggestion in 1913 when she attended a dinner party hosted by her friends, the prominent industrial chemist George Beilby (1850–1924) and his wife, at their house in Glasgow. Also present were the Beilby's daughter Winifred and her husband Frederick Soddy (1877–1956), lecturer in physical chemistry and radioactivity at the

Figure 1. Margaret Todd by Walter Stoneman (© National Portrait Gallery, London).

University. Soddy had in 1900–1903 worked at McGill University in Montreal with Ernest Rutherford (1871–1937), and after a year working with William Ramsay (1852–1916) in London had moved to Glasgow. It is not known if there were any other guests at the dinner party.

The conversation must have turned to some of the astonishing recent developments in radiochemistry. In the period 1898–1900, four previously unknown elements had been discovered — polonium, radium, radon and actinium. But then, in the first decade of the twentieth century, an additional 35 radioactive disintegration products were found, which were initially thought to be new elements. This situation presented a challenge to the periodic table — at least one version was published in which the supposedly new elements could be accommodated [Van Spronsen, 1969, p.188]. But it was soon demonstrated that all these disintegration products, resulting from either alpha (α-) or beta (β-) particle emission, were inseparable from previously known elements. Alexander Fleck (1889–1968), who at the time of the dinner party was conducting research under Soddy's supervision, showed that for each of the 13 beta emitting

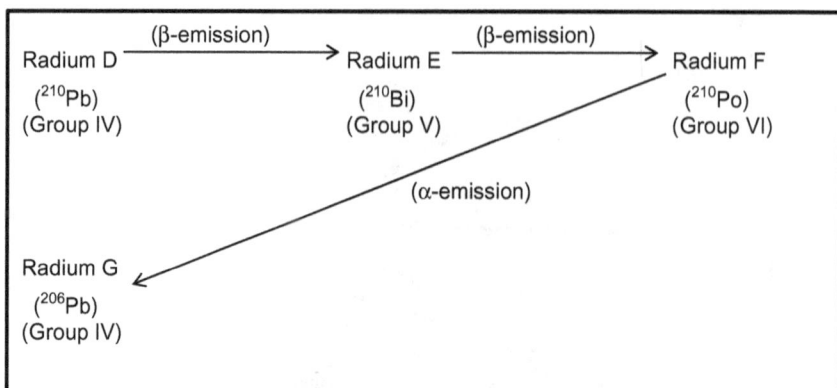

Figure 2.   The Group Displacement Law, exemplified by part of the ^{238}U decay series. The new radioelements were initially designated as being varieties of radium. It was Fleck who identified them as being identical to known elements, indicated underneath by their modern symbols and periodic groups.

radioelements he investigated the product was a new element occupying one place to the right in the periodic table [Holroyd, 1971]. Soddy had earlier established that when an element decays by alpha particle emission, the new element occupies a place two spaces to the left in the periodic table. Other workers, notably Kasimir Fajans (1887–1975), were obtaining similar results. These rules describing the consequences of alpha and beta particle emissions were summarised as the Group Displacement Law (Figure 2). This work demonstrated that elements could exist as more than one kind of atom exhibiting identical chemical properties but possessing different atomic weights. It follows that atomic weight cannot be the fundamental constant determining all the chemical properties of an element.

Thus at this time two long-held beliefs had to be abandoned. John Dalton's (1766–1844) concept that all the atoms of a particular element were identical, especially in weight, was now disproved. Furthermore, since each element could have more than one atomic weight this could no longer be the determining factor of chemical periodicity. Soddy had recently summed up the position in a paper in which, after reviewing Fleck's results, he wrote: "These results prove that almost every vacant place in the Periodic Table between thallium and uranium is crowded with non-separable elements of atomic weight varying over several units, and leads inevitably to the presumption that the same may be true in other parts of the table" [Soddy, 1913a, p. 99].

Fleck described how, up to this time, he and Soddy were using the expression "radio elements chemically non-separable" to refer to these chemically identical species of an element, the atoms differing only in mass [Fleck, 1957, p. 208]. Clearly a less cumbersome designation was called for, and presumably the dinner party discussion turned to the topic of a suitable name. The word "isotope" proposed by Todd emphasised that these identical versions of an element should all be accommodated in the same place (or box) in the periodic table, thereby reinforcing the concept that the fundamental property responsible for chemical periodicity was not atomic weight and had yet to be discovered.

Dr Margaret Todd

Margaret Todd, the proposer of the new term, was not a scientist but held medical qualifications. As well as receiving recognition as an early female member of the medical profession she acquired a reputation as a novelist of some distinction. When she died in 1918 the announcements of her death in the newspapers gave brief accounts of her career, but there is no mention of the term "isotope" [N. N., 1918a, 1918b, 1918c]. She was born in Fife, Scotland to James Cameron Todd, described as being of Glasgow and Rangoon, and his wife Jeannie McBain. She was educated at Edinburgh, Glasgow and Berlin and then worked as a school teacher in Glasgow. She left teaching when the Edinburgh School of Medicine for Women was opened in 1886, enrolling as one of its first students.

The Edinburgh School of Medicine for Women was founded by Dr Sophia Jex-Blake (1840–1912), one of the principal campaigners for the admission of women into the medical profession in Britain [Roberts, 2004]. Whilst still studying medicine Todd wrote her first book, a three-volume novel, published under the pseudonym Graham Travers, entitled *Mona Maclean, Medical Student* [Travers, 1892]. The book describes the heroine's medical and romantic adventures whilst a student at the London School of Medicine. It was published in 1892, and was clearly intended to win increased acceptance for women doctors amongst both the medical profession and the general public. It was well received, reaching 15 editions by 1900. As a result of her literary endeavours, Todd took longer than usual to complete her studies, but by 1894 she had acquired medical qualifications from Edinburgh and a medical degree (M.D.) from Brussels. Todd was to produce several other novels over the coming years. After qualification Todd took the post of Assistant Medical Officer at the Edinburgh Hospital and Dispensary for Women and Children, which had been founded by Jex-Blake. Todd had by this time formed a close friendship with Jex-Blake, and in 1899, when the latter retired, the two women moved to a small farm at Rotherfield, a village in the south of England. Jex-Blake died there in 1912, and it was during a return visit to her native Scotland in the following year that Todd proposed the name "isotope."

Todd's role in the introduction of this new term remained unknown for over 30 years. It was Fleck who, in an obituary of Soddy, revealed that Todd had made the proposal, writing: "From this [the Group Displacement Law] also came the introduction into our language of the word 'isotope'. It was suggested by Dr Margaret Todd, arising from a discussion that Soddy had in the house of his father-in-law, Sir George Beilby, in Glasgow" [Fleck, 1956, p. 893]. It is not known if Todd had taken science courses in her earlier education, but she would certainly have studied some chemistry whilst a medical student. Although never a practising scientist she would have been able to understand the revolutionary nature of the recent discoveries in the field of radioactivity and hence to propose the term "isotope." She was also proficient in Greek. Fleck, in another obituary of Soddy, commented: "The readiness and etymological accuracy with which she produced 'isotopes' is a standing testimony to her practical knowledge of the Greek tongue" [Fleck, 1957, p. 208].

The Term Gains Acceptance

It is perhaps hard to for us today to appreciate the revolutionary nature of the concept that more than one version of an element could exist with atoms of different weights but displaying identical chemical properties. So the question arose as to what caused an element to exhibit its characteristic chemical properties if it was not the weight of its atom. It was known at the time that the atom contained a positively charged nucleus. It was also thought that the nucleus additionally contained negatively charged particles, fewer in number than the positive particles, thus giving the nucleus its net positive charge. Soddy first used Todd's term "isotope" in a paper in which he suggested that the different isotopes of an element would have the same net positive nuclear charge, but different numbers of positive and negative particles. The relevant passage reads: "The algebraic sum of positive and negative charges in the nucleus, when the arithmetic sum is different, gives what I call "isotopes" or isotopic elements, because they occupy the same place in the periodic table" [Soddy, 1913b, p. 400].

Soddy's phrasing is interesting. He says he calls the differing forms of the same element isotopes without actually claiming he invented the name. But seven years later he wrote: "I have termed them *isotopes* or *isotopic elements*" [Soddy, 1920, p. 229], giving no credit to Todd as the originator of the new word. By contrast, he acknowledged Fleck's work in establishing the chemical inseparability of the isotopes. It is entirely possible that Todd did not wish her name to be mentioned. Maybe she wanted to be remembered both as an author (when Soddy first used the term she had published five novels) and for her medical work, rather than the originator of the name for a phenomenon which she had played no part in discovering. Whatever the reason, if Soddy had undertaken not to reveal Todd's role, it is understandable that Fleck, who would certainly have learned about the dinner party discussion immediately afterwards, chose not to disclose the information during Soddy's lifetime.

The word "isotope" soon entered the scientific vocabulary, the only competitor being "pleiad," coined by Fajans [Scerri, 2007, p. 178]. The Pleiades in classical mythology were the seven daughters of Atlas, and the name is now applied to a star cluster. The term "pleiad" in the singular has at times been used for a member of a group of similar persons or things, especially when seven in number. But although a collection of elemental pleiads would be expected to have similar properties, the term makes no reference to the periodic table. Todd's term "isotope" conveyed the meaning that the different forms of the same element were not just closely similar in properties, but identical, and thus justified occupying the same place, or box, in the periodic table. The superiority of the term "isotope" for the new concept was soon recognised.

Fleck continued his researches at Glasgow, publishing nine papers by 1915 on his work on radioactive disintegration, and being awarded the doctor of science (DSc) degree at Glasgow in the following year. He soon left academia to become an industrial chemist, and by 1956 he was Chairman of the British chemical giant Imperial Chemical Industries. In 1921 Soddy received the Nobel Prize for Chemistry, the citation reading: "For his contributions to our knowledge of the chemistry of radioactive substances, and his investigations into the origin and nature of isotopes" [N. N., 1921].

Isotopy Explained

By 1913 the periodic table was so well established that the recent experimental work on radioactive disintegration never seriously threatened its position as the great unifying system of inorganic chemistry. Todd's term "isotope" reaffirmed the validity of the table, even though the theoretical basis of periodicity had yet to be discovered. The key to the solution of the problem was soon provided when Henry Gwyn Jeffreys Moseley (1887–1915) published his work on the X-ray spectra of the elements. From the measured frequency of the principal line in the spectrum of each element, Moseley derived a quantity which increased by a constant amount between consecutive elements when ordered according to the periodic table. Crucially this quantity placed cobalt before nickel in the table in accordance with their chemical properties, whereas the atomic weight order was the reverse. Moseley realised that the quantity he had derived was related to the positive charge on the nucleus, and hence that the magnitude of this charge was responsible for each element's position in the table [Moseley, 1913]. Later, when it was realised that most non-radioactive elements existed as mixtures of isotopes, atomic weight inversions such as cobalt/nickel were explained by the isotopic compositions of the elements yielding atomic weight averages out of sequence with the periodic table order. The discovery of isotopes had never posed a serious threat to the periodic law, but now the concept of isotopy was used to explain an anomaly that had existed ever since Mendeleev first proposed the table.

The numerical position of an element in the sequence derived by Moseley is now known as its atomic number. Isotopes of an element have the same atomic number and all the isotopic atoms have the same nuclear positive charge. In 1920 Rutherford was to identify the nuclear positive particles as hydrogen nuclei, coining the name "proton," and in 1932 Chadwick was to identify the neutron as an electrically neutral nuclear particle [Brock, 1985, p. 217–219]. Thus the nuclear positive charge is simply the number of protons (not the algebraic sum of nuclear positive and negative charges as Soddy and others had thought). Isotopes of an element have nuclei with the same number of protons but different numbers of neutrons. In beta particle emission a neutron decomposes to a proton (which is retained in the nucleus) and an electron (which is ejected), thereby yielding an atom with one more proton in the nucleus, and hence one place to the right in the periodic table. An alpha particle consists of two protons and two neutrons, so alpha emission yields an atom with two fewer protons in the nucleus, and hence two places to the left in the periodic table. Figure 2 illustrates some of changes resulting from alpha and beta emission which were investigated by Fleck when working with Soddy.

Todd's Life after the Proposal of the Name

Todd continued to live at Rotherfield, producing her sixth and final book in 1918, which was a biography of Jex-Blake, her deceased friend [Todd, 1918]. At this time her health was apparently poor; one source refers to "periods of recuperation" spent at Tankerton in Kent [N. N., 1918c]. Three months after the publication of the biography of Jex-Blake, Todd died at a nursing home in London. During her lifetime Margaret Todd achieved significant recognition as an author, and although her works of fiction are largely forgotten today, her biography of Jex-Blake provides a detailed account of the difficulties faced by women in Britain who wished to enter the medical profession.

A plaque on the wall of the Beilby's former house in Glasgow commemorates the dinner party, simply stating that Soddy introduced the "concept" of isotopes (Figure 3). That statement is strictly correct, but no mention is made of Margaret Todd, or that she proposed the name "isotope." The word "isotope" is familiar to most people, although its precise meaning may not be properly understood by many non-scientists. It is Margaret Todd's most enduring legacy, but since 1957, apart from occasional very brief references in profiles of Soddy [e.g. Nagel, 1982; James, 1993], there

Figure 3. The plaque on the wall of the Beilby's former residence, 11 University Gardens Glasgow (photo by the author).

has been little mention of her and the fact that the word "isotope" was her suggestion. But hopefully the situation is changing; recently an online article has appeared about her [Wikipedia, 2018]. Perhaps the time is ripe for a replacement plaque giving Dr Margaret Todd her due recognition?

Acknowledgements

The author wishes to thank Peter Reed for his comments on an earlier version of this chapter.

References

Brock, W. H. (1985). *From Protyle to Proton* (Adam Hilger, Bristol).
Fleck, A. (1956). Obituary, Prof. Frederick Soddy, F. R. S., *Nature*, 178, p. 893.
Fleck, A. (1957). Frederick Soddy 1877–1956, *Biogr. Mems Fell. R. Soc.*, 3, pp. 203–216.
Holroyd, R. (1971). Alexander Fleck, Baron Fleck of Saltcoats 1889–1968, *Biogr. Mems Fell. R. Soc.*, 17, pp. 242–254.
James, L. K. (1993). *Nobel Laureates in Chemistry, 1901–1992* (American Chemical Society and The Chemical Heritage Foundation, Washington, DC).

Moseley, H. G. J. (1913). The High-Frequency Spectra of the Elements, *Phil. Mag.*, 27, pp. 1024–1034.

Nagel, M. C. (1982). Frederick Soddy: From Alchemy to Isotopes, *J. Chem. Ed.*, 59, pp. 739–740.

N. N. (1918a). Obituary, Dr Margaret Todd, *The Times (London)*, 5 September 1918, p. 9.

N. N. (1918b). Obituary, *The Scotsman*, 7 September 1918, p. 4.

N. N. (1918c). Obituary, *The Whitstable and Herne Bay Herald*, 14 September 1918, p. 8.

N. N. (1921). (https://www.nobelprize.org/prizes/chemistry/1921/summary/, last accessed 28 December 2018)

Roberts, S. (2004). Blake, Sophia Louisa Jex-, in *Oxford Dictionary of National Biography*, ed. Harrison, B. (Oxford University Press, Oxford) (http://www.oxforddnb.com/view/10.1093/ref:odnb/9780198614128.001.0001/odnb-9780198614128-e-34189?rskey=oc2C6Y&result=4, last accessed 04 August 2018).

Scerri, E. R. (2007). *The Periodic Table* (Oxford University Press, Oxford).

Soddy, F. (1913a). The Radio-Elements and the Periodic Law, *Chem. News*, 107, pp. 97–99.

Soddy, F. (1913b). Intra-Atomic Charge, *Nature*, 92, pp. 399–400.

Soddy, F. (1920). *The Interpretation of Radium and the Structure of the Atom*, 4th ed. (John Murray, London).

Van Spronsen, J.W. (1969). *The Periodic System of Chemical Elements* (Elsevier, Amsterdam).

Todd, M. G. (1918). *The Life of Sophia Jex-Blake* (Macmillan, London).

Travers, G. (1892). *Mona Maclean, Medical Student* (Blackwood, Edinburgh).

Wikipedia (2018). Margaret Todd (doctor) (https://en.wikipedia.org/w/index.php?title+Margaret_Todd_(doctor)&oldid=846120717, last accessed 1 August 2018).

Select Bibliography

Fleck, A. (1957). Frederick Soddy 1877–1956, *Biogr. Mems Fell. R. Soc.*, 3, pp. 203–216.

Holroyd, R. (1971). Alexander Fleck, Baron Fleck of Saltcoats 1889–1968, *Biogr. Mems Fell. R. Soc.*, 17, pp. 242–254.

Wikipedia (2018). Margaret Todd (doctor) (https://en.wikipedia.org/w/index.php?title+Margaret_Todd_(doctor)&oldid=846120717, last accessed 1 August 2018).

Stefanie Horovitz: A Crucial Role in the Discovery of Isotopes

Marelene Rayner-Canham* and Geoffrey Rayner-Canham[†]

*Department of Physics, Grenfell Campus,
Memorial University, Corner Brook, NL, Canada
[†]Department of Chemistry, Grenfell Campus,
Memorial University, Corner Brook, NL, Canada

The discovery of chemical elements cannot be separated from the discovery of isotopes. And one of the most important contributions to the discovery of the existence of isotopes was made by a long-forgotten woman chemist, Stefanie Horovitz (1887–1942) (Figure 1). It was her incredibly precise and accurate gravimetric measurements which confirmed that lead from uranium-rich minerals had a definite different atomic weight (mass)[1] to that of "ordinary" lead. Later, she and her supervisor, Otto Hönigschmid (1878–1945), disproved the existence of the claimed element ionium.

Horovitz and the Conclusive Evidence for Isotopes

The British chemist Frederick Soddy (1877–1956) had proposed that the radioactive elements, such as uranium and thorium, disintegrated

---

[1]We will use "weight" rather than "mass" in this chapter in keeping with the historical terminology.

Figure 1. Stefanie Horovitz, November 1915 (Austrian Central Library for Physics, reproduced by kind permission of the Austrian Central Library for Physics).

step-wise to ultimately form lead. However, according to his Group Displacement Law, the atomic weight of the lead produced from the decay of uranium should be 206.5 rather than the 207.1 found for common lead [Kauffmann, 1982; Badash, 1979a]. Soddy had proposed the name "isotope" for these forms of the same element that differed only in their masses (the word "isotope" was actually devised by Soddy's friend, Dr Margaret Todd)[2] [Fleck, 1963]. This strange concept of atoms being identical in everything but weight acquired few adherents, and Soddy's own measurements had been regarded with scepticism. If it could be shown that lead from radioactive sources really did have a different atomic weight than that of common lead, it would be a triumph for his Law and for the concept of isotopes.

It was this task to which Stefanie Horovitz was assigned by her research supervisor, Otto Hönigschmid: the determination of the atomic

---

[2] See Hudson in this volume.

weight of the lead present in radioactive minerals. Horovitz and Hönigschmid were in the right place at the right time. The major source of radioactive uranium and radium ores was the St. Joachimsthal mine in Bohemia, part of the Austro-Hungarian Empire. The ore from the St. Joachimsthal mine was shipped to the Radium Institute at Vienna. In Vienna the valuable radium was separated out (most of it going from there to Marie Curie (1867–1934) in Paris and to Ernest Rutherford (1871–1937) in Manchester), leaving lead in the residues. Thus the first task for Horovitz was to separate out very pure samples of lead from these residues, not an easy task in itself. These careful extraction procedures were followed by the vital mass measurements that could be used to determine the atomic weights. Each of the measurements was made to the nearest one hundred thousandth of a gram — a level of precision that is unusual even today.

Horovitz's results were startling. In fact, the atomic weight of the lead from the disintegration of uranium and radium was found to be 206.736 compared to 207.190 for 'normal' lead [Hönigschmid and Horovitz, 1914a; 1914b]. The difference of 0.4 units, although seemingly small, was enough to demolish the belief that atomic weights were invariant.[3] It was the first authoritative evidence for the concept of isotopes, and it provided strong support for Soddy's Group Displacement Law. As a result, this paper by Horovitz and Hönigschmid is regarded as one of the crucial publications in chemistry in the first half of the twentieth century [Leicester, 1968]. This discovery of atoms of the same chemical identity but different weights caused astonishment and almost consternation among many chemists of the time [Badash, 1979b, p. 206].

Following this first result, Horovitz analysed new samples from St. Joachimsthal as well as samples from two other mines: "pitchblende" from German East Africa; and "bröggerite" from Norway, the latter supplied by the Norwegian chemist Ellen Gleditsch (1879–1968) [Kubanek, 2010]. Horovitz's results this time were even more convincing, giving values as low as 206.046 [Hönigschmid and Horovitz, 1915]. The difference of more than one weight unit could not be explained by experimental error, and this definitely meant that atomic weights were not necessarily

---

[3] See also Lykknes on Gleditsch in this volume.

invariant, as had been believed until that date. During this period Hönigschmid mentioned Horovitz in a letter to Lise Meitner (1878–1968):

> Miss Horovitz and I worked like coolies. On this beautiful Sunday we are still sitting in the laboratory at 6 o'clock. ... I am sending you greetings from Miss Horovitz, who does not believe that you still remember her. I have just argued with her about that [Hönigschmid, 1914a].

Horovitz and Meitner had probably met in 1907, the year that Horovitz started at the University of Vienna and that Meitner departed from Vienna for Berlin.

Disproving the Existence of "Ionium"

Horovitz's second claim to fame was disproving the existence of the radioactive element "ionium."[4] It was in 1907 that the American chemist Bertram Boltwood (1870–1927) announced that he had discovered a new element, ionium, to which he assigned the chemical symbol, Io [Boltwood, 1907]. This element and its symbol became accepted amongst researchers in radioactivity, and Boltwood (and others) continued to publish papers on the behaviour of this element [Boltwood, 1911].

Between 1914 and 1916 Horovitz determined the precise atomic weight of "ionium" [Hönigschmid and Horovitz, 1916]. She showed that ionium had the same spectroscopic and chemical properties as thorium, the only difference being the atomic weight. Thus ionium was not a separate element at all but, as we now know, simply an isotope of thorium, thorium-230. So, in one experiment, she had disproved the existence of an element and found the second established case of isotopes.

The Institute for Radium Research

It was at the Institute for Radium Research where Horovitz performed her research. Had this Institute not been so hospitable to women scientists,

---

[4]The radioactive "ionium" claimed as a discovery by Boltwood should not be confused with the claimed stable element of the same name "discovered" by Sir William Crookes (1832–1919) in 1886. See Fontani *et al.*, 2015, pp. 202–206 and 473.

Horovitz would have never been able to accomplish her ground-breaking discoveries. As we mentioned earlier, the major source of radioactive uranium and radium was the St. Joachimsthal mine in Bohemia [Paneth, 1964]. It was from this mine that, in the late nineteenth century, Pierre (1859–1906) and Marie Curie obtained the samples used in their researches. Many Austrians considered it a national disgrace that foreigners were making great scientific advances using Austrian resources while there was nowhere in the Empire undertaking such important work.

It was the industrialist Karl Kupelwieser (1872–1930) who offered to finance a multidisciplinary Institute for Radium Research [Rentetzi, 2005, p. 279]. The building was to be of the most up-to-date design. Maria Rentetzi has observed that the plans indicate the presence of female toilets in addition to those for men, anticipating the presence of women researchers. In our studies we have shown that radioactivity research attracted a remarkably high proportion of women scientists. We hypothesized that there were probably several contributing factors: female-supportive mentors; the greater willingness of women to take on the tedious and dangerous task of working with radioactive materials; and the newness of the field which lacked established male hierarchies [Rayner-Canham and Rayner-Canham, 1997]. Rentetzi [2004, p. 366] added another factor:

> At this intersection of a number of disciplines, women who entered radioactivity research could choose different paths and take advantage of its multiple applications. Indeed, the fact that radioactivity stood on the border of physics, chemistry, biology, and medicine played an enormous role in sustaining women's positions in the field, as they could pursue work in hospitals, medical laboratories, and technical companies.

Whatever the cause, under the acting Director, Stefan Meyer (1872–1949), there was a supportive atmosphere for women scientists at the Institute for Radium Research.

Hönigschmid and Atomic Weights

It was her research supervisor, Hönigschmid, who assigned the research challenge to Horovitz. She commenced work at the Radium Institute of Vienna with Hönigschmid in 1913 or early 1914. Hönigschmid was actually affiliated with the Technical University of Prague from 1911 to 1918, but he maintained research facilities in Vienna. It is unclear whether Horovitz stayed in Vienna, where she was listed as a research worker, or whether she travelled regularly with Hönigschmid to Prague. It would seem that she visited Prague on at least one occasion, for Hönigschmid, in a letter from Prague to his friend Max Lembert closed with the comment (English translation provided here): "With best wishes also from Fräulein Doctor Horovitz, the beautiful graduate" [Hönigschmid, 1914b].

Hönigschmid had a particular interest in the measurement of the precise atomic weights of the chemical elements. This was very important work, for the determination of the formulas of new chemical compounds relied upon the knowledge of the atomic weights of the combining elements. The first table of atomic weights had been prepared by the American Chemical Society in 1893, after which the German Chemical Society published its own table in 1898. It became obvious that to obtain consistent and reliable data, chemists needed to collaborate in the development of an accurate, unified table of atomic weights. To accomplish this goal, the German Chemical Society invited chemical societies from around the world to participate in this endeavour [Oesper, 1941].

Many of the measurements of the period were of very dubious accuracy. Hönigschmid had spent a year studying precise techniques under one of the few experts, Theodore William Richards (1868–1928) at Harvard University. Using extremely careful chemical procedures, Richards had developed the most precise and reliable values for atomic weights at that time. Hence the subsequent work of any of Richard's associates had instant credibility.

Horovitz — Her Early Life

Horovitz was born in Warsaw on 17 April 1887, her father, Leopold Horovitz (1838–1917) being a well-known artist [Rentetzi, 2011]. The family moved to Vienna about 1890, and it was there that Horovitz completed the university entrance requirements in 1907. She entered the faculty of philosophy at the University of Vienna that fall. In 1914, Horovitz received a PhD in chemistry under the supervision of Guido Goldschmiedt (1850–1915) with a rating of "very good." Two publications resulted from the thesis research, both on the rearrangement of quinone using sulfuric acid [Böttcher and Horovitz, 1913; 1914]. According to Rentetzi [2011], in January 1914 Hönigschmid had contacted Meitner to ask whether she knew of anyone in Vienna who could assist him with his work on atomic weight determinations. It was Meitner, then in Berlin, who had suggested the name of Horovitz, and Hönigschmid offered her the position. There was also a direct link between Hönigschmid and Goldschmiedt, for Hönigschmid himself had been a former student of Goldschmiedt [Zintl, 1938].

Horovitz — The Confusing Later Years

The publication in 1916 on the atomic weight of the claimed "ionium" was Horovitz's last. Then she disappeared from scientific research. Unfortunately, Hönigschmid's correspondence was destroyed in World War II, so no letters between Horovitz and Hönigschmid survive [Hönigschmid, 1991].[5]

After the end of World War I, family matters and the political upheavals of the time disrupted her life [Rentetzi, 2011]. In the interwar years her interests shifted to psychology and she worked as an educational consultant with the Association for Individual Psychology. In 1924, together with psychologist Alice Friedmann (1897–1980), she set up an individual psychology-oriented home for children and adolescents. The ultimate fate of Horovitz was discussed in an exchange of letters between Kasimir Fajans (1887–1975) and Elizabeth Róna (1890–1981).[6] Fajans, who knew that Róna had also worked at the Radium Institute, asked whether she knew of the fate of Stefanie Horovitz [Fajans, 1963a]. Róna [1963] replied that Horovitz had been at the Institute before her time, but that she would try to find out from Berta Karlik (1904–1990), Marietta Blau (1894–1970), or Professor Karl Przibram (1878–1973). One assumes that she was unsuccessful, for it was Fajans who revealed the tragic news:

> You probably have not received any information from Vienna about the fate of Dr. Stephanie Horovitz. I learned about it from a mutual relative at Warzawa. Stephanie moved there after World War I and after her parents

---

[5] According to Dr Wolfgang Hönigschmid, the nephew of Otto Hönigschmid, all of his uncle's papers were destroyed during an air raid in 1944.
[6] See Rentetzi in this volume.

had died in Vienna to join her married sister. She was not active in chemistry and both were liquidated by the Nazis in 1940 [Fajans, 1963b].

However, in recent years more details about Horovitz's final years have been published [N. N., 1917]. Apparently, in 1937 she travelled to her sister's home in Warsaw. After the establishment of the Warsaw ghetto, the two Jewish sisters initially escaped. However, in order not to endanger those who were sheltering them, they reported to the Nazi authorities in 1942 when the deportations from the ghettos were ordered. Both Stefanie Horovitz and her sister were sent to the Treblinka extermination camp where they died.

***

So ended the life of a women who, in her brief career, had made a crucial discovery in the study of the atom. Although the conclusive proof of the lower atomic weight of lead from radioactive lead is ascribed to "Hönigschmid and Horovitz," [Soddy, 1966] it was Hönigschmid who received the honours. This was not the fault of Hönigschmid, for he had a very positive attitude towards his women students [Hönigschmid, 1991].[7] Rentetzi [2011] uncovered a letter from Hönigschmid to Meitner which showed that he clearly saw his work with Horovitz as a team effort: "We now isolate lead from pure Joachimsthal pitchblende … We hope that in the next two weeks before the holidays we will analyze these preparations of lead …." As Rentetzi [2011] comments, it is very difficult to untangle roles and contributions in research partnerships. She does, however, consider that, while Hönigschmid was the professor and the originator of the project, Horovitz was an active colleague in the research process.

Acknowledgements

The authors would like to thank K. Mühlberger, University Archivist, University of Vienna (1996) for some of the biographical information; and Jytte Selno, German Department, Memorial University, for translations of Hönigschmid's correspondence into English.

---

[7]According to Dr Wolfgang Hönigschmid, the encouragement of women researchers by Otto Hönigschmid was confirmed by his last woman student, Dr Luitgard Görnhardt.

References

Badash, L. (1979a). The Suicidal Success of Radiochemistry, *Brit. J. Hist. Sci.*, 12, pp. 245–256.

Badash, L. (1979b). *Radioactivity in America: Growth and Decay of a Science* (Johns Hopkins University Press, Baltimore, MD).

Boltwood, B. (1907). Note on a New Radioactive Element, *Amer. J. Sci.*, 24, pp. 370–372.

Boltwood, B. (1911). Report on the Separation of Ionium and Actinium from Certain Residues and on the Production of Helium by Ionium, *Proc. Roy. Soc. A*, 85, pp. 77–81.

Böttcher, B. and Horovitz, S. (1913). Über die Umlagerung von Chinin durch Schwefelsäure I, *Monatshefte für Chemie und verwandte Teile anderer Wissenschaften*, 32, pp. 793–796.

Böttcher, B. and Horovitz, S. (1914). Über die Umlagerung von Chinin durch Schwefelsäure II, *Monatshefte für Chemie und verwandte Teile anderer Wissenschaften*, 33, pp. 567–582.

Fajans, K. (1963a). Letter to Elizabeth Róna, 3 January 1963, Bentley Historical Library (BHL), University of Michigan, call number: 85327 Aa/2.

Fajans, K. (1963b). Letter to Elizabeth Róna, 31 August, 1963, BHL, call number: 85327 Aa/2.

Fleck, A. (1963). Early Work in the Radioactive Elements, *Proc. Chem. Soc.*, p. 330.

Fontani, M., Costa, M. and Orna, M. V., eds. (2015). *The Lost Elements: The Periodic Table's Shadow Side*. (Oxford University Press, Oxford).

Hönigschmid, O. (1914a). Letter to Lise Meitner, 27 June 1914, Meitner Collection, Churchill College, Cambridge, collection: GBR/0014/MTNR.

Hönigschmid, O. (1914b). Letter to Max Lembert, 13 June, 1914, Deutsche Staatsbibliothek.

Hönigschmid, O. and Horovitz, S. (1914a). Sur le poids atomique du plomb de la pechblende, *Comptes. Rendus.*, 158, pp. 1796–1798.

Hönigschmid. O. and Horovitz, S. (1914b). Über das Atomgewicht des 'Uranbleis' I, *Monatshefte für Chemie*, 35, pp. 1557–1560.

Hönigschmid, O. and Horovitz, S. (1915). Über das Atomgewicht des 'Uranbleis' II, *Monatshefte für Chemie*, 36, pp. 335–380.

Hönigschmid, O and Horovitz, S. (1916). Zur Kenntnis des Atomgewichtes des Ioniums, *Monatshefte für Chemie*, 37, pp. 305–334.

Hönigschmid, W. (1991). Letter to G. Rayner-Canham, 24 March 1991, author's collection.

Kaufmann, G. B. (1982). The Atomic Weight of Lead of Radioactive Origin: A Confirmation of the Concept of Isotopy and the Group Displacement Laws, *J. Chem. Educ.,* 59, part I: pp. 3–8; part II: pp. 119–123.

Kubanek, A.-M. W. (2010). *Nothing Less Than an Adventure: Ellen Gleditsch and Her Life in Science* (CreateSpace Independent Publishing Platform, Montreal).

Leicester, H. M., ed. (1968). *Source Book in Chemistry 1900–1950* (Harvard University Press, Cambridge, MA), pp. 82–84.

N. N. (2017). Stefanie Horovitz, *Wien Geschichte Wiki*, 20 December 2017 (https://www.geschichtewiki.wien.gv.at/Stefanie_Horovitz, last accessed 7 October 2018).

Oesper, R. E. (1941). Otto Hönigschmid, *J. Chem. Educ.,* 18, p. 562.

Paneth, F. A. S. (1964). Joachimstal and the History of the Chemical Elements, in *Chemistry and Beyond: A Selection from the Writings of the Late Professor F. A. Paneth,* eds. Dingle, H. and Martin, G. R. (Interscience, New York, NY), pp. 20–28.

Rayner-Canham, M. F. and Rayner-Canham, G. W. (1997). Pioneer Women of Radioactivity, in *A Devotion to Their Science: Pioneer Women of Radioactivity,* eds. Rayner-Canham, M. F. and Rayner-Canham, G. W. (McGill-Queen's University Press, Montreal, and Chemical Heritage Foundation, Philadelphia, PA), pp. 12–28.

Rentetzi, M. (2004). Gender, Politics, and Radioactivity Research in Interwar Vienna: The Case of the Institute for Radium Research, *Isis,* 95, pp. 359–393.

Rentetzi, M. (2005). Designing (for) a New Scientific Discipline: The Location and Architecture of the Institut für Radiumforschung in Early Twentieth-Century Vienna, *Brit. J. Hist. Sci.,* 38, pp. 275–306.

Rentetzi, M. (2011). Stephanie Horovitz (1887–1942), in *European Women in Chemistry,* eds. Apotheker, J. and Sarkadi, L. S. (Wiley-VCH, Weinheim), pp. 75–79.

Róna, E. (1963). Letter to Kasimir Fajans, 6 January, 1963, BHL, call number: 85327 Aa/2.

Soddy, F. (1966). The Origins of the Conceptions of Isotopes, in *Nobel Lectures: Chemistry 1901–1921* (Elsevier, Amsterdam), p. 396.

Zintl, E. (1938). Otto Hönigschmid zum 60. Geburtstag, *Zeitschrift für anorganische und allgemeine Chemie,* 236, pp. 3–11.

Select Bibliography

Apotheker, J. and Sarkadi, L. S., eds. (2011). *European Women in Chemistry* (Wiley-VCH, Weinheim).

Fontani, M., Costa, M. and Orna, M. V., eds. (2015). *The Lost Elements: The Periodic Table's Shadow Side* (Oxford University Press, Oxford).

Rayner-Canham, M. F. and Rayner-Canham, G. W., eds. (1997). *A Devotion to Their Science: Pioneer Women of Radioactivity* (McGill-Queen's University Press, Montreal, and Chemical Heritage Foundation, Philadelphia, PA).

Rentetzi, M. (2009). *Trafficking Materials and Gendered Experimental Practices: Radium Research in Early Twentieth Century Vienna* (Columbia University Press, New York, NY).

Ellen Gleditsch and Research on Radium, Chlorine and Potassium

Annette Lykknes

Department of Teacher Education, NTNU — Norwegian
University of Science and Technology, 7491 Trondheim, Norway

The years following the discovery of radioactivity were exciting times for chemists and physicists. Scientists sought to find out why certain substances emitted this newly discovered radiation, what characterized it, and what the relation between the radioactive substances was. In the decades leading up to the 1920s concepts and phenomena such as the theory of atomic disintegration (the transformation theory), isotopes, half-lives, and decay series displaying parenthood among the radioactive elements helped unveil the cloud of mystery that had surrounded the field of radioactivity from the very beginning.

To this exciting new field ambitious women from all over Europe were drawn to establish themselves in the world of science. One of them was the Norwegian pharmacist and chemist Ellen Gleditsch (1879–1968), who arrived in Paris in 1907 to work with Marie Curie (1867–1934).[1] During

---

[1] Two women who came to the Curie laboratory in Paris just before or after Gleditsch, were the Canadian physicist Harriet Brooks (1876–1933) (see Rayner-Canham and Rayner-Canham on Brooks in this volume) and the British chemist May-Sybil Leslie (1887–1937) (see Rae on Leslie in this volume). See also Roqué on Curie in this volume.

Figure 1. Ellen Gleditsch at the university in the old quarters facing Oslo's main street, Karl Johans gate (private collection, reproduced by kind permission of Nils Petter Gleditsch).

her five years at the Curie laboratory Gleditsch developed into a full-fledged researcher in radiochemistry. In 1916, back in Norway, she became the country's first associate professor of radiochemistry at the University of Oslo (Figure 1),[2] and in 1929 was appointed full professor, this time in

---

[2] At the time, the university now known as the University of Oslo was the Royal Frederick University in Kristiania. The city changed its name to Oslo in 1925, and the university became University of Oslo in 1939.

inorganic chemistry. Thus she became the second female full professor of Norway, after Kristine Bonnevie (1872–1948) who had been appointed to professor in zoology in 1912. Once tenured part of the faculty Gleditsch, among other things, determined the half-life of radium (while visiting at Yale during 1913–1914); with her co-workers she determined the atomic weight of chlorine; and towards the end of her career, she worked out the role of the potassium-40 isotope in the heating of the earth at the beginning of earth's history.

Early Research on Radium

As one of the few chemists in Curie's laboratory, Gleditsch was given special responsibility for crystallizations of radium and barium salts. For this work Curie waived her laboratory fee [Lykknes, Kragh *et al.*, 2004]. In 1908–1909, upon Curie's request, Gleditsch took up investigations related to the radium (Ra) and uranium (U) ratio in minerals. In 1902, the New Zealand-born British physicist Ernest Rutherford (1871–1937) and his colleague, the British chemist Frederick Soddy (1877–1956) had introduced the so-called transformation theory, or theory of radioactive disintegration, which explained the phenomenon of radioactivity as atomic disintegration: radioactive atoms decaying spontaneously into other atoms through the emission of rays. Eager to add empirical evidence to this theory, from 1904 on the American radiochemist Bertram Boltwood (1870–1927) set out to determine the Ra-U ratio in a large number of radioactive minerals and to confirm that the amount of radium was proportional to the amount of uranium in minerals. Gleditsch's work was a response to Boltwood's initial investigations.

Uranium decayed very slowly (half-life of the magnitude of 1 billion years), and was still present in ancient minerals. However, since radium was known to decay much more rapidly than uranium (i.e., had a considerably shorter half-life), the radium found in minerals at the beginning of the twentieth century could not stem from ancient times — it had to be produced continually. If it turned out that the Ra-U ratio was constant in minerals, that would support the hypothesis that uranium was radium's mother substance, since it showed that radium was produced from uranium at a constant rate [Lykknes, Kragh *et al.*, 2004].

In 1906 Rutherford and Boltwood established that the ratio was, indeed, constant. It had, however, proved impossible to "grow" radium from uranium directly, which led Boltwood to conclude that there existed an intermediate element between uranium and radium which had a long half-life. This element was named "ionium" (later recognized to be the thorium isotope Th-230).[3]

Gleditsch objected to some of Boltwood's procedures, especially for the collection of the so-called radium emanation (radon gas; radon-222 produced from radium-226) used to determine the amount of radium in the minerals. For three years she worked on the ratio, investigating no fewer than twenty-one different uranium minerals from different parts of the world. At the Curie lab comparative analysis of radioactive ores was considered important, not just of scientific interest, but for the benefit of companies who needed an assessment of the commercial viability of particular sites. Gleditsch's work was part of this tradition of updating tables of radioactive constants and characteristics of minerals [Roqué, 1997]. Her ratio for, e.g., pitchblende from Cornwall agreed well with the ratio found by Rutherford and Boltwood, but that for others, e.g. chalcocite from Saxony, was considerably lower. Her conclusion that the ratio was not constant created some stir, as this implied that uranium was not radium's mother substance after all [Gleditsch, 1909; 1911]. Today, the Ra-U ratio is known to vary in younger minerals, while it is constant in older minerals. Although Gleditsch also found variations among the older minerals, her results were noted in the most important English, German and French textbooks on radioactivity of the time, testifying to the impact of her investigations [Lykknes, Kragh *et al.*, 2004].

Gleditsch's and Boltwood's paths crossed in 1913, when a fellowship from the American-Scandinavian Foundation gave her the opportunity to work for one year in the United States. She first turned to Theodore Lyman (1874–1954) at Harvard University, but was rejected, apparently since no woman had ever set foot in the Harvard physics laboratory [Lykknes, Kragh *et al.*, 2004]. However, Boltwood at Yale accepted her, not without reluctance. Gleditsch would later explain that she had once worked with a man "who was reputed to hate women" [N. N. 1930].

---

[3] See Rayner-Canham and Rayner-Canham on Horovitz in this volume.

However, he thought of her as a "rare exception" and came to appreciate her work. At Yale, Gleditsch set out to resolve the discrepancy between the existing experimental values for the half-life of radium (radium-226).

Radium was of industrial interest, and furthermore it was regarded the standard substance in radioactivity research; for example, the half-life of uranium was determined on the basis of the value for radium's half-life and the Ra-U ratio [Badash, 1979, p. 153]. These two half-lives were also reckoned important for geological age determinations using the ratio of lead to uranium (the Pb-U ratio), which would also interest Gleditsch. Boltwood had already conducted experiments to determine the half-life of radium, which led him to estimate 2000 years for its value. He had compared the amount of radium produced by ionium to the amount in equilibrium with ionium in uranium minerals. In Manchester, however, Rutherford and Hans Geiger (1882–1945) had found the half-life to be 1760 years (later adjusted to 1690 years using the new radium standard) by means of counting the number of alpha (α-) particles emitted by a salt with a known amount of radium.

Gleditsch preferred Boltwood's method over Rutherford's and Geiger's, since it was independent of the radium standard. But it was a demanding chemical task involving the complete separation of ionium from uranium minerals. She managed to isolate ionium by precipitating it with fluoride [Lykknes, Kragh *et al.*, 2004]. She arrived at values of 1642 and 1674 years, which largely agreed with Rutherford and Geiger's results. Many years later Gleditsch repeated the experiments with the Norwegian minerals broeggerite and cleveite, and obtained values of 1691 and 1686 years. The consensus today is a value of 1600 years based on a weighted average of five experimental values [Bé *et al.*, 2008, p. 607].[4]

The determination of the half-life of radium was amongst Gleditsch's most noted and celebrated early achievements. When she applied for the full professorship in inorganic chemistry in 1928, all of the referees highlighted this work. In his book *Radioactivity in America*, the historian Lawrence Badash noted that even though the half-life of radium would later

---

[4]The report by Bé *et al.* includes Gleditsch's results from 1935 in an overview of previous results, but this value is not included among the five selected for a weighted average [Bé *et al.*, 2008, p. 607]. The selected values are of more recent origin.

be revised, Gleditsch's work "had developed assurance that future changes would be small" [Badash, 1979, p. 158; Lykknes, Kvittingen *et al.*, 2004].

The Isotopic Composition of Chlorine

In 1919, at the Cavendish Laboratory in Cambridge, Francis W. Aston (1877–1945) announced that he had detected two distinct atomic weights for chlorine, 35 and 37, with his newly invented mass spectrograph.[5] The concept of isotopes — that the same element can have more than one atomic weight (or mass, in modern terminology) — had been introduced six years earlier by Frederick Soddy (and coined by Margaret Todd).[6] It solved the problem of how to accommodate the many newly discovered "radioelements" in the periodic system.[7] As many of them turned out to be isotopes of already known elements, there was room after all. That lead, which is not radioactive, could have different atomic weights depending on its origin was relatively easy to accept, since lead was known to be produced in radioactive decay.[8] However, that elements of non-radioactive origin, too, consisted of more than one isotope, was harder to swallow. As Gleditsch explained:

> The discovery of isotopes among the common elements caused general surprise although it had been anticipated, and in spite of the fact that it does not contradict present day ideas on the constitution of matter. On the whole it was less readily accepted than had been the theory of isotopy put forward a few years earlier. This was mainly due to the difficulty of admitting that an element can have a complex constitution and at the same time a perfectly constant atomic weight [Gleditsch, 1925, p. 25].

Together with her collaborators Bjarne Samdahl (1896–1969), Margot Dorenfeldt (Holtan) (1895–1986) and Liv Gleditsch[9] (1895–1977),

---

[5]The mass spectrograph could identify chemical substances by sorting gaseous ions into electric and magnetic fields according to their mass-to-charge ratios, allowing particles of the same ratio to concentrate at fixed points. This historical mass spectrograph eventually developed into the modern mass spectrometer.

[6]See Hudson in this volume.

[7]See Lykknes and Van Tiggelen, Introduction, pp. 27–29, in this volume.

[8]See Rayner-Canham and Rayner-Canham on Horovitz in this volume.

[9]Liv Gleditsch was Ellen Gleditsch's sister and the first woman to obtain a master's degree in chemistry in Norway (1923).

Figure 2. Ellen Gleditsch in her laboratory with her assistants Ernst Føyn (1904–1984) and Ruth Bakken (1909–1998), ca. 1930s (Norsk Farmasihistorisk Museum, reproduced by kind permission of Norsk Folkemuseum).

Gleditsch set out to investigate whether chlorine from different sources always had the same isotopic composition (Figure 2). A lot was at stake, since a varying isotopic composition could lead to a reconsideration of the constancy of atomic weight. Gleditsch and her team investigated chlorine from sources other than of the most common marine origin, and concluded that the isotopic composition of chlorine did not vary from source to source [Lykknes *et al.*, 2005].

At about the same time, in Paris Irène Curie (1897–1956)[10] was conducting similar experiments with terrestrial and ancient minerals from Canada, Norway and Central Africa, and compared them with chlorine from sea salts, in keeping with the tradition of comparative

---

[10] See Jacquemond in this volume.

analyses of ores at the Curie laboratory. Their correspondence reveals that the two women were not even aware of each other's work in the beginning [Lykknes *et al.*, 2005]. For the sample from Africa, Curie found differences in atomic weight beyond experimental error. Careful at first, Gleditsch did not initially comment on Curie's results, but eventually suggested that Curie's experiments were tinted with experimental errors and that her samples were contaminated.

To resolve the discrepancy, Marie Curie asked Gleditsch to investigate a sample of chlorine-bearing water from the United States. This water was old and probably unaltered, and her investigations agreed with her previous results. Gleditsch and her team maintained their conclusion: chlorine's (average value for) atomic weight did not vary, and the proportion of the isotope was always the same for chlorine in different products. The constant isotopic composition was thought to be due to the "relative stability of the isotopes, and to the abundance of the atomic species from which they are formed" [Harkins and Stone, 1926, p. 947].

Potassium and the Heat Production in Rocks

Aston's mass spectrograph stimulated isotope research. More than a decade earlier, in 1907, potassium had been shown to be radioactive. But which isotopes were responsible for this property? Aston's mass spectrum revealed the existence of two different isotopes: potassium-39 and potassium-41. Almost thirty years later, potassium-40 was detected by the American physicist and mass spectrometrist Alfred Nier (1911–1994). At about the same time it was established that potassium-40 constituted a very small fraction of the naturally occurring potassium isotopes. In fact, the Hungarian radiochemist George Hevesy (1885–1966) proposed that all radioactivity of potassium was due to this isotope.

For Gleditsch research on potassium-40 was initially a minor project. Many publications on the radioactivity of potassium had appeared after 1930, but her research interests lay mainly in radio-geology as she had coined it, i.e. the investigation of the radioactivity of rocks. In 1939/40 she was given opportunity to do precise measurements using a Geiger-Müller (G-M) counter, which was much faster and easier than the electroscope she had used for previous studies. In that year the Hungarian Tibor Gráf came to work with her and constructed a G-M counter which they used for joint investigations of uranium and thorium in rocks. The problem was that any potassium present would influence the measurements, so the determination of the radioactivity from potassium became a separate line of research.

Gleditsch and Gráf found indications of a secondary, penetrating ray that turned out to be much more intense than previously reported. They realized that decaying potassium was giving off enormous amounts of heat. This sensational finding helped resolve the discrepancy between the theoretical and the experimental values for the ionization of air emitted by certain rocks, and implied a much higher heat production from potassium-containing rocks than previously accounted for. In the early history of the earth, Gleditsch and Gráf claimed, "the heat produced by ^{40}K alone was about 200 times that generated at present by all the radioactive elements in the earth together" [Gleditsch and Gráf, 1947]. Therefore, the heat caused a considerable slowing down of the cooling process on the earth's surface, and this was consistent with an age of the earth of ca. 1 billion years.

***

Ellen Gleditsch was one of the many young, ambitious women and men who were attracted to the new research field of radioactivity — to the rays one could not see, but the effects of which were nevertheless measurable. As an assistant to Marie Curie at the beginning of the century, she learned how to handle radioactive substances and how to conduct research in radioactivity, as well as acquiring a broad network that would help her pursue her research later on. The path to establishing her research and becoming a full professor in Norway (Figure 3) was not easy — Gleditsch was belittled as a diligent but not very innovative female scientist, at the very time when her research approach, which was rooted in wet chemistry was declining and being taken over by the new fields of nuclear chemistry and physics [Lykknes, Kvittingen *et al.*, 2004]. Yet she was privileged to be part of a pioneering field and to play an important role at the very beginning of radioactivity research in the first third of the twentieth century — including research on the elements radium, chlorine and potassium that was noticed by the scientific community. Gleditsch was internationally acknowledged as a researcher, and she raised new generations of chemists to work in the

Figure 3.   Gleditsch at her desk (Norsk Farmasihistorisk Museum, reproduced by kind permission of Norsk Folkemuseum).

field of radiochemistry and nuclear chemistry. For some time, she also served as the president of the International Federation of University Women (1926–1929), helping women to enter university careers worldwide.

Acknowledgements

The author wishes to thank Xavier Roqué for valuable comments to an earlier draft of this article.

References

Badash, L. (1979). *Radioactivity in America: Growth and Decay of a Science* (The John Hopkins University Press, Baltimore, MD, and London).
Bé, M.-M., Chisté, V., Dulieu, C., Browne, E., Baglin, C., Chechev, V. P., Egorov, A., Kuzmenko, N. K., Sergeev, V. O., Kondev, F. G., Luca, A., Galán, M., Huang, X., Wang, B., Helmer, R. G., Schönfeld, E., Dersch, R., Vanin, V. R.,

de Castro, R. M., Nichols, A. L., MacMahon, T. D., Pearce, A., Lee, K. B. and Wu, S. C. (2008). *Table of Nuclides (Comments on evaluation)* (Bureau International des Poids et Mesures, Sèvres).

Gleditsch, E. (1909). Sur le radium et l'uranium contenus dans les minéraux radioactifs, *Comptes rendus*, 148, pp. 1451–1453.

Gleditsch, E. (1911). Sur le rapport entre l'uranium et le radium dans les minéraux actifs, *Le Radium*, 8, pp. 256–273.

Gleditsch, E. (1925). *Contribution to the Study of Isotopes* (Skrifter Vitenskapsakademiet, klasse I, Oslo).

Gleditsch, E. and Gráf, T. (1947). Significance of the Radioactivity of Potassium in Geophysics, *Phys. Rev.*, 72, p. 641.

Harkins, W. D. and Stone, S. B. (1926). The Isotopic Composition and Atomic Weight of Chlorine from Meteorites and from Minerals of Non-Marine Origin, *Journal of the American Chemical Society*, 48, pp. 938–949.

Lykknes, A., Kragh, H. and Kvittingen, L. (2004). Ellen Gleditsch: Pioneer Woman in Radiochemistry, *Physics in Perspective*, 6, pp. 126–155.

Lykknes, A., Kvittingen, L. and Børresen, A. K. (2004). Appreciated Abroad, Depreciated at Home. The Career of a Radiochemist in Norway: Ellen Gleditsch (1879–1968), *Isis*, 95, pp. 576–609.

Lykknes, A., Kvittingen, L. and Børresen, A. K. (2005). Ellen Gleditsch: Duty and Responsibility in a Research and Teaching Career, 1916–1946, *HSPS*, 36, pp. 131–188.

N. N. (1930). Mannen som Kollega: Hvad Professor Ellen Gleditsch Mener, *Adresseavisen*, 10 May 1930.

Roqué, X. (1997). Marie Curie and the Radium Industry: A Preliminary Sketch, *History and Technology*, 13, pp. 267–291.

Select Bibliography

Badash, L. (1979). *Radioactivity in America. Growth and decay of a science* (The John Hopkins University Press, Baltimore, MD, and London).

Boudia, S. (2001). *Marie Curie et son laboratoire* (Paris: Éditions des archives contemporaines).

Kubanek, A.-M. W. (2010). *Nothing Less Than an Adventure: Ellen Gleditsch and Her Life in Science* (Crossfield Publishing, Ontario).

Lykknes, A., Kvittingen, L. and Børresen, A. K. (2004). Appreciated Abroad, Depreciated at Home. The Career of a Radiochemist in Norway: Ellen Gleditsch (1879–1968), *Isis*, 95, pp. 576–609.

Pigeard-Micault, N. (2013). *Les femmes du laboratoire de Marie Curie* (Éditions Glyphe, Paris).

Rayner-Canham, G. W. and Rayner-Canham, M. F., eds. (1997). *A Devotion to Their Science: Pioneer Women of Radioactivity* (McGill-Queen's University Press, Montreal and Chemical Heritage Foundation, Philadelphia).

May Sybil Leslie and the Disintegration of her Element — Thorium

Ian D. Rae

School of Chemistry, University of Melbourne,
Victoria 3010, Australia

Thorium, today a member of the actinide group of elements, was first identified by the Swede Jöns Jacob Berzelius (1779–1848) [1829], and within a few weeks of each other, first the German chemist Gerhard Carl Schmidt (1865–1949) [1898] and then Marie Curie (1867–1934)[1] [1898] announced that it was radioactive. The major source of thorium is the mineral monazite, the major constituent of which is cerium phosphate, but other lanthanides and uranium are also present in low concentrations. The English chemist May Sybil Leslie (1887–1937) made several contributions on the emanation of radioactive thorium, and her minute measurement practices allowed her to provide robust data on the half-life, atomic weight and activity of the products of the disintegration series of thorium ("the thorium series") and actinium ("the actinium series"). These products would later be identified as isotopes of both known and newly

---

[1] See Roqué on Curie in this volume.

discovered elements, but at that time her research was part of exciting new developments in science that revolved around the phenomenon of radio-activity, and the nomenclature that was far from settled.

***

In the early years of the twentieth century, two institutions were following the activity of thorium very closely: in her laboratory in the Department of Physics at the Sorbonne, Paris, Marie Curie concentrated on the radioactivity of radium and polonium (derived from thorium and uranium minerals), while Ernest Rutherford (1871–1937) took up the work on thorium in his laboratory at McGill University in Montreal, Canada. Rutherford [1900] reported that thorium emitted a radioactive substance that induced a current to flow between charged plates even if these were placed some distance from the sample (thorium oxide), showing that the emanation could travel on a stream of air. Adjusting the flow-rate enabled determination of the half-life of the emanation as "about one minute" [Rutherford 1900, p. 7]. With Miss Harriet Brooks (1876–1933)[2] Rutherford [Rutherford and Brooks, 1902] showed that the emanation was not deflected by a magnetic field, but behaved more like a radioactive gas, which Rutherford and Soddy [1902, p. 581] showed was chemically inert and "analogous in nature to the members of the argon family." May Sybil Leslie, who graduated from the University of Leeds in 1908 with First Class Honours in Chemistry, was awarded a so-called "Exhibition of 1851 Scholarship" (from money left over from the Great Exhibition in Crystal Palace in 1851) that enabled her to work, first, for two years with Marie Curie, and then for a further year with Ernest Rutherford (now in Manchester), studying the radioactivity of thorium [Dawson, 1938]. After her scholarship expired, Leslie did not continue with research in radioactivity. Her subsequent career was disrupted by the lack of opportunities for women in university research, by munitions work during World War I, and by personal circumstances.

Leslie in Paris

At the Radium Institute in Paris, Leslie began by exploring the products of disintegration of thorium [Leslie, 1911a], a family identified by the

---

[2] See Rayner-Canham and Rayner-Canham on Brooks in this volume.

German chemist Otto Hahn (1879–1968) as consisting of nuclei formed by successive losses of alpha (α-) and beta (β-)particles, and the "emanation" thoron (now identified as radon) [Rutherford and Hahn, 1906; Hahn, 1907]. The chain of disintegrations begins with the loss of an α-particle from the nucleus of naturally occurring thorium (conf. Table 1).[3]

Leslie began her research with the mineral thorite, thorium phosphate, which did not contain significant quantities of radium or uranium. It was dissolved in acid, freed of base metals by treatment with hydrogen sulphide, and subjected to repetitive precipitations from ammoniacal solutions as oxalate salts, before calcination and redissolution in "aqua regia."[4] The activities of the various fractions were measured with a Laborde

Table 1. Nuclear species produced by successive decay from thorium (adapted from [Rutherford, 1913, p. 552]).[5]

Thorium series	Atomic weight	Half-life period	Radiation
Thorium	232	1.3×10^{10} years	α
Mesothorium 1	228	5.5 years	[practically] rayless
Mesothorium 2	228	6.2 hours	$\beta + \gamma$
Radiothorium	228	2 years	α
Thorium X	224	3.65 days	$\alpha + \beta$
Th emanation	220	54 seconds	$\alpha + \beta$
Thorium A	216	0.14 seconds	$\alpha + \beta$
Thorium B	212	10.6 hours	$\beta + \gamma$
Thorium C	212	60 minutes	$\alpha + \beta$
Thorium C_2	212	very short	α
Thorium D	208	3.1 minutes	$\beta + \gamma$

[3] The process can be described by the equation $_{90}\text{Th}^{232} \rightarrow {}_{88}\text{Ra}^{228} + {}_2\text{He}^4$, where the subscripts are the numbers of protons in the nucleus (= atomic number) and the superscripts are the total nuclear masses (protons plus neutrons). It can be seen that the superscripts on either side of the arrow balance.

[4] "Aqua regia," literally "royal water," is a mixture of concentrated nitric and hydrochloric acids that has greater power of solution that either of the acids alone.

[5] Subsequent work led to refinement of the series and identification of "thorium" species as isotopes of other elements such as radium and actinium. See e.g. [Shea, 1983].

electroscope in which the magnitude of the charge was indicated by the deviation of a strip of aluminium foil [Cheneveau and Laborde, 1909]. In this instrument, repulsion between the electrical charges on the foil and the main frame of the electroscope keeps the foil away from the frame. The distance between them decreases as ionizing radiation from the substance under investigation dissipates the electrical charge. The rate of change is a measure of the radioactivity. Black oxide of uranium was used as a standard. The results showed that thorite contained thorium and its transformation products. The major one of these products was called mesothorium at the time, and it was later shown to be an isotope of radium (see Table 1). A pure preparation of thorium had not been achieved, but this material was found in several of the fractions from the acid treatment. Thorium itself is only weakly radioactive, but the disintegration products are more active, and they produce up to four α-particles. Beyond mesothorium was radiothorium that had a period (half-life)[6] of two years and produced α-particles with a range of 2 cm.[7]

This minutely detailed experimental work was followed with a further publication on the thorium disintegration products [Leslie, 1912a]. Working with carefully prepared material from several samples of thorite, Leslie revised her estimate of the period of radiothorium to 710 days, bringing it into accord with the result — 737 days — obtained by Blanc [1907]. Rutherford [1913] gave a value of two years for the "half-value period" of radiothorium in the thorium series. Successive disintegrations in this series produce a chain of daughter products as α- and β-particles are lost from the nuclei and γ-rays are emitted, as summarized in Table 1.

While she was in Paris, Leslie also made measurements of the mass of the entities in the thorium emanation [Leslie 1911b]. André-Louis Debierne (1874–1949), described as the "chef de travaux" in Marie Curie's laboratory [Quinn, 1995, p. 401] — especially after the death of her husband, Pierre — had constructed an apparatus in which a radioactive

---

[6]The term used at the time was "period," but later usage for this quantity is "half life," consistent with an exponential (first order) kinetic decay process.

[7]The energy with which an α-particle is released from the nucleus determines the distance it will travel in air before collisions with molecules of nitrogen and oxygen will bring it to a stop.

gas at very low pressure could be diffused through a small hole in a thin platinum foil (0.1 mm thickness) and its presence detected by means of an electroscope [Debierne, 1910] — the method referred to as the method of gaseous diffusion.[8] The apparatus was modified slightly to take account of the shorter lifetime and low intensity of the thorium emission compared to that from radium, and experiments were allowed to continue over several days. Although Debierne's publication was entitled "atomic weight," in the body of his article he referred to "molecular weight," as did Leslie, suggesting that there was still uncertainty about the nature of the emanation that we now understand is a monatomic gas (radon).

In the interpretation of her measurements, Leslie had to allow for the rapid decay of the radioactivity of the thorium emanation, for which a period (half-life) of 54 seconds had been determined by Hahn [1905], 53.3 seconds by Bronson [1905] and 51.2 seconds by Le Rossignol and Gimingham [1904]. The accuracy of the calculated molecular weight depends on the value taken for the period, as Leslie observed, because "a difference of 1 in 100 in the period results in a difference of 5 in 100 in the calculated molecular weight," and the accuracy of the counting is also critical. Her measurements as reported are shown in Table 2.

Table 2.   Leslie's measurements of the molecular weight of thorium emanation (adapted from [Leslie, 1911b, p. 330]).

Assumed period	Days of counting	Molecular weight
54 seconds	One	210
54 seconds	Two	201
53.3 seconds	One	203
53.3	Two	194

---

[8]The kinetic theory of gases provides a basis for the understanding of the diffusion phenomenon, and its rate is described by Graham's Law, which states that the rate of diffusion is proportional to the square root of the gas density or, because of the relationship between gas density and atomic/molecular weight, the square root of that quantity. The method thus involves measurement of diffusion rates of gases of known atomic or molecular weight, and then of the substance under investigation, from which its atomic or molecular weight rate can then be calculated.

Leslie was unable to extend her stay in Paris beyond the two years allowed by her scholarship, but she was able to gain a one-year extension to work with Ernest Rutherford who had moved from McGill to the University of Manchester in 1907.

Leslie in Manchester

In Manchester, Leslie continued her investigation of the thorium emanation and the relationship between diffusion coefficients and molecular weights. Instead of the electroscope used in Paris, the activity (and hence the concentration) of the emanation was measured by the extent to which it caused a current to flow between charged plates. She began her report of the experiments [Leslie, 1912b] by noting that the values determined in Paris for the emanations of radium [Debierne, 1910] and thorium [Leslie, 1911b] were "higher than would be expected from the probable molecular weights calculated on the basis of disintegration theory." She also noted the differences between the values for the molecular weights and half-lives of actinium emanation reported by different laboratories, in which the measurements had been made under different conditions of pressure and temperature. She reasoned that a comparison of the diffusion data for thorium and actinium emanations "under strictly identical conditions" would provide the ratio of their molecular weights, following which (based on that of the thorium emanation being 220) that for actinium could be calculated. The course of true radioactivity research did not run smooth, however, because

> Unfortunately, through the discovery of an unexpected source of error, no very definite conclusions could be drawn from the result, but as these throw some light on the cause of the discrepancies between the observed values for the diffusion coefficients, it will be useful to record them. [Leslie, 1912b, p. 638].

The source of error identified by Leslie was radiation arising from the "active deposit" that was left on surfaces such as those of the plates in the detector by the disintegration of the gaseous emanation. Referring to Rutherford's listing of the thorium series (Table 1), we can see that this solid deposit would consist of several radioactive daughter products. The extent of such deposits would depend on the quantity of emanation used in the measurement, and its location would be dependent on the configuration of the apparatus — both points made by Leslie.

Turning to a gold-leaf electroscope for measuring activity in the diffusion chamber, Leslie employed a model that incorporated a pendulum that enabled a timed recharging of the electroscope, and also provided a basis for time measurements, an important point for the short lifetimes that were expected for the emanations. With this combination, the half-life of the actinium emanation (average of several values) was observed to be 3.92 seconds, in good agreement with Debierne [Debierne, 1904], and that of thorium emanation 54.3 seconds, also in good agreement with values reported earlier by Leslie and others. With her thorough work, therefore, Leslie contributed to the resolution of discrepancies in values from different laboratories.

After Radioactivity

Leslie's career never approached the heights that one might have expected for someone who held a prestigious scholarship, worked in two world-leading laboratories, and contributed to the development of an exciting new area of science during the long first decade of the twentieth century. At the conclusion of her year with Rutherford she left radioactivity forever, taking employment as a school teacher and then an Assistant Lecturer at University College, Bangor, Wales. During the war she worked on the industrial production of explosives at government munitions factories at Litherland (near Liverpool) and Penrhyndeudraeth (North Wales), where she was chemist-in-charge. The skills she exhibited there, together with her work on radioactivity, earned her a Doctor of Science degree from the University of Leeds. Retrenched from government service in 1918, she was appointed to an Assistant Lectureship at Leeds (Figure 1), where she lectured and was in charge of laboratories — typical of the role assigned to women in universities — until she was appointed to a full Lectureship in Physical Chemistry. She married another chemist, Alfred Hamilton Burr (1885–1933), in 1923, interrupted her career in the early 1930s to

Figure 1.   May Sybil Leslie with a colleague at the University of Leeds (University of Leeds, reproduced with the permission of Special Collections, Leeds University Library).

care for him in his terminal illness, and resumed work at Leeds until her own death at age 49 in 1937.

***

Leslie's experiments with thorium emanation were a valuable contribution to this field of science at a time when the process of radioactive decay and the nature of the decay products were not fully understood. She was working in an extremely active field of research, one that lay in the shared borderland between chemistry and physics. This led to her most significant published work, although the results of her subsequent research in solution chemistry at Leeds were also published [Dawson and Leslie, 1909; 1911]. Many years later she joined her husband in research on the dyeing of artificial fibres [Burr and Burr, 1934] and resumed working with her mentor at Leeds, Henry Medforth Dawson[9] on solution chemistry [Burr and Dawson, 1937]. Her high standing in the chemical profession was recognised by her admission in 1918 as Associate to the Institute of Chemistry of Great Britain and Northern Ireland, and as a Fellow of the Chemical Society of London in 1920 [Rayner-Canham and Rayner-Canham, 1997].

References

Berzelius, J. J. (1829). Untersuchung eines neuen Minerals und einer darin erhalten zuvor unbekannten Erde, *Annalen der Physik*, 92:3, pp. 385–415.

Blanc, G. A. (1907). Die Zerfallskonstante des Radiothoriums, *Physikalische Zeitschrift*, 8, pp. 321–324.

Bronson, H. L. (1905). Radio-Active Measurements by a Constant Deflection Method, *Amer. J. Sci.*, series 4:19, pp. 185–187.

Burr, A. H. (deceased) and Burr, M. S. (1934). The Nature of the Dyeing Process in Cellulose Acetate Rayon and the Hydrolytic Constant of Chrysodine G, *J. Soc. Dyers Colourists*, 50, pp. 42–47.

Burr, M. S. and Dawson, H. M. (1937). Analysis of the Complex Group of Reactions Involved in the Final Stages of the Idealized Hydrolysis of Aqueous Solutions of Sodium Bromo-Acetate, *Proc. Leeds Phil. Lit. Soc.*, 3, pp. 293–299.

Chéneveau, C. and Laborde, A. (1909). Appareils pour la mesure de la radioactivité, après la method électroscopique, *J. phys.*, 4th series, 8, pp. 161–174.

Cobb, J. W. (1939). Prof. H. M. Dawson, F.R.S., *Nature*, 143, p. 628.

Curie, M. (1898). Rayons émis par les composes l'uranium et le thorium, *Compt. Rend.*, 126, pp. 1101–1103.

Dawson, H. M. (1938). May Sybil Burr (1887–1937), *J. Chem. Soc.*, pp. 151–152.

Dawson, H. M. and Leslie, M. S. (1909). Dynamics of the Reaction Between Iodine and Acetone, *J. Chem. Soc. Trans.*, 95, pp. 1860–1870.

---

[9]For biographical information on Dawson, see [Gray and Smith, 1940; Cobb 1939].

Dawson, H. M. and Leslie, M. S. (1911). Ionisation in Non-Aqueous Solvents, part 1, *J. Chem. Soc. Trans.*, 99, pp. 1601–1609.

Debierne, A. (1904). Sur l'émanation de l'actinium, *Comptes rendus*, 88, pp. 411–414.

Debierne, A. (1910). Sur le poids atomique de l'émanation du radium, *Comptes rendus*, 150, pp. 1740–1743.

Gray, R. W. and Smith, G. F. (1940). Harry Medforth Dawson 1876–1939, *Obit, Notices Fellows Roy. Soc.*, 3:8, pp. 139–154.

Hahn, O. (1905). A New Radio-Active Element which Evolves Thorium Emanation, preliminary communication, *Proc. Roy. Soc.* A, 76, pp. 115–117.

Hahn, O. (1907). Ein neues Zwischenprodukt im Thorium, *Ber. Deut. Chem. Gesell*, 40:2, pp. 1462–1469.

Le Rossignol, C. and Gimingham, C. T. (1904). The Rate of Decay of Thorium Emanation, *Phil. Mag.*, series 6, 8, pp. 107–110.

Leslie, M. S. (1911a). Le thorium et ses produits de désagrégation, *Le Radium*, 8:10, pp. 356–363.

Leslie, M. S. (1911b). Sur le poids moléculaire de l'émanation du thorium, *Comptes rendus*, 153, pp. 328–330.

Leslie, M. S. (1912a). Sur la période du radiothorium et le nombre des particules α données par le thorium et ses produits, *Le Radium*, 9:8, pp. 276–277.

Leslie, M. S. (1912b). A Comparison of the Coefficients of Diffusion of Thorium and Actinium Emanations, with a Note on their Periods of Transformation, *Phil. Mag.*, series 6, 61, pp. 637–647.

Quinn, S. (1995). *Marie Curie, a Life* (Heineman, London).

Rayner-Canham, M. F. and Rayner-Canham, G. W. (1997). May Sybil Leslie: From Radioactivity to Industrial Chemistry, in *A Devotion to Their Science: Pioneer Women of Radioactivity*, eds. Rayner-Canham, M. F. and Rayner-Canham, G. W., McGill-Queens University Press, Montreal), pp. 76–81.

Rutherford, E. (1900). A Radio-Active Substance Emitted from Thorium Compounds, *Phil. Mag.*, series 5, 49, pp. 1–14.

Rutherford, E. (1913). *Radioactive Substances and their Radiations* (Cambridge University Press, Cambridge).

Rutherford, E. and Brooks, H. T. (1902). Comparison of the Radiations from Radioactive Substances, *Phil. Mag.*, series 6, 4, pp. 1–23.

Rutherford, E. and Hahn, O. (1906). Mass of the α Particles from Thorium, *Phil. Mag.*, series 6, 12, pp. 371–378.

Rutherford, E. and Soddy, F. (1902). A Comparative Study of the Radioactivity of Radium and Thorium, *Phil Mag.*, series 5, 28, pp. 445-457.

Schmidt, G. C. (1898). Über die vom Thorium und den Thoriumverbindungen ausgehende Strahlung, *Verhandl. Physik. Ges. Berlin*, 17, pp. 14–16.

Shea, W. R., ed. (1983). *Otto Hahn and the Rise of Nuclear Physics* (Reidel, Dordrecht, Boston, MA and Lancaster).

Select Bibliography

Badash, L. (1964). Radioactivity before the Curies, *Am. J. Phys.*, 33, pp. 128–135.

Badash, L. (1966). The Discovery of Thorium's Radioactivity, *J. Chem. Ed.*, 43:4, pp. 219–220.

Campbell, J. (1999). *Rutherford: Scientist Supreme* (AAS Publications, Christchurch).

Curie, M. (1935). *Radioactivité* (Hermann & Co., Paris).

Dawson, H. M. (1938). May Sybil Barr (1887–1937), *J. Chem. Soc.*, pp. 151–152.

Quinn, S. (1995). *Marie Curie, A Life* (Heineman, London).

Rayner-Canham, M. F. and Rayner-Canham, G. W. (1997). May Sybil Leslie from Radioactivity to Industrial Chemistry, in *Devotion to Their Science: Pioneer Women of Radioactivity*, eds. Rayner-Canham, M. F. and Rayner-Canham, G. W. (McGill-Queens University Press, Kingston, Ontario), pp. 76–81.

Rutherford, E. (1913). *Radioactive Substances and their Radiations* (Cambridge University Press, Cambridge).

Shea, W. R., ed. (1983). *Otto Hahn and the Rise of Nuclear Physics* (Reidel, Dordrecht, Boston, MA, and Lancaster).

Lise Meitner and Protactinium

Xavier Roqué

Centre for the History of Science,
Universitat Autònoma de Barcelona, 08193 Bellaterra, Spain

When the Austrian physicist Lise Meitner (1878–1968) (Figure 1) moved from Vienna to Berlin in 1907 to do research on radioactivity, the field was full of both promise and uncertainty. Not a decade had passed since the discovery of polonium and radium by Pierre and Marie Curie,[1] and the complex relations between radioactive species were being investigated by the radioactivists, physicists and chemists with access to those rare, energy-emitting substances [Hughes, 1993]. More than twenty radioactive species had been described and grouped into three main series, named after the longest-lived precursors, as in the uranium and thorium series, or the species' presumed parent element, as in the actinium series. Elements in each series decayed into one another emitting alpha (α-) and beta (β-)radiations, whose nature was not yet entirely clear.

***

Born in Vienna in 1878, Meitner's family provided her with an advanced education, even though women were excluded from secondary and higher

---

[1] See Roqué on Curie in this volume.

Figure 1. Lise Meitner at the Kaiser-Wilhelm Institute for Chemistry, Berlin, ca. 1931, with a number of electroscopes as those used in the discovery of protactinium (reproduced by kind permission of the Archiv der Max-Planck-Gesellschaft, Berlin-Dahlem).

education in Austria by law. From 1897 women were admitted to a number of faculties at the University of Vienna. Meitner took the entrance exam to the University of Vienna in 1901, and completed her PhD in physics in 1906. Prospects for a professional career were rather limited, and Meitner left for Berlin with the help of an allowance from her parents, in order to attend lectures by Max Planck (1858–1947), a physicist she greatly admired. She also conducted experiments in the basement of the Chemical Institute of the University of Berlin. As a woman she was not allowed to work or even be in the main building, nor was she paid. As with Marie Skłodowska Curie and other women scientists, family support was a crucial element of Meitner's scientific education and her career.

At the Institute Meitner met the radiochemist Otto Hahn (1879–1868), with whom she would work, sometimes on the same project, for the next three decades. Their early jointly written papers dealt with the absorption of beta rays (high-energy electrons) by matter. Absorption curves representing energy as a function of distance provided crucial information about beta-emitters. By chemically separating known substances from unknown ones and tracking their radiations, Meitner and Hahn were able to identify new radioactive substances.

However, the identification was difficult because the radiations emitted by the substances were not unique to each of them, and also because chemically identical radioelements radiated differently from each other. In 1913 the radiochemist Frederick Soddy (1877–1956) proposed that chemical elements did not necessarily always have the same mass, and introduced the term "isotope" ("same place") to designate chemically identical elements of different weights.[2] Also, the discovery of the so-called Group Displacement Law clarified the genetic relation between elements. The law correlated the positions of parent and daughter elements in radioactive decay: an alpha-emitter decayed into an element two places to the left in a horizontal row of the periodic table, a beta-emitter decayed into an element one place to the right.

Searching for Actinium's Mother Substance

The Group Displacement Law introduced order in the uranium and thorium series, but posed a challenge for the actinium series. Actinium (atomic number 89) was the least explored of the main radioactive elements. With a half-life estimated to be 25 years, it could not be the parent element of its series but had to derive from a longer-lived element, but which one? The actinium precursor had to be a beta-emitting isotope of radium (88) or an alpha-emitting isotope of element 91, between thorium (90) and uranium (92) (Figure 2). The chemical properties of this as-yet-unknown element would resemble those of tantalum (73), a pentavalent element, and the name eka-tantalum was used for it ("eka" was the sanskrit prefix introduced by Dmitri Mendeleev (1834–1907) to designate "one below").

In 1913 Meitner and Hahn set out "to find that substance which ... forms the starting point for the actinium series, and to determine whether

---

[2] See Hudson in this volume.

Tabelle 2. Die Elemente der Actiniumreihe.

Element	Symbol	T	τ	λ	Kern-strahlung	Gleich-gewichts-mengen (U = 1)
Uran Y	UY	24,6 h	35,5 h	$7{,}8 \cdot 10^{-6} s^{-1}$	β	$1{,}9 \cdot 10^{-14}$
Protactinium . . .	Pa	$3{,}2 \cdot 10^4$ a	$4{,}6 \cdot 10^4$ a	$6{,}86 \cdot 10^{-13} s^{-1}$	$\alpha\gamma$	$0{,}21 \cdot 10^{-6}$
Actinium	Ac	13,5 a	19,4 a	$1{,}63 \cdot 10^{-9} s^{-1}$	β	$0{,}8 \cdot 10^{-10}$
Radioactinium . .	RdAc	18,9 d	27,3 d	$4{,}24 \cdot 10^{-7} s^{-1}$	$\alpha\gamma$	$3{,}3 \cdot 10^{-13}$
Actinium X	AcX	11,2 d	16,2 d	$7{,}14 \cdot 10^{-7} s^{-1}$	$\alpha\gamma$	$1{,}9 \cdot 10^{-13}$
Actinon . . . . . (Actiniumemanation)	An	3,92 s	5,66 s	$0{,}177 s^{-1}$	α	$7{,}6 \cdot 10^{-19}$
Actinium A	AcA	$2 \cdot 10^{-3}$ s	$2{,}88 \cdot 10^{-3}$ s	$347 s^{-1}$	α	$3{,}8 \cdot 10^{-22}$
Actinium B	AcB	36,0 m	51,9 m	$3{,}21 \cdot 10^{-4} s^{-1}$	β	$4 \cdot 10^{-16}$
Actinium C	AcC	2,16 m	3,12 m	$5{,}35 \cdot 10^{-3} s^{-1}$	$\alpha\beta$	$2{,}4 \cdot 10^{-17}$
Actinium C' . . .	AcC'	ca. $5 \cdot 10^{-3}$ s	ca. $7 \cdot 10^{-3}$ s	ca. $140 s^{-1}$	α	ca. 10^{-24}
Actinium C'' . . .	AcC''	4,76 m	6,87 m	$2{,}43 \cdot 10^{-3} s^{-1}$	$\beta\gamma$	$5{,}2 \cdot 10^{-17}$
Actinium D . . .	AcD		stabil		—	—

Figure 2. The Actinium series in 1933, from G. Kirsch's contribution on Radioactivity in the *Handwörterbuch der Naturwissenschaften* (Jena, 1933), p. 157 (public domain).

and through which intermediates actinium is derived" [Hahn and Meitner, 1918, p. 208; English translation quoted from Sime, 1996, p. 49; see also Sime, 1986]. Their search took place in far better material and professional conditions than at the Chemical Institute of the University. In 1912 the Kaiser Wilhelm Institute for Chemistry in Berlin-Dahlem, a newly established research institution protected by the German state and supported by German industry, was opened by the Kaiser-Wilhelm-Society for the Advancement of the Sciences. As Meitner moved to the Institute, still as a guest, Planck provided her with her first paid job by appointing her as his assistant. Within a year she would become a full member of the Institute, working together with Hahn in the radioactivity section, the "Laboratorium Hahn-Meitner," which would eventually become the institute's most important laboratory [Kant, 2002; Vogt, 2007].

A radium isotope had been dismissed as the actinium precursor on the grounds that actinium was not found in radium preparations. Meitner and Hahn continued looking for the precursor in uranium minerals. They settled for a sample of uranium salts and for the rich uranium ore pitchblende. They knew how to treat the salts for tantalum-like substances, and they developed a new concentration procedure for pitchblende. Treating the ore with nitric acid left a residue that contained "most of the tantalumlike substances and relatively little extraneous activity" [Sime,

1996, p. 52]. Regardless of the source, the challenge was twofold: first, to distinguish the weak radiation of the mother substance from that of the highly radioactive substances that followed it in the concentration procedure; and second, to establish the existence of actinium by tracking the radiation of its decay products. Both challenges required year-long measurements with fixed electroscopes that monitored radiation. Meitner and Hahn expected the short-lived, intense activities "to fade away while the actinium, steadily supplied by its mother substance, would gradually increase" [Sime, 1996, p. 51].

Work on the identification and isolation of the actinium precursor continued throughout World War I. Hahn was drafted, and from January 1915 he was assigned to a unit developing and deploying toxic gases. Meitner spent the period from July 1915 to October 1916 as a nurse to the Austrian army. Upon returning to Berlin she continued the experiments largely on her own, with Hahn occasionally joining in when on leave. Meitner's letters to Hahn offer extraordinary insights into the difficulties of the work and the nature of their cooperation [Ernst, 1992; Sime, 1996]. The Institute was largely devoted to the war research led by chemist Fritz Haber (1868–1934), and Meitner feared for their samples:

> The Haber people treat us of course like captured territory; they do not take what they *need* but what they *like* … Who will guarantee that they won't come over here, and then everything will really be lost. I will do everything to prevent it; we have measurements here that have been in progress for so long [Meitner, 1916; transcribed in Ernst, 1992, p. 63; translated by Sime, 1996, p. 62 (emphasis original)].

Experiments had begun with a sample of 21 grams of pitchblende which, once treated with nitric acid, left a sample of 2 grams that contained the actinium precursor. Meitner set 1.5 grams aside and submitted the remaining 0.5 grams to further treatment, until a tiny residue was obtained. Larger amounts of pitchblende were needed to confirm the existence of the decay products of actinium, and she asked Friedrich Giesel (1852–1927), an expert radiochemist who worked for

Buchler & Co in Braunschweig, to do the process. In 1902 Giesel had discovered actinium independently from André Debierne (1874–1949), a co-worker of the Curies in Paris. Late in 1917, Giesel was able to treat 1 kilogram of pitchblende, which allowed Meitner and Hahn to prepare samples up to thirty times stronger than the ones they had obtained before.

The last stage of the experiments was performed by Meitner alone, but there was never any question that the results would be jointly published. In March 1918 Meitner and Hahn submitted a paper claiming the discovery of a new element, which they also named: "The supposition that pitchblende was the suitable starting material was indeed justified. We have succeeded in discovering a new radioactive element, and demonstrating that it is the mother substance of actinium. We propose, therefore, the name protactinium" [Hahn and Meitner 1918, p. 211, quoted in Sime, 1996, p. 70; see also Hahn and Meitner 1919a-d]. Both "protactinium" and "protoactinium" (Soddy's proposal) were used until 1949, when the International Union of Pure and Applied Chemistry (IUPAC) settled on protactinium.

The relation with previous discoveries remained to be clarified, and a symbol for the new element needed to be chosen. Soddy had discovered a short-lived decay product of thorium that decayed into actinium, a discovery substantiated by Kasimir Fajans (1887–1975) and Ostwald Göhring (1889–ca. 1915), who called the element "brevium" on account of its two-minute half-life. But such a short half-life excluded it from being the mother substance of actinium, and therefore no one challenged Meitner and Hahn's priority in discovering the mother substance.

Meitner consulted the Austrian radioactivist Stefan Meyer (1872–1949) on the question of the symbol, and he clearly expressed to whom he attributed the discovery: "You pose a terribly difficult question. I would prefer the names Lisonium, Lisottonium, etc., and I therefore propose the symbol Lo, but unfortunately these are unsuitable if one wants general acceptance" [Meyer, 1918; transcribed in Ernst, 1992; English translation quoted from Sime, 1996, p. 71]. The symbol Pa was finally adopted.

Recognition and Further Work

In 1917, even before the experiments were completed, Meitner was appointed to Head of the Physical section of the Radioactive Unit at the Kaiser Wilhelm Institute for Chemistry, and was awarded the title of full professor in 1919. Both the discovery of protactinium and her new position strengthened her confidence as a researcher. In 1922 she received the *venia legendi* (her "Habilitation"), which qualified her for teaching at the University, and in 1926 she was appointed to extraordinary professor. Throughout the interwar years Meitner led the development of research on radioactivity and was a pioneer in the field of nuclear physics in Germany. She was an expert in the use of both the cloud chamber and the Geiger counter in nuclear research, and specialised in the study of the thorium series. "Mesothorium" (MsTh, radium-228), a radium isotope discovered by Hahn in 1906, was a cost-effective substitute for "radium" (radium-226), and it was also widely used in the production of gas mantles. German companies such as the "Deutsche Gasglühlicht Gesellschaft" or the "Auergesellschaft" were leaders in the international market and could be relied upon to supply mesothorium to the Institute.

Between 1924 and 1948 Meitner was nominated 19 times for the Nobel Prize in Chemistry (mostly together with Hahn), and between 1937 and 1965 she was nominated 29 times for the Prize in Physics (of which nine times together with Hahn), for the discovery of protactinium and their work on radioactive elements and nuclear physics [N. N., 2019]. In 1933, after the racial laws of the German Reich had been passed, she gave up her independent line of research and resumed collaboration with Hahn, convincing him to take on the transuranes, elements beyond uranium allegedly formed by the bombardment of uranium with neutrons. She was a key member of the interdisciplinary team that discovered and named nuclear fission in December 1938, but she had been exiled in Sweden since July 1938. Even though she kept in touch with the Berlin team and kept providing crucial insights, Hahn did not include Meitner as an author in his first publication on the subject. Racial policies may have prevented him from doing so at the time, but he failed to acknowledge Meitner's contribution to the discovery of fission after the war. She was also dismissed by the Nobel Committee in 1945, who awarded the Nobel Prize of Chemistry for 1944 to Hahn alone [Crawford, Sime and Walker, 1996].

***

Meitner is one of the greatest women scientists of the twentieth century. Her long career bears witness to the difficulties of a career in science, but also to the ways in which women sought and managed to overcome them. Meitner is also connected with the periodic table in another way. In 1994 IUPAC approved the name "meitnerium" for element 109. Meitnerium is a radioactive synthetic element — one of the superheavy elements[3] — whose most stable known isotope has a half-life of 7.6 seconds.

References

Crawford, E., Sime, R. L. and Walker, M. (1996). A Nobel Tale of Wartime Injustice, *Nature*, 382, pp. 393–395.

Ernst, S. (1992). *Lise Meitner an Otto Hahn: Briefe aus den Jahren 1912 bis 1924* (Wissenschaftliche Verlagsgesellschaft, Stuttgart).

Hahn, O. and Meitner, L. (1918). Die Muttersubstanz des Actiniums, ein neues radioaktives Element von langer Lebensdauer, *Phys. Z.*, 19, pp. 208–218.

Hahn, O. and Meitner, L. (1919a). Über das Protactinium und die Lebensdauer des Actiniums, *Phys. Z.*, 20, pp. 127–130.

Hahn, O. and Meitner, L. (1919b). Der Ursprung des Actiniums, *Phys. Z.*, 20, pp. 529–533.

Hahn, O. and Meitner, L. (1919c). Über das Protactinium und die Frage nach der Möglichkeit seiner Herstellung als chemisches Element, *Naturwiss.*, 7, pp. 611–612.

Hahn, O. and Meitner, L. (1919d). Über die chemischen Eigenschaften des Protactiniums, *Ber. Dt. Chem. Ges.*, 52, pp. 1812–1838.

Hughes, J. A. (1993). *The Radioactivists: Community, Controversy, and the Rise of Nuclear Physics* (PhD dissertation, University of Cambridge).

Kant, H. (2002). Vom KWI für Chemie zur KWI für Radioaktivität: Die Abteilung(en) Hahn/Meitner am Kaiser-Wilhelm-Institut für Chemie, *Dahlemer Archivgespräche*, 8, pp. 57–92.

Kirsch, G. (1933). Radioactivity, in *Handwörterbuch der Naturwissenschaften*, eds. Dittler, R., Joos, G., Korschelt, E., Linck, G., Oltmanns, F., and Schaum, K., 2nd ed. (Gustav Fischer, Jena).

---

[3] See Murray and Wade in this volume.

Meitner, L., (1916). Meitner to Hahn, 16 November 1916, Archiv der Max-Planck-Gesellschaft, Berlin, III. Abteilung, Repositur 14 B.

Meyer, S. (1918). Meyer to Meitner, 5 June 1918, Churchill College Archives Centre, Cambridge, MTNR 5/12.

N. N. (2019). The Nobel Prize Nomination Archive (https://www.nobelprize.org/nomination/redirector/?redir=archive/, last accessed 5 January 2019).

Sime, R. L. (1986). The Discovery of Protactinium, *J. Chem. Educ.*, 63, pp. 653–657.

Sime, R. L. (1996). *Lise Meitner: A Life in Physics* (California University Press, Berkeley and Los Angeles, CA, and London).

Vogt, A. (2007). *Vom Hintereingang zum Hauptportal? Lise Meitner und ihre Kolleginnen an der Berliner Universität und in der Kaiser-Wilhelm-Gesellschaft* (Franz Steiner Verlag, Stuttgart).

Select Bibliography

Badash, L. (1979). *Radioactivity in America: Growth and Decay of a Science* (John Hopkins University Press, Baltimore, MD and London).

Johnson, J. (1990). *The Kaiser's Chemists: Science and Modernization in Imperial Germany* (The University of North Carolina Press, Chapel Hill, NC).

Sime, R. L. (1996). *Lise Meitner: A Life in Physics* (California University Press, Berkeley and Los Angeles, CA, and London).

"She is in the Next Room": Elizabeth Róna and Polonium

Maria Rentetzi

National Technical University of Athens, Department of
Humanities, Social Sciences, and Law, School of
Applied Mathematical and Physical Sciences, Greece

"She is in the next room!" This is how one of Elizabeth Róna's colleagues responded to a biologist of the Medical Division of the Argonne National Laboratory who, in need of a polonium source, was trying to locate Róna in Europe in 1947. An amount of polonium as small as the head of a pin was radioactive enough to be used as an alpha (α-)particle source in biological experiments of single cells. At the time, Elizabeth Róna (1890–1991), a Hungarian radiochemist, known also as the "polonium woman," was the leading expert in preparing polonium sources. She had been trained by Irène Curie (1897–1956) at the Radium Institute in Paris;[1] she worked for 13 years at the Institute for Radium Research in Vienna preparing polonium for medical and laboratory use; and she got involved in one of the most important scientific controversies of the

---

[1] See Jacquemond in this volume.

interwar period over the artificial disintegration of light elements as part of the Viennese group.

Like most of her female colleagues at the Vienna Radium Institute, Róna had been part of a strong network of "radioactivists" [Hughes, 1993], men and women of science who devoted their lives to radioactivity research. Threatened by the political upheavals Róna left Vienna in 1938, immigrated first to Sweden and then to Norway, returned to Budapest, and finally, in 1941, she fled to the USA. The biologist at Argonne had not realized that the Jewish Róna was now working literally next door to him, for the US Atomic Energy Commission on projects related to isotope separation. Her rare expertise on polonium helped her to overcome issues of US secrecy and provided her with clearance to work on war-related, highly confidential projects. Indeed, polonium — the radioactive material — secured Róna's professional identity and, at the same time, embodied the transformations of laboratory culture after the end of the war.

Elizabeth Róna

Róna was born into a prosperous Jewish family in Budapest in 1890. Her father, Samuel Róna (1857–1910), a physician who had close contacts with the founders of radium therapy in Paris, Louis Wickham (1861–1913) and Henri-August Dominici (1867–1919), was influential in introducing the field into the academic community in Budapest. After growing up in a stimulating environment, Róna studied physics at the University of Budapest and pursued further graduate studies at the University of Karlsruhe, Germany, with a focus on physical chemistry. There she was introduced to radioactivity research by Kasimir Fajans (1887–1975), a Polish radiochemist who was working on radioactive isotopes.

Mentors have, always, played a key role in supporting women's careers in science by introducing them to their networks, assigning laboratory space to them, and encouraging them to succeed. Fajans was one of them. In 1911, right after she received her PhD in chemistry, physics and geophysics from the University of Budapest, Róna chose to work with Fajans on radioactivity rather than with Georg Bredig (1868–1944) on physical chemistry. Although devoted to his students, Bredig was known to be an authoritative German professor, a model that Fajans, luckily, did not subscribe to. Instead, Fajans introduced Róna to a new and exciting field, encouraged her to devote her creativity and energy to it, and made her feel at home. In addition, Fajans created a social and professional network of both men and women. In contrast to Bredig, at whose work parties at home Róna was expected to join the ladies, "Fajans gave many small parties in the laboratory with no discrimination against women" [Róna, 1978, p. 4].

Róna's second experience of working in the field, an "exciting and pleasant" one, was with Georg Hevesy (1885–1966) in Budapest a few months later. Róna was free to use her own imagination in her research, and described Hevesy as someone who "did not feel the need to keep his students in their place" [Róna, 1978, p. 3]. During World War I Róna returned to Hungary to work with Hevesy, who had just left the Institute for Radium Research in Vienna to accept the position of lecturer at the University of Budapest. His work on radioactive elements as tracers of chemical reactions attracted Róna's interest. Her collaboration with Hevesy placed Róna among the key figures of the radioactivity community, such as Ernest Rutherford (1871–1937), Frederick Soddy (1877–1956), Alexander Fleck (1889–1968), Lise Meitner (1878–1968), and Otto Hahn (1879–1968).[2]

At the time Hevesy got involved in a controversy concerning the production of isotopes and their relationship to the periodic table. While Soddy was involved in a dispute with Fajans concerning the Group Displacement Law that defined the production of radioactive isotopes, a second, related dispute arose between Soddy and Rutherford's group in Manchester. George Antonoff, a Russian research student working with Rutherford, claimed to have isolated uranium Y, an unknown element in 1911.[3] Soddy and Fleck, unable to repeat his experiments, engaged in a public and fierce dispute with Rutherford's student. Hoping to apply his radioactive tracer method to the problem, Hevesy asked Róna to repeat the controversial experiment. She succeeded in separating the uranium Y from all the interfering elements and proved that it was a beta (β-)emitter with a half-life of twenty-five hours. "Soon

---

[2] See Hudson (on Todd) and Roqué (on Meitner) in this volume.

[3] This element was later identified as Thorium-231.

after my paper was published by the Hungarian Academy of Science," Róna recalls, "Soddy, Hahn, and Meitner also verified Antonoff's results" [Róna, 1978, p. 8]. Impressed with her work, Otto Hahn offered Róna a fellowship to work at the radioactivity department of the Kaiser Wilhelm Institute in Berlin-Dahlem.

In the early 1920s it was Stefan Meyer (1872–1949), the director of the Institute for Radium Research in Vienna, who asked Róna to join his Institute. "What a windfall for me," Róna recalled with surprise, "to be able to work full time in my chosen field in well-equipped laboratories, with scientists expert in the field of radioactivity!" [Róna, 1978, p. 15]. She remembered the pleasant atmosphere, the collaborative research, and her feeling of being part of the "family" of the Institute, all mainly thanks to Meyer's personality and his personal interest in the work of each of his students and staff [Rentetzi, 2008]. It was there that she was introduced to research on polonium, and became the leading polonium expert who crossed the borders between physics and biology, and science and medicine, for the rest of her career (Figure 1), passing on her knowledge in polonium preparation to her young colleague, Ewald Schmidt [Róna and Schmidt, 1927; 1928].

Polonium

In July 1898, after using a combination of electrometric methods and chemical analyses, Pierre (1859–1906) and Marie Curie (1867–1934) discovered polonium.[4] As the French Academy insisted on the isolation and spectroscopic corroboration of polonium, the Curies needed to process enormous amounts of uranium pitchblende in order to extract a few milligrams of polonium. The only available uranium mine was the one in St. Joachimsthal, Bohemia, then part of the Austro-Hungarian Empire. It was Meyer's Institute that finally offered them the required pitchblende, building a strong and long-lasting collaboration between the Paris and Vienna Radium Institutes. Thus, when Meyer's colleagues later needed to develop expertise in preparing polonium sources, they turned to Curie.

---

[4] See Roqué on Curie in this volume.

Figure 1. Elisabeth Róna in the Institute for Radium Research, ca. 1925 (the author's archive, donated by Agnes Rodhe).

As an alpha emitter, polonium is a strong radioactive metal, hazardous to handle, and highly dangerous to human health. It appears rarely in nature, all of its unstable isotopes have short half-lives, and it poses great challenges in its handling. During the 1920s polonium proved to be the ideal radioactive source for experiments on artificial disintegration. A group working under the leadership of the Swedish physicist Hans Pettersson (1888–1966) in Vienna conducted experiments in parallel with Rutherford's group at the Cavendish Laboratory. The first place in what proved to be a fierce controversy was to be judged on the use of strong radioactive sources [Stuewer, 1985]. Since polonium did not emit the beta particles that usually interfered in scintillation counting, its use as a radioactive source was most advantageous.

The preparation of polonium sources did not simply involve its separation from other elements. It was a tedious process of concentrating radioactively pure polonium on a small surface. This involved the tiresome task of the chemical separation of the element, its purification, and the final concentration. In May 1926 Pettersson reported to his father that "I have now managed to get Meyer to write a letter to Curie asking to send one of our scientists, Frau Doktor Róna, chemist and specialist in polonium, to her lab for three weeks in order to learn the art from Irène Curie ... If she is allowed to go, we have no problems next year and can make our own polonium samples" [Petterson, 1926a].[5] Róna was finally accepted into Curie's laboratory in 1928, and was assigned to working with Irène on her own preparation method. On Róna's return to the Radium Institute a few weeks later, Curie was generous enough to donate to the Viennese a strong polonium source concentrated on a small silver disc of 12 mm. Most of the subsequent studies on artificial disintegration performed at the Institute were done either using Curie's source or Róna's preparations [Petterson, 1926b; 1928].

Róna also studied the use of polonium in leukaemia therapy. By doing so, she constantly crossed the disciplinary boundaries between physics and medicine. Medical doctors urgently needed radium supplies and measurements of radium preparations for clinical use, while the physicists were glad to use the spent radon needles used in medicine for the polonium preparations that became Róna's expertise [Meyer, 1950]. It was this expertise that later gave Róna the chance to join the war effort in the big US laboratories, and to ensure contracts with several key research institutions such as the Argonne and the Oak Ridge National Laboratory. She even provided for free her polonium preparation method that led to mass production of polonium sources by the Canadian Radium and Uranium Company, which was contracted by the US Office of Scientific Research and Development.

***

Róna represents a generation of accomplished European female scientists who were able to travel to other European institutes in order to receive training in new experimental techniques, who supervised young male students, and were deeply involved in the design, production, and evaluation of experiments. Yet, World War II interrupted their careers and

---

[5] Letter in Swedish, translated by Agnes Rodhe.

rendered them invisible in the big US laboratories. Like many others, Róna had to flee the continent and start a new career across the Atlantic. She was 51 years old when she obtained a three-month visitor's visa to enter the US. Along with her few belongings, Róna carried her polonium expertise developed at the Radium Institute in Vienna, as well as a working laboratory culture that favoured team work and was distinctively different from the one flourishing at the time in the big national US laboratories. The unusual intertwinements of two biographies — that of Róna and that of polonium — provide a glimpse into women's contributions to chemistry in the first part of the twentieth century. Being a woman in science was inseparable from political history and the history of the discipline.

References

Hughes, J. (1993). The Radioactivists: Community, Controversy and the Rise of Nuclear Physics, unpublished PhD thesis (University of Cambridge).

Meyer, S. (1950). Die Vorgeschichte der Gründung und das erste Jahrzehnt des Institutes für Radiumforschung, in *Festschrift des Institutes für Radiumforschung anlässlich seines 40jährigen Bestandes (1910–1950)*, Sitzungsberichte der Österreichischen Akademie der Wissenschaften, 159 (Vienna), pp. 1–26.

Petterson, H. (1926a). Letter to Otto Pettersson, 24 May 1926, Agnes Rodhe Papers, author's collection.

Pettersson, H. (1926b). Report on the Investigations Regarding Artificial Disintegration, Archives of the Gothenburg University Library.

Pettersson, H. (1928). Report to the International Education Board, April 1928, Archives of the Gothenburg University Library.

Rentetzi, M. (2008). *Trafficking Materials and Gendered Experimental Practices: Radium Research in Early Twentieth Century Vienna* (Columbia University Press, New York, NY) (http://www.gutenberg-e.org/rentetzi (2007), last accessed 27 January 2019).

Róna, E. (1978). *How It Came About: Radioactivity, Nuclear Physics, Atomic Energy* (Oak Ridge Associated Univ. Press, Oak Ridge, TN).

Róna, E. and Schmidt, E. (1927). Untersuchungen über das Eindringen des Poloniums in Metale, *Sitzungsberichte Akademie der Wissenschaften in Wien, mathematisch-naturwissenschaftliche Klasse,* Abteilung IIa, 136, pp. 65–73.

Róna, E. and Schmidt, E. (1928). Eine Methode zur Herstellung von hochkonzentrierten Poloniumpräparaten, *Sitzungsberichte Akademie der Wissenschaften in Wien, mathematisch-naturwissenschaftliche Klasse,* Abteilung IIa, 137, pp.103–105.

Stuewer, R. (1985). Artificial Disintegration and the Vienna-Cambridge Controversy, in *Observation, Experiment, and Hypothesis in Modern Physical Science,* ed. Achinstein, P. and Hannaway, O. (MIT Press, Cambridge, MA), pp. 239–307.

Select Bibliography

Des Jardins, J. (2010). *The Madame Curie Complex: The Hidden History of Women in Science* (Feminist Press, New York, NY).

Meitner, L. (1964). Looking Back, *Bulletin of the Atomic Scientists,* 20, pp. 2–7.

Merricks, L. (1996). *The World Made New: Frederick Soddy, Science, Politics, and Environment* (Oxford University Press, Oxford).

Rayner-Canham, M. and Rayner-Canham, G. (1997). Elizabeth Rona: The Polonium Woman, in *A Devotion to their Science: Pioneer Women of Radioactivity,* eds. Rayner-Canham, M. and Rayner-Canham, G. (McGill-Queen's University Press, Montreal and Chemical Heritage Foundation, Philadelphia, PA), pp. 209–216.

Rentetzi, M. (2004). Gender, Politics, and Radioactivity Research in Interwar Vienna: The Case of the Institute for Radium Research, *Isis,* 95, pp. 359–393.

Rentetzi, M. (2008). *Trafficking Materials and Gendered Experimental Practices: Radium Research in Early Twentieth Century Vienna* (Columbia University Press, New York, NY) (http://www.gutenberg-e.org/rentetzi (2007), last accessed 27 January 2019).

Rentetzi, M. (2014). Creating a Niche for Women Chemists in Cosmetics Industry, in *Women in Industry,* eds. Tobies, R. and Vogt, A., Wissenschaftskultur um 1900, series ed. Breidbach, O. (Steiner, Stuttgart), pp. 115–118.

Rona, E. (1978). *How it Came About: Radioactivity, Nuclear Physics, Atomic Energy* (Oak Ridge Associated Universities, Oak Ridge, TN).

Rona, E. (1979). Laboratory Contamination in the Early Period of Radiation Research, *Health Physics,* 37, pp. 723–727.

Sobel, D. (2016). *The Glass Universe: How the Ladies of the Harvard Observatory Took the Measure of the Stars* (Penguin Random House, New York, NY).

Marguerite Perey: The Discoverer of Francium

Marelene Rayner-Canham* and Geoffrey Rayner-Canham[†]

*Department of Physics, Grenfell Campus,
Memorial University, Corner Brook, NL, Canada
[†]Department of Chemistry, Grenfell Campus,
Memorial University, Corner Brook, NL, Canada

For decades, chemists had been searching for the missing heaviest alkali metal. It was to be the French scientist, Marguerite Perey (1909–1975) (Figure 1) who conclusively identified this elusive short-lived radioactive element and named it after her country: francium. The discovery was to bring fame to this researcher at the Institut du Radium (the Radium Institute), a centre for radioactivity studies founded by Marie Curie (1867–1934).

The Search for Element 87

Element 87 had been incredibly elusive [Adloff and Kauffmann, 2005b]. Element 84, polonium; element 86, radon; element 88, radium; and element 89, actinium, had all been identified by the beginning of the twentieth century. Only element 85, the bottom member of the halogen family, and element 87, the lowest member of the alkali metals, were missing. We now know that

Figure 1.   Marguerite Perey, ca. 1938 (Musée Curie, coll. Institut du Radium, reproduced by kind permission of the Musée Curie).

element 85 (astatine) only exists in nature in such infinitesimally low concentrations it cannot be considered a naturally occurring element [Thornton and Burdette, 2010].[1] Thus our story here is of the discovery of the last element to be identified which can be extracted chemically from minerals.

In the 1920s and 1930s there was still hope that a stable isotope of element 87, also called *Eka*-caesium (or *eka*-Cs), existed, even though the neighbouring elements could only be found as radioactive isotopes. Such

---

[1] See also Forstner in this volume.

hope spawned many false element discoveries, including: russium (1925); alkalinium (1930); virginium (1930); and moldavium (1937); each of which was disproven [Fontani *et al.*, 2015].

The search for a possible radioactive isotope began in earnest in 1913, following the discovery of the Group Displacement Law, a rule governing the transmutation of elements during radioactive decay. It is also known as the Fajans and Soddy Law, after Kasimir Fajans (1887–1975) and Frederick Soddy (1877–1956), who independently proposed it in 1913. This law stated that an isotope decay could occur as a result of emission of an alpha (α-) particle (a helium nucleus), in which case the daughter nucleus would have two fewer protons and two fewer neutrons.[2] Alternatively, emission of a beta (β-)particle (electron) would result in the number of protons increasing by one, while the number of neutrons would decrease by one. In the specific case of forming element 87, there could be only two possible synthetic pathways: that beginning with an α-emitting isotope of actinium; or that beginning with a β-emitting isotope of radon. Displaying just the atomic number of each (as this would apply to any feasible number of neutrons, and hence mass number), the generic pathways are shown below:

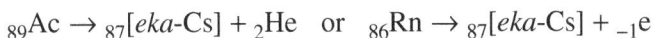

$$_{89}\text{Ac} \rightarrow {}_{87}[eka\text{-Cs}] + {}_2\text{He} \quad \text{or} \quad _{86}\text{Rn} \rightarrow {}_{87}[eka\text{-Cs}] + {}_{-1}\text{e}$$

Unfortunately, the known isotopes of actinium were β-emitters while the known isotopes of radon were α-emitters. Some radioactive nuclei were known for which a proportion of the atoms decayed by the α-route while the remainder decayed by the β-route. In the search for element 87 it was hoped that perhaps one of the isotopes of actinium, or one of radon, exhibited this branching behaviour resulting in some atoms of element 87. Radon, being a radioactive gas,[3] was not easy to work with safely thus attention turned to isotopes of actinium.

Marguerite Perey and the Discovery of Actinium K

In 1929, a young chemical technician, Marguerite Perey, was hired at the Radium Institute in Paris [Adloff and Kauffmann, 2005a] (Figure 2). Her

---

[2] The proton and neutron were not known at the time, so this explanation is a modern one.

[3] See Rayner-Canham and Rayner-Canham on Brooks in this volume.

Figure 2. At the Radium Institute in Paris, 1930. From left to right, front row: Marguerite Perey, Léonie Razet, Isabelle Archinard, Sonia Cotelle; back row: André Regnier, Alexis Jakimach, Raymond Grégoire, Renée Galabert, Tchang Da Tcheng and Frédéric Joliot-Curie (Musée Curie, coll. Institut du Radium, reproduced by kind permission of the Musée Curie).

role was to be Marie Curie's personal laboratory assistant with the task of working with the radioactive actinium series. She had two tasks to learn and perform. The first was the concentration of actinium itself by means of dozens of crystallizations, evaporations, and fractional precipitations (Figure 3). The second was to rapidly separate out the short-lived daughter isotopes which were being formed. Only when she was sure of the purity of the actinium could she search for any uncatalogued radiation.

The death of Marie Curie intervened. Perey continued her work under the direction of André-Louis Debierne (1874–1949), the new Director of the Radium Institute, and Marie Curie's daughter, Irène Joliot-Curie (1897–1956).[4] Independently, and unbeknownst to each other, they asked

---

[4] See Jacquemond in this volume.

Figure 3.   At the "salle de chimie" of the Curie laboratory, Radium Institute in Paris, July 1930. From left to right: Sonia Cotelle, Marguerite Perey, Alexis Jakimach and Tchang Da Tcheng (Musée Curie, coll. ACJC, reproduced by kind permission of the Musée Curie).

Perey to determine a precise value for the half-life on actinium-227. Actinium-227 was known to undergo β-emission to give thorium-227 with a half-life of 22 years, followed by an α-emission with a half-life of 19 days. Displaying the mass number of each, the specific pathway for this isotope is shown below:

$$^{227}Ac \rightarrow {}^{227}Th + {}^{0}e$$
$$^{227}Th \rightarrow {}^{223}Ra + {}^{4}He$$

In her careful measurements, Perey made the crucial observation that there was some β-radiation which was unaccounted for. This radiation, with a short half-life of 22 minutes, seemed to correspond to the appearance of a new element which she called actinium K (symbol AcK). After numerous chemical tests, actinium K gave one conclusive result: when perchlorate ion was added, actinium K co-precipitated with caesium ion (caesium perchlorate is one of the very few insoluble compounds of the "heavier" alkali metals). The inescapable conclusion was that Perey had discovered an alkali metal, the missing *eka*-Cs. This pathway, undergone by only about 1% of the actinium-227 atoms, had the converse order of first, the α-emission, then second, the β-emission, as shown below:

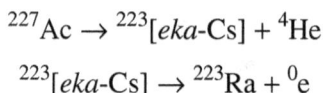

$$^{227}Ac \rightarrow {}^{223}[eka\text{-}Cs] + {}^{4}He$$

$$^{223}[eka\text{-}Cs] \rightarrow {}^{223}Ra + {}^{0}e$$

The first of a series of publications which Perey authored on "element 87, derived from actinium" appeared in 1939 [Perey, 1939]. It has been surmised by Jean-Pierre Adloff and George B. Kauffman [2005b] that it was her name alone on the paper due to friction between Debierne and Joliot-Curie over which one of them she had been working for.[5] As the discoverer, Perey was entitled to propose a name for the element. Her first choice, "catium" due to its readiness to form a cation, was favoured by Debierne but opposed by Joliot-Curie. Joliot-Curie's preference and that of her husband, Frédéric Joliot-Curie (1900–1958), was for Perey's second choice of "francium" after her native land, France.

Though originally given the symbol "Fa," Friedrich Paneth (1887–1958) of the International Committee of Inorganic Nomenclature (established 1919), pointed out that the general rule was that the first and second letters of the name were to be used, where possible, thus the symbol was to be "Fr" [F. Paneth, letter to M. Perey, 27 November 1946, cited in Adloff and Kauffman, 2005b, p. 392]. Perey [1957] devoted the remainder of her working life to the study of the properties of francium and its possible medical applications, authoring the definitive work on the element which was published in 1957.

Marguerite Perey: A Life

Marguerite Catherine Perey was born 19 October 1909 in Villemomble, a suburb of Paris, the youngest of five children of a middle-class Protestant

---

[5] Adloff, Perey's first research student and later Directeur of the Department of Nuclear Chemistry at Strasbourg, was Perey's close friend and kept in contact with her until her health declined.

industrialist [Adloff and Kauffmann, 2005a]. She had hoped to have a career in medicine but the financial difficulties of her family made it impossible to have the funds to do so. Instead, she studied chemistry at the "École Féminine d'Enseignement Technique," a private but state-approved school for laboratory technicians in Paris.

It was a custom of Marie Curie that, funds permitting, she would ask the Directeur of the School for permission to hire the best chemistry laboratory technician of the graduating class. In June 1929, Perey received her "Diplôme d'Etat de Chimiste" and being top of the class was invited to an interview with Marie Curie. Despite Perey's belief that the interview had gone badly, Curie hired her. As described above, following Curie's death, Debierne and Joliot-Curie independently assumed they were her supervisor.

As a result of her success, Debierne and Joliot-Curie encouraged Perey to obtain a formal qualification in science, and in 1946 she received a doctorate from the University of Paris based upon her discovery of element 87. Having completed her doctorate, Perey was now able to pursue her work as an independent qualified researcher at the "Centre National de la Recherche Scientifique" (CNRS) while training a new generation of laboratory technicians. She stayed there from 1946 to 1949, but Adloff and Kauffman [2005a, p. 380] state that it was not a happy time for Perey:

> The initial enthusiasm accorded the discovery of francium had faded with the rapid development of new topics following the discoveries of nuclear fission and the first of the transuranium elements. Perey felt progressively isolated at the Institut du Radium and had no opportunity to assume responsibilities for supervising young research students.

Fortunately, in 1949 Perey was appointed to the Chair of Nuclear Chemistry at the University of Strasbourg. This was a position created especially for her and, at the same time, she became "Directeur" of the newly created associated laboratory. Perey's interest shifted to the potential use of francium for the early diagnosis of cancer, as the element concentrated in cancerous tissues. However, the project was abandoned due to the Radium Institute in Paris holding most of the supplies of the parent actinium (which had been prepared by Perey), as well as a lack of interest among physicians.

To the surprise of many, in 1962, Perey was elected to the physics section of the "Académie des Sciences" — the French Academy of

Sciences [Adloff and Kauffmann, 2005c]. She was the first woman ever elected to the Académie, the honour denied to her research supervisors, Marie Curie and Irène Joliot-Curie. At the time of Perey's discovery in 1939, there was very little interest, but election to the Academy resulted in a "media frenzy" [Adloff and Kauffmann, 2005c, p. 396]. Adloff and Kauffman [2005c] have, however, commented that she was not actually elected a full member (the first woman elected as full member was the physicist Yvonne Choquet-Bruhat (b. 1923), in 1979):

> Perey was elected as a corresponding (*correspondant*) member, not a full member. Such second-rank members have no "academician seats" or other prerogatives of full members and do not bear the official title of *académicien* [Adloff and Kauffman, 2005c, p. 396].

Like many members of the Curie group, Perey developed cancer from her work with such highly radioactive sources [Róna, 1979]. Her last recognition was that of "Commandeur dans l'Ordre National du Mérite," awarded in 1974. Perey's fifteen years of suffering from radiation-related diseases finally ended with her death on 13 May 1975. Thus ended the unconventional story of a scientist who rose from being a lowly technician through to a university professor on the basis of her dedication to the discovery of a missing chemical element.

References

Adloff, J.-P. and Kauffman, G. B. (2005a). Marguerite Perey (1909–1975): A Personal Retrospective Tribute on the 30th Anniversary of her Death, *Chem. Educator*, 10, pp. 378–386.

Adloff, J.-P. and Kauffman, G. B. (2005b). Francium (Atomic Number 87), the Last Discovered Natural Element, *Chem. Educator*, 10, pp. 387–394.

Adloff, J.-P. and Kauffman, G. B. (2005c). Triumph Over Prejudice: The Election of Radiochemist Marguerite Perey (1909–1975) to the French Académie des Sciences, *Chem. Educator*, 10, pp. 395–399.

Fontani, M., Costa, M. and Orna, M. V., eds. (2015). *The Lost Elements: The Periodic Table's Shadow Side* (Oxford University Press, Oxford).

Perey, M. (1939). Sur un élément 87, dérivé de l'actinium, *Comptes Rendus.*, 208, pp. 97–99.

Perey, M. (1957). Francium, in *Nouveau traité de chimie minérale*, vol. 3, ed. Pascal, P. (Masson, Paris), pp. 131–141.

Róna, E. (1979). Laboratory Contamination in the Early Period of Radiation Research, *Health Physics*, 37, pp. 723–727.

Thornton, B. F. and Burdette, S. C. (2010). Finding Eka-Iodine: Discovery Priority in Modern Times, *Bull. Hist. Chem.*, 35, pp. 86–96.

Select Bibiography

Adloff, J.-P. and Kauffman, G. B. (2005a). Marguerite Perey (1909–1975): A Personal Retrospective Tribute on the 30th Anniversary of her Death, *Chem. Educator*, 10, pp. 378–386.

Adloff, J.-P. and Kauffman, G. B. (2005b). Francium (Atomic Number 87), the Last Discovered Natural Element, *Chem. Educator*, 10, pp. 387–394.

Adloff, J.-P. and Kauffman, G. B. (2005c). Triumph Over Prejudice: The Election of Radiochemist Marguerite Perey (1909–1975) to the French Académie des Sciences, *Chem. Educator*, 10, pp. 395–399.

Kauffman, G. B. and Adloff, J. P. (1989). Marguerite Perey and the Discovery of Francium, *Educ. Chem.*, 26, pp. 135–137.

Berta Karlik and Traude Bernert: The Natural Occurring Astatine Isotopes 215, 216, and 218

Christian Forstner

Department of History — History of Science, Goethe-University Frankfurt, IG-Farben-Haus, Postfach 28, Norbert-Wollheim-Platz 1, 60629 Frankfurt am Main, Germany

During World War II the Austrian physicist Berta Karlik (1904–1990), together with her assistant Traude Bernert (1915–1998), discovered isotopes 215, 216, and 218 of element 85, "eka-iodine." Element 85 is known today as astatine. It is part of two of the natural decay chains, namely the uranium or radium series, and the actinium series. Astatine is very rare in nature, and therefore several erroneous discoveries were made. Astatine naturally appears for just a short time during the decay of radium, thorium and actinium. However, in 1940 Emilio Segré (1905–1989) and his team at the University of California in Berkeley managed to produce the isotope astatine-211 artificially by bombarding bismuth-209 with alpha (α-)particles. This was 3 years before Karlik and Bernert were finally able to detect it in the natural decay chains.

The Institutional Context of the Discovery

In the early twentieth century there were two centres of nuclear research in Vienna, one at the Second Institute for Physics of Vienna University, and the other at the Institute for Radium Research (hereafter: Radium Institute) of the Austrian Academy of Science. The Radium Institute opened in 1910 and, due to Austria's monopoly on pitchblende, the raw material for radium production, became a major centre within the international network of the so-called "radioactivists." One astonishing feature of the Institute at that time was the high proportion of female staff, which the historian Maria Rentetzi traced back to the social and political milieu of the "Red Vienna" of the 1920s and 1930s [Rentetzi, 2004].

Berta Karlik and Traude Bernert were two of the women who were carrying out their research at the Radium Institute. However, before they made their discovery, the "Anschluss" (annexation) of Austria to Nazi-Germany in 1938 led to major changes at the institute. About a fourth of all Austrian nuclear researchers lost their jobs, principally due to anti-Jewish sanctions, and the number of women, which were employed at the Radium Institute plummeted by half within the course of a year. Two positions for full professors and two for associate professors at the physics department of the University of Vienna were subsequently filled by scholars who followed the Nazis out of conviction or opportunism. One of them, Gustav Ortner (1900–1984) replaced Stefan Meyer (1872–1949), the former director of the Radium Institute, who had had to leave due to his Jewish heritage. During World War II the Viennese institutes were formally reorganised: the Second Institute for Physics and some parts of the Radium Institute were merged to create the Four-Year-Plan Institute for Neutron Research in 1943. However, even before the foundation of the Four-Year-Plan-Institute the discovery of nuclear fission attracted the interest of Austrian physicists, and the German so-called "uranium club" opened up new possibilities for their research, which were embraced by most of the Austrian researchers. All German research projects dealing with uranium fission were organised in the uranium club from 1939 to 1945. It was the central German project for building a nuclear reactor during the war [Walker, 1995]. The Austrian part of the research, which was carried out in the uranium club, was mainly basic research which was to determine different cross sections in scattering processes [Fengler, 2014; Forstner, 2017; Reiter, 2017].

Berta Karlik and Traude Bernert

Berta Karlik finished her PhD at the University of Vienna in 1928, started her research at the Radium Institute in 1928/29, and became a graduate assistant at the Institute in 1933. With the help of a fellowship of the International Federation of University Women she studied under William Bragg (1862–1942) at the Royal Institution in London for one year, between November 1930 and December 1931. In 1935 she was invited to Sweden to undertake research there for several months. After she finished her "Habilitation" the University of Vienna awarded her the "venia leg-endi" in 1937. She received several fellowships until she was appointed a lecturer with remuneration in 1942. Karlik never took part in the research programme of the German "Uranverein" (the German nuclear weapons programme), but tried to develop her own line of research within the insti-tute. After the Nazis had seized power in Austria the future of her work was uncertain. Her request for an extension of her fellowship was denied by the NS-curator for the Viennese University with the argument that academia did not offer any opportunities for women. Thanks the Institute director Gustav Ortner's intervention she was allowed to stay at the Radium Institute on a regular salary. In a report written by the "NS-Dozentenführer" ("leader of the National Socialist organisation for university lecturers") she is described as politically disinterested. Overall it seems that she pursued her own scientific path without attracting any political attention — neither positive nor negative — from the NS-government [Marchet, 1941]. It was this unobtrusive behaviour that paved the way for her post-war career.

Karlik's assistant Traude (Cless-)Bernert was born in Vienna in 1915. She started to study physics at the University of Vienna in 1933 and was awarded her PhD in 1939. In the following years she worked as a volun-teer and assistant at the Radium Institute.

The Discovery

Astatine is the rarest naturally occurring element on earth. Discoveries of element 85 were announced from the 1920s onwards. The element, which is in the periodic table below iodine, the "eka-iodine" in Dmitri Mendeleev's (1834–1907) terminology, belongs to the group of halogens and should have similar chemical properties. However, natural astatine has a very short half-life and only exists for a few fractions of a second before it decays further. Therefore the chemical detection of astatine is impossible, and its detection needs to start in the decay chain with a known element, so that the type of decay that follows (α- or β- (beta) decay) can be determined and then the energy of the radiation particles. When a known element is reached in the decay chain a reverse calculation on the basis of the observed radiation particles is used to determine the unknown elements in the chain.

The first discovery of element 85 was achieved by Emilio Segré, Dale Corson (1914–2012), and Kenneth MacKenzie (1912–2002). They bombarded bismuth-209 with α-particles and produced the isotope 211 of element 85 in 1940 [Corson, MacKenzie; Segré, 1940]. In the same year the Swiss chemist Walter Minder (1905–1992) claimed that he had discovered the natural element 85 as a β-decay product of polonium-218 (then called "radium A") [Minder 1940]. At first, Karlik and Bernert tried to reconstruct Minder's measurements. However, Karlik and Bernert were not able to reproduce Minder's results, and concluded that his measurements resulted from impurities in the radium emanation (radon-222), which is the parent element of radium A (polonium-218) [Karlik and Bernert, 1942]. Hence, the race for the discovery of the new element continued.

Half a year later, in April 1943, Karlik and Bernert (Figure 1) sent a letter to the editor of the journal *Die Naturwissenschaften* [Karlik and Bernert, 1943]. In this short paper they repeated their critique on Minder's measurements and presented their own results using radium emanation, which was immediately analysed after its production. They used a modified vacuum tube, which had been developed by the Austrian physicists Georg Stetter (1895–1988) and Willibald Jentschke (1911–2002). Using this "vacuum tube electrometer" they were able to measure exactly the range of the emitted radiation. From the results they calculated the energy of the α-radiation following M. Stanley Livingston's (1905–1986) and Hans A. Bethe's (1906–2005) theory of 1937 [Livingston and Bethe, 1937]. Karlik and Bernert found a maximum range of 5,53 cm and a corresponding energy of 6,6 MeV. This was definitely the radiation of a new element, that had been previously unknown. Putting the energy in relation to the mass number of the element they found that the new α-radiation belonged to a new element with an atomic number between 84 and 86: the element 85 was discovered.

Figure 1.   Berta Karlik (right) and Traude Bernert (left) with their experimental setup in 1944 (bpk/Liselotte Orgel-Köhne, reproduced with permission of bpk).

In the following months Karlik and Bernert repeated their experiments also using thorium emanation (radon-220) and actinium emanation (radon-219). Again, they used the vacuum tube electrometers to determine the maximum range and the energy of the α-particles. As in their first experiment, they then plotted the decay energy in relation to the mass number. The connecting line was exactly between the line of element 84 and element 86: Karlik and Bernert had discovered the isotopes 215, 216, and 218 of element 85. The half-lives of these isotopes are: 0.2 ms, 0.3 ms, and 1.5 s [Karlik and Bernert, 1944].

After the war Fritz Paneth (1887–1958), the chairman of the committee of the International Union of Pure and Applied Chemistry (IUPAC), who was responsible for the naming of new elements, once more drew the attention to element 85. Upon his request, Segré, Corson, and MacKenzie published a letter to the editor in *Nature* [Corson *et al.*, 1947], suggesting the name astatine from the Greek *astatos*, which means "unstable" [Davis, 1959, p. 29]. Segré and his colleagues were the first to produce astatine isotopes. Following the convention it was their task to choose the name of the new element. Karlik and Bernert did not discover the natural isotopes until three years after the Californian group had done so.

After the War

Following World War II Gustav Ortner lost his directorship of the Radium Institute as a result of the denazification of the Austrian universities. Berta Karlik initially took over the provisional leadership of the Radium Institute, before Stefan Meyer, who survived the Nazi dictatorship in Bad Ischl, returned to the Institute and was again appointed as its formal director. In 1947 Meyer resigned, and Karlik continued to manage the Radium Institute as its full director until 1974. She was appointed a "titular professor" in 1946, and extraordinary professor in 1950. In 1956 she became the first woman in Austria to be appointed to a full professorship when a chair in experimental nuclear physics was established for her at the University of Vienna. In the 1950s she became one of the central figures in physics in Austria. Together with her colleagues at the University of Vienna she led the Austrian accession to CERN (the European Organisation for Nuclear Research). In the context of the American Atoms for Peace Program, in which the USA made research reactors available to allied nations, she took the leading role in Austrian science. She was the only scientist in the Austrian Atomic Commission and wrote the central memoranda for the construction of the two reactor centres in Vienna and Seibersdorf, about 40km south of Vienna. She also wrote the central memorandum for the Austrian delegation at the Geneva Conference on the Peaceful Use of Nuclear Energy in August 1955. Karlik was awarded numerous national and international prizes. In 1998 a memorial gate at the University of Vienna (one of its "gates of remembrance," "Tore der Erinnerungen"), which is located near the "Old General Hospital," was named after her. In 2016 a memorial was erected for Berta Karlik and six other female scientists in the arcade courtyard of the main building of the University of Vienna. Until then only one of the 154 busts had represented a woman.

The Radium Institute also became the Austrian distribution centre for radioactive isotopes. It controlled the import and distribution of radioactive material in Austria from Harwell (UK) from 1949 onwards, and from the US from 1952 onwards [Karlik, 1950]. Traude Bernert became the manager of this centre. In the Biennium 1948/49 she went to Sweden for research. In 1959 she became the scientific secretary at the second "Atoms for Peace" conference in Geneva. She continued her professional career as the organiser of the department for industrial consulting for isotope applications in the Seibersdorf reactor centre [Angetter and Martischnig 2005].

References

Angetter D. and Martischnig M. (2005). Biografien österreichischer (Physiker) innen: Eine Auswahl (Österreichisches Staatsarchiv, Vienna).

Corson, D. R., MacKenzie, K. R., and Segré, E. (1940). Possible Production of Radioactive Isotopes of Element 85, *Phys. Rev.*, 57, p. 459.

Corson, D. R., MacKenzie, K. R., and Segré, E. (1947). Astatine: The Element of Atomic Number 85, *Nature*, 159, p. 24.

Davis, H. M. (1959). *The Chemical Elements*, 2nd ed. (Ballantine Books Inc., New York, NY).

Fengler, S. (2014). *Kerne, Kooperation und Konkurrenz: Kernforschung in Österreich im internationalen Kontext (1900–1950)* (Böhlau, Cologne, Weimar, Vienna).

Forstner, C. (2017). Alltagsphysik statt Atombomben: Ein erneuter Blick auf den deutschen Uranverein, in *Physik, Militär und Frieden: Physiker zwischen Rüstungsforschung und Friedensbewegung*, eds. Forstner, C. and Neuneck, G. (Springer Spektrum, Wiesbaden), pp. 51–68.

Karlik, B. (1950). 1938–1950, in *Festschrift des Institutes für Radiumforschung anlässlich seines 40jährigen Bestandes (1910–1950)* (Akademie der Wissenschaften, Vienna), pp. 35–53.

Karlik, B. and Bernert T. (1942). Über eine vermutete α-Strahlung des Ra A und die natürliche Existenz des Elementes 85, *Die Naturwissenschaften*, 30, p. 685.

Karlik B. and Bernert T. (1943). Eine neue natürliche α-Strahlung, *Die Naturwissenschaften*, 31, p. 298.

Karlik B. and Bernert T. (1944). Das Element 85 in den natürlichen Zerfallsreihen, *Zeitschrift für Physik*, 123, pp. 51–72.

Livingston, M. S. and Bethe, H. A. (1937). Nuclear Physics C. Nuclear Dynamics, Experimental, *Rev. Mod. Phys.*, 9, pp. 245–390.

Marchet, A. (1941). Letter to the Dekanat der Philosophischen Fakultät of the University of Vienna, 8 April 1941, Archiv der Universität Wien, Postgasse 9, A-1010 Wien, Personalakte Berta Karlik, Aktnr. 2152.

Minder, W. (1940). Element 85, *Nature*, 146, p. 225.

Reiter, W. L. (2017). Aufbruch und Zerstörung: Zur Geschichte der Naturwissenschaften in Österreich 1850 bis 1950 (LIT, Münster).

Rentetzi, M. (2004). Gender, Politics, and Radioactivity Research in Interwar Vienna: The Case of the Institute for Radium Research, *Isis*, 95, pp. 359–393.

Walker, M. (1995). *Nazi Science: Myth, Truth, and the German Atomic Bomb* (Plenum Press, New York, NY, London).

Select Bibliography

Fengler, S. (2014). *Kerne, Kooperation und Konkurrenz: Kernforschung in Österreich im internationalen Kontext (1900–1950)* (Böhlau, Cologne, Weimar, Vienna).

Forstner, C. (2019). *Kernphysik, Forschungsreaktoren und Atomenergie: Transnationale Wissensströme und das Scheitern einer Innovation in Österreich* (Springer, Heidelberg).

Reiter, W. L. (2017). Aufbruch und Zerstörung: Zur Geschichte der Naturwissenschaften in Österreich 1850 bis 1950 (LIT, Münster).

Rentetzi, M. (2008). Trafficking Materials and Gendered Research Practices: Radium Research in Early 20th Century Vienna (Columbia University Press, New York, NY).

Part 5

Manufacturing Elements:
From Artificial Radioactivity
to Big Science

Irène Joliot-Curie and the Discovery of "Artificial Radioactivity"

Louis-Pascal Jacquemond

IEP-Sciences Politiques Paris, 27 Rue Saint-Guillaume,
75007 Paris, France

14 November 1935. A postman rings the doorbell of the Curie laboratory in the Institut du Radium (the Radium Institute), and delivers a telegram to the French scientist Irène Joliot-Curie (1897–1956) (Figure 1).[1] She reads aloud to her husband and colleague Frédéric Joliot-Curie (1900–1958), the recipient of the telegram:

> I HAVE THE HONOUR OF INFORMING YOU THAT THE SWEDISH ACADEMY OF SCIENCES HAS AWARDED THE NOBEL PRIZE IN CHEMISTRY OF 1935 TO YOU AND TO MADAME CURIE-JOLIOT. LETTER TO FOLLOW. SIGNED: PLEIJEL SECRÉTAIRE PERPÉTUEL [Pleijel, 1934].[2]

---

[1] This article is based mainly on the author's biography of Irène Joliot-Curie and its references [Jacquemond, 2014].

[2] All translations from the French by the author/editorial team unless otherwise indicated. Henning Bernhard Mathias Pleijel (1873–1962) was a Swedish physicist and electrical

Figure 1. Irène Joliot-Curie in the laboratory of the Institut du Radium, 1932 (André Kertesz/Musée Curie; coll. ACJC, reproduced by kind permission of the Musée Curie).

The Nobel! Here was the official recognition of the work of the two researchers who had discovered "artificial" radioactivity on 11 January 1934. The Joliot-Curies had immediately informed the scientific community via a notice ("Un nouveau type de radioactivité") read by Jean Perrin (1870–1942) before the French Academy of Sciences [Joliot-Curie and Joliot-Curie, 1961], which appeared in the *Comptes rendus de l'Académie des Sciences* on 15 January [Curie and Joliot, 1934a] and in *Nature* on the 19 January 1934 [Curie and Joliot, 1934b]. Back in 1903, Irène's parents, Marie (1867–1934) and Pierre Curie (1859–1906), had

---

engineer, and the "secrétaire perpétuel" (lifetime secretary) of the Royal Swedish Academy of Sciences between 1933 and 1943.

(with Henri Becquerel, 1852–1908) received the Nobel prize in Physics for their 1898 discovery of natural radioactivity.[3] Now, thirty-two years later, and two years after their own discovery of artificial radioactive processes, Irène and her husband were likewise honoured with the most prestigious prize in the world of science.

The discovery of artificial radioactivity was an important step in the understanding of the atom and its nuclear processes, and provided scientists with information about the production of elements beyond uranium. Irène Joliot-Curie's role in this ground-breaking discovery is best understood by looking at the ways in which the Joliot-Curies arrived at their conclusions.

***

The Joliot-Curies' discovery has its origins in the mid-1920s, when it was known that polonium was produced in the process of radium decay. Thanks to the amount of radium that had been accumulated by Marie Curie, who at the time was the head of the Radium Institute in Paris,[4] Irène was able to experiment with the alpha (α-)rays emitted by polonium. She would dedicate her doctoral thesis to this topic. After defending her thesis in 1925, in parallel with her own research, Irène worked with the young physicist Frédéric Joliot, whom she married one year later (Figure 2). Irène and Frédéric were eager to understand the structure of the atom. They were not alone: scientists in Rome, Copenhagen, Cambridge, Berlin and the United States pursued the same goal.

As part of this fruitful international competition, Irène Curie, a chemist and an imaginative theoretician, and Frédéric Joliot, a physicist and an exceptional experimenter, used alpha particles of radioactive polonium as projectiles to explore the nuclei of beryllium and boron. In 1932 they concluded that beryllium nuclei bombarded with highly penetrating alpha particles had to emit very high energy gamma rays (photons) to be able to

---

[3] See Roqué on Curie in this volume.

[4] The Radium Institute was created for Marie Curie in 1909 by the University of Paris and the Pasteur Institute. In 1914 the Institute moved into new buildings comprised of the Curie laboratory, which was responsible for theoretical research on radium, and the Pasteur laboratory, led by professor Claudius Regaud (1870–1940), who carried out applied research on radioactivity (e.g. brachytherapy).

Figure 2. Irène and Frédéric Joliot-Curie in the laboratory, 1935 (Henri Manuel/Musée Curie, coll. ACJC, reproduced by kind permission of the Musée Curie).

eject protons from hydrogen-rich substances such as paraffin or cellophane. The British physicist James Chadwick (1891–1974), who was aware of their work and unconvinced by their interpretation, proposed the existence of a neutron that was formed in the nuclear reaction between beryllium and alpha particles [Chadwick, 1932]:

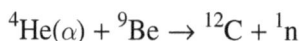

$$^{4}He(\alpha) + {}^{9}Be \rightarrow {}^{12}C + {}^{1}n$$

Electrically neutral, with a mass similar to that of the proton, the neutron is — along with the proton — the constituent of the nuclei [Pais, 1995, pp. 119–120].

Disappointed not to have discovered the neutrons themselves, Irène and Frédéric focussed on radioelements from then on. They conducted more and more varied experiments and discovered that, when they covered their sources with a foil of aluminium or boron, traces of positive electrons (or "positrons"),[5] and some protons, ejected by the neutrons, could be detected. From November 1933, advised by the renowned physicists Niels Bohr (1885–1962) and Wolfgang Pauli (1900–1958), the Joliot-Curies measured the appearance threshold of neutrons by varying the energy of the alpha particles. To count the positive electrons individually, they added a Geiger-Müller counter and an amplifier to their existing equipment. This technique had been perfected by Wolfgang Gentner (1906–1980), a German postdoctoral researcher who had arrived at the Curie laboratory two years previously.[6]

On Thursday, 11 January 1934, Irène and Frédéric realised that the emission of positrons continued even after the bombardment with alpha rays had stopped completely. They noted that the initial intensity of positrons increased with the duration of exposure to alpha rays, and decreased exponentially when exposure ceased, which is the typical presentation of radioactive disintegration. The conclusion was obvious: neutrons and positrons were not emitted at the same time but in two stages, which made them suspect the formation of an intermediate that would subsequently undergo radioactive decay. Gentner verified that the Geiger counter had functioned correctly, and the Joliot-Curies repeated the experiment with the same results.

On the following day, with Gentner's assurance that all was in order, Frédéric repeated the experiment once more, while Irène tried different irradiated elements. If Frédéric had just provided physical

---

[5]The positive electron, or positron, was discovered in 1932 in cosmic radiation by the American physicist Carl D. Anderson (1905–1991), who would be the Nobel laureate in Physics in 1936. This was the first evidence supporting the existence of anti-particles: the positron is an anti-electron, as it has an electric charge of +1e (opposite to the electron), but the same spin (1/2) and the same mass as the electron. Because annihilation occurs when an electron and a positron collide, and this produces highly energetic photons, it is the anti-matter counterpart of the electron.

[6]Gentner worked at the Radium Institute before joining the Berlin Institute of Physics in 1936. As an opponent of Nazism he was enlisted as officer for the "Abwehr," the German occupation army in Paris. He was in charge of the sector including the Sorbonne and the Collège de France. Suspected of collusion with Frédéric Joliot, to whom he provided information, Gentner was relieved of his position and posted to Hamburg in 1941. After 1945, he renewed contact with the Joliot-Curies.

proof for the existence of artificial radioactivity, it was Irène who assigned its radiochemical significance by demonstrating the presence of the isotope phosphorus-30 when aluminium was irradiated. She added hydrochloric acid to the test tube containing the irradiated aluminium which Frédéric gave to her, which then reacted with the aluminium to produce hydrogen gas and aluminium salts. She transferred the gases swiftly to another test tube, closed the tube and held it up to the Geiger counter. The Joliot-Curies heard the typical "tic-tic" sound rapidly decreasing. All activity from the irradiated foil had concentrated in the gaseous residue they had collected in the test tube. They called this new radioelement "radiophosphorus." The phenomenon lasted for three minutes, which is the half-life of this element, i.e. the period during which half of the radioactivity disappears and the element turns into ordinary and stable silicon [Pais, 1995, p. 121]:

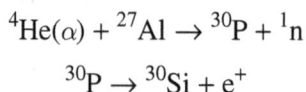

$$^{4}He(\alpha) + {}^{27}Al \rightarrow {}^{30}P + {}^{1}n$$
$$^{30}P \rightarrow {}^{30}Si + e^{+}$$

In the late afternoon of 12 January 1934 (Figure 3), Frédéric, Irène, Wolfgang and their collaborators were euphoric. In the spirit of the moment, they called Frédéric's friend and accomplice, the engineer Pierre Biquard (1901–1992), who later explained:

> I will never forget that afternoon in January 1934, when a telephone call urged me to leave my laboratory in the Rue Vauquelin and go to the Rue Pierre Curie and the Radium Institute... the experiment that had been hastily set up, and by way of demonstration reproduced the discovery that had been made a few hours earlier with Irène. ... In that moment the laboratory door behind the experimenting scientist opened to admit Marie Curie and Paul Langevin to the room. The demonstration resumed with the same precision and simplicity. The scene just described was over within thirty minutes [Biquard, 1961, pp. 41–42].

Thus, not even six months before her death on 4 July 1934, Marie Curie was so fortunate as to witness her daughter's and son-in-law's discovery.

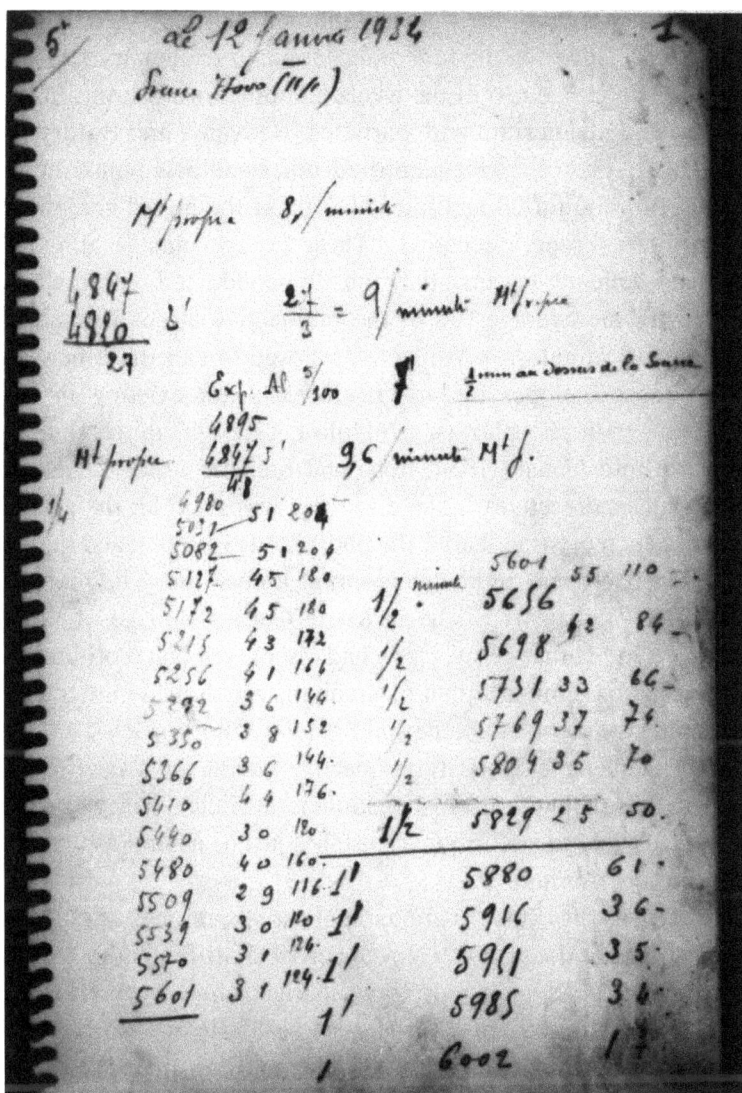

Figure 3. Laboratory notebook of Irène et Frédéric Joliot-Curie devoted to the discovery of artificial radioactivity, entry for 12 January 1934 (Musée Curie, coll. ACJC, reproduced by kind permission of the Musée Curie).

In the preparation of their presentation of their results to the French Academy of Sciences on the following Monday, 15 January 1934, Irène and Frédéric spent the rest of the weekend bombarding atoms of boron, aluminium and magnesium with particles [Radvanyi and Bordry, 1984, pp. 106–116]. They thereby identified other radioelements including silicon-27, aluminium-28 and nitrogen-30. At the award ceremony for the Nobel Prize, Frédéric recalled: "These experiments, relating to aluminium, are delicate, because they must be conducted within about six minutes, with the average life of the radioactive atoms formed being fewer than five minutes. ... We have proposed to call these new radioelements, these isotopes of known elements not existing in nature, 'radioazote,' 'radiophosphorus,' ..." [Joliot, 1935b] (Figure 4).

Thus, in mid-January 1934, Irène and Frédéric Joliot-Curie discovered "artificial radioactivity." The discovery was twofold: the artificially induced radioactivity (i.e. that of the phosphorus isotope), and the subsequent radioactive decay by the emission of a positron.[7] The roots of the term "artificial radioactivity" lie in Marie Curie's coinage of "natural" radioactivity: in February 1898 Curie had discovered that thorium emitted ionizing radiation similar to that of uranium, which made the traditional term "uranic rays" obsolete. Irène and Frédéric Joliot-Curie's 1934 feat of making an element temporarily radioactive called for a new term: the process of this temporary transmutation is artificial, hence "artificial" radioactivity, while the radioactivity itself is just as natural as that of uranium, radium or thorium.

Radioactivity has found various applications in chemistry, biology, medicine and other fields.[8] On 9 November 1935, its first biological application was announced in the prestigious journal *Nature* by the Hungarian-born Swedish chemist Georg von Hevesy (1885–1966) and his medical assistant Otto Chiewitz, who had studied the metabolism of phosphorus in

---

[7]This nuclear reaction would eventually be called "β + radioactivity."

[8]In biology and medicine, a little radioactive isotope (called "tracer" or "indicator") is included in an organic molecule (a labelled molecule) to make the observation of the metabolism (fixation, elimination, concentration) of the element in question possible in a living body, thanks to the element's radiation. This can be used in diagnostic identification, and if used in controlled quantities, also for therapeutic purposes, by emitting radiation onto the neighbouring cells.

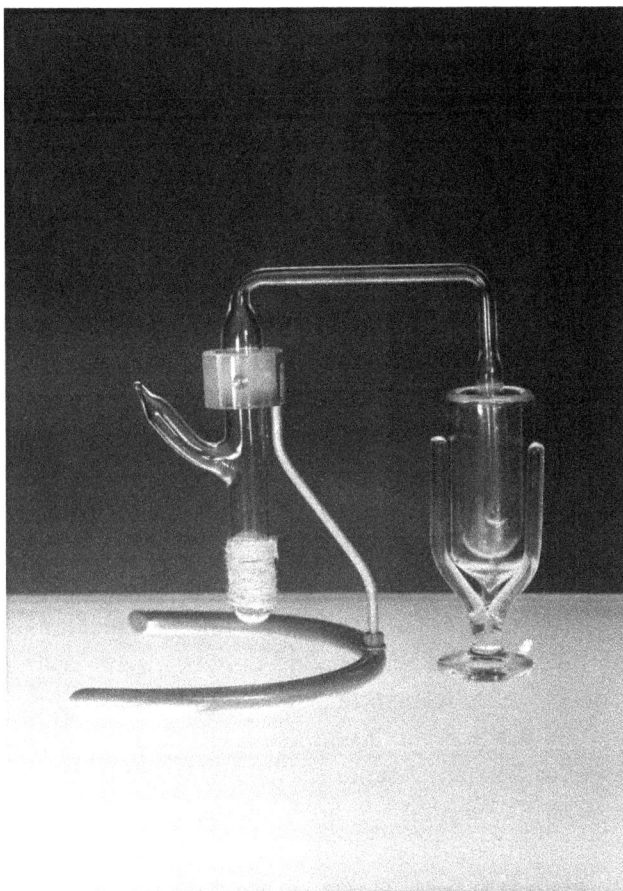

Figure 4. Device used by Irène et Frédéric Joliot-Curie to isolate radioazote, one of the earliest radioisotopes created in 1934 (replica from 1964) (Musée Curie, coll. ACJC, reproduced by kind permission of the Musée Curie).

rats with a radioactive element, phosphorus-32,[9] as indicator [Von Hevesy and Chiewitz, 1935].

---

[9] Georg von Hevesy won the Nobel Prize in Chemistry in 1943 "for his work on the use of isotopes as tracers in the study of chemical processes" [N. N., 2019]. A Swedish chemist born in Budapest, he spent most of his life in Copenhagen. In 1923 he filled void 72 in the periodic table, which at that time was still empty, with his discovery (with Dirk Coster) of hafnium, whose chemical properties he studied (1927).

When their joint Nobel Prize in Chemistry was awarded to the Joliot-Curies in Stockholm in December 1935, Irène described the history of radioactivity and the experimental part of the joint investigation and Frédéric Joliot followed up with the conclusions that could be drawn. He stated: "The diversity of the chemical properties, [that is, the] diversity of the average lives of these synthetic radio-elements will without doubt enable further advances in research in biology and in physical chemistry. In order to undertake this work properly, fairly large quantities will be required," which justified the use of particle accelerators, the cyclotrons, to produce these radioactive bodies in appreciable quantities. Frédéric ended his Nobel lecture by predicting new nuclear processes: "we are correct in thinking that scientists, building up or shattering elements at will, will be able to bring about transmutations of an explosive type, true chemical chain reactions" [Joliot, 1935a, p. 373]. One such chain reaction is the process at work in nuclear bombs and in nuclear reactors.

In June 1936, when French women had not yet secured the right to vote (which they would not be granted until April 1944), Irène Joliot-Curie was one of three women ministers of the Popular Front government chaired by the Socialist Leon Blum (1872–1950); the other two were the socialist Suzanne Lacore (1875–1975) and the Radical Cécile Brunschvicg (1877–1946). After three months of ministerial experience with the Popular Front, Irène Joliot-Curie returned to the laboratory[10] and devoted the following two years to investigating the bombardment of uranium with neutrons together with her Yugoslavian assistant Pavel Savitch.

In their production of new radioelements via uranium bombardment they noticed something unprecedented: one of the radioelements which

---

[10] Irène herself provided two reasons for the short duration of her involvement with the Popular Front: her lungs requiring rest and a mountain cure (Haute-Savoie), and her candidacy for a vacant position of professor at the Sorbonne. Two actual political, reasons for her disengagement were Frédéric's prior agreement with Léon Blum to be replaced by Jean Perrin, the Nobel laureate in Physics of 1926 and the advisor to the President of the Council; and Frédéric's disagreement with the non-interventionist position towards Spain, which had begun a civil war (Francoists against Republicans) in July 1936.

they referred to "R-3,5h" appeared to be similar to lanthanum and they speculated that if it was, indeed, lanthanum, it could only be obtained by splitting the uranium nucleus [Curie and Savitch, 1938] — the very principle of nuclear fission "invented" and defined in early 1939.

***

As the chair of research, and then sole "master" of the Curie Laboratory from 1946, Irène Joliot-Curie continued to study isotopes, especially the natural and stable isotopes that aided her radioactivity measurements and mineral dating. Some isotopes were linked to uranium, including actinium-227[11], or thorium-230, and others to carbon, e.g. carbon-11 and carbon-14. As chair of General Physics and Radioactivity at the Sorbonne, she also trained highly competent researchers including Pierre Radvanyi (b. 1926), the future research director at the Centre National de la Recherche Scientifique (CNRS), and Georges Charpak (1924–2010), the Nobel laureate in Physics of 1992.

When her book on natural radioelements produced during World War II was published in 1946, Irène was forced to look after her health, which was affected by tuberculosis, and her exposure to X-rays and polonium. But she relied on her collaborators to make the Curie laboratory a great research centre for the calibration of radioelements sources.

Finally, with the guidance of Frédéric Joliot, who had been named High Commissioner for Atomic Energy (CEA) in 1945, and as a member of the CEA herself, Irène developed Zoé, the first French atomic reactor, in December 1948. She then oversaw the construction of the new Orsay's Institute of Radium, which would later be the Higher Institute of Nuclear Physics, on the Plateau de Saclay. After her passing on 17 March 1956, Frédéric Joliot took over her work and followed the institute's development until it opened before he, too, died two years later. Irène Joliot-Curie undoubtedly advanced our knowledge of the atom and gave a new dimension to French nuclear research while, interestingly, proscribing the military use of such discoveries.

---

[11] Actinium was discovered in 1899 by the chemist Andre Debierne (1874–1949), the student and collaborator of Pierre Curie, who was Irène's tutor from 1906 and Marie Curie's successor as the head of the Radium Institute from 1934 to 1946.

Acknowledgements

The author wishes to acknowledge Annette Lykknes, Brigitte Van Tiggelen and Anke Timmermann for their careful editorial work on this article. He expresses his gratitude to them.

References

Biquard, P. (1961). *Frédéric Joliot-Curie et l'énergie atomique* (Éditions Seghers, Paris).

Chadwick, J. (1932), Possible Existence of a Neutron, *Nature*, 129, pp. 312.

Curie, I. and Joliot, F. (1934a). Un nouveau type de radioactivité, *Comptes rendus des séances de l'Académie des Sciences*, 198, pp. 254–256.

Curie, I. and Joliot, F. (1934b). Artificial Production of a New Kind of Radio-Element, *Nature*, 133, pp. 201–202.

Curie, I. and Savitch, P. (1938). Sur les radioéléments formés dans l'uranium irradié par les neutrons, II, *J. Phys. Radium*, 9:9, pp. 355–359.

Jacquemond, L.-P. (2014). *Irène Joliot-Curie: Biographie* (Éditions Odile Jacob, Paris).

Joliot, F. (1935a). Chemical Evidence of the Transmutation of Elements, Nobel lecture, 12 December 1935 (https://www.nobelprize.org/uploads/2018/06/joliot-fred-lecture.pdf, last accessed 8 February 2019).

Joliot, F. (1935b). Nobel Lecture in Stockholm, 12 December 1935, Archives Curie, file JC-6, bundle 50 (https://musee.curie.fr/decouvrir/archives-et-collections/archives, last accessed 8 February 2019).

Joliot-Curie, F. and Joliot-Curie, I. (1961). *Œuvres scientifiques complètes* (Presses universitaires de France, Paris).

N. N. (2019). The Nobel Prize in Chemistry 1943 (https://www.nobelprize.org/prizes/chemistry/1943/summary/, last accessed 8 February 2019).

Pais, A. (1995). Introducing Atoms and their Nuclei, in *Twentieth Century Physics*, eds. Brown, L.M., Pais, A. and Sir Pippard, (American Institute of Physics, New York, NY, and London), vol. 1, pp. 103–129.

Pleijel, H. B. M. (1934). Telegram to Frédéric Joliot-Curie, 14 November 1935 (https://musee.curie.fr/decouvrir/zoom-sur/prix-nobel-joliot-curie, last accessed 2 February 2019).

Radvanyi, P. et Bordry M. (1984). *La radioactivité artificielle et son histoire* (Éditions du Seuil (Points sciences), Paris).

Von Hevesy, G. and Chiewitz, O. (1935). Indicateurs radioactifs dans l'étude du métabolisme du phosphore chez le rat, *Nature*, 136, p. 754.

Select Bibliography

Conkling, W. (2016). *Radioactive! How Irène Curie & Lise Meitner Revolutionized Science and Changed the World* (Algonquin Young Readers, Chapel Hill, NC).

Curie, E. (1938). *Madame Curie* (Éditions Gallimard, Paris). (Also available in paperback, Folio, n°1336, 1981).

Emling, S. (2012). *Marie Curie and Her Daughters: The Private Lives of Science's First Family* (Palgrave Macmillan, Hampshire).

Fernandez, B. (2006). *De l'atome au noyau: Une approche historique de la physique atomique et de la physique nucléaire* (Éditions Ellipses, Paris).

Gilmer, P. J. (2011). Irène Joliot-Curie, a Nobel Laureate in Artificial Radioactivity, in *Celebrating the 100th anniversary of Madame Maria Sklodowska Curie's Nobel Prize in Chemistry*, eds. Chiu, M.-H., Gilmer, P. J. and Treagust, D. F., (Sense publishers, Rotterdam), pp. 41–57.

Guerra, F., Leone, M. and Robotti, N. (2012). The Discovery of Artificial Radioactivity, *Physics in Perspective*, 14, pp. 33–58.

Jacquemond, L.-P. (2014). *Irène Joliot-Curie: Biographie* (Éditions Odile Jacob, Paris).

Joliot-Curie, I. (1946). *Les radioéléments naturels: Propriétés chimiques, préparation, dosage* (Hermann, Paris).

Pflaum, R. (1989). *Grand Obsession: Madame Curie and Her World* (Doubleday, New York, NY); French translation: de Martinoir, F. (1992). *Marie Curie et sa fille Irène* (Éditions Belfond, Paris).

Pigeard-Micault, N. (2013). *Les femmes du laboratoire de Marie Curie* (Éditions Glyphe, Paris).

Pinault, M. (2000). *Frédéric Joliot-Curie* (Éditions Odile Jacob, Paris).

Radvanyi, P. (2005). *Les Curie, pionniers de l'atome* (Éditions Belin, Paris).

Weart, S. R. (1980). *La grande aventure des atomistes français* (Éditions Fayard, Paris).

Isabella L. Karle and the Synthesis of Plutonium Chloride

Ann E. Robinson

Independent Scholar, Cambridge, MA

Isabella L. Karle (1921–2017) (Figure 1) has been described as "the most distinguished female chemist" to be employed at the Manhattan Project's Metallurgical Laboratory (Met Lab) [Howes and Herzenberg, 1999, p. 74]. She spent six months there, working on the chemistry of plutonium. This period is generally overlooked in sketches of her career, the majority of which was spent at the United States Naval Research Laboratory. Karle was a well-known and highly regarded crystallographer. She developed the symbolic addition procedure that demonstrated the usefulness of the mathematical technique for determining the structure of crystals, known as direct method. The direct method was pioneered by her husband, the chemist Jerome Karle (1918–2013) (Figure 1), and the mathematician Herbert A. Hauptman (1917–2011), who won the Nobel Prize in Chemistry in 1985.

Early Life and Education

Born Isabella Helen Lugoski in Detroit, Michigan, Karle was the daughter of Polish immigrants. Her mother had taught her how to read and write in

374

Figure 1. Dr Jerome Karle with his wife Dr Isabella Karle (U.S. Naval Research Laboratory; public domain).

Polish and how to do basic maths, but she spoke no English until she began the first grade. She first encountered chemistry in high school and found it fascinating. When she told her teacher that she wanted to pursue a career in chemistry, "he tried to discourage me by saying that chemistry was not a proper field for girls" [Karle, 1973, p. 11]. Karle was awarded a full scholarship to attend the University of Michigan, where she finished her bachelor's degree with the highest honours. The University declined to give her funding for graduate studies, but she was able to continue with a fellowship from the American Association of University Women.

Karle met her husband in the laboratory section of a graduate physical chemistry class. They were lab partners and married in 1942. Their advisor was Lawrence O. Brockway (1907–1979), whose main research was in the field of electron diffraction, a technique used to study matter by firing electrons at a sample of the material being studied, and observing the interference pattern caused by the electrons interacting with each other. This process is used to study the crystal structures of solids. In response to queries about his students and their usefulness to the war effort, Brockway wrote that Karle and her husband had "the best records" at the University of Michigan "of any recent graduates in chemistry" [Seaborg, 1978, p. 326]. She finished her PhD in 1943 at the age of 22. Her husband moved to Chicago in 1943 to work at the Met Lab, and Karle followed several months later. The day she began work at the Met Lab, Seaborg [1978, p. 326] wrote in his journal: "Those that know Mrs. Karle praise her as 'brilliant' and having a 'most pleasing personality'".

Plutonium

Plutonium is a synthetic element, with an atomic number 94 and twenty isotopes.[1] Plutonium-239 is the primary element used in the production of nuclear weapons. Plutonium was first produced at the University of California, Berkeley in December 1940 by Glenn T. Seaborg (1912–1999), Arthur Wahl (1917–2006), and Joseph W. Kennedy (1916–1957). They were continuing work begun by Edwin McMillan (1907–1991) who, with Philip H. Abelson (1913–2004) in May 1940, produced neptunium (no. 93), the first element beyond uranium on the periodic table. Although the United States was not yet an active participant in World War II, the discovery of this new element was kept secret due to its potential use in a weapon. Research on the new element continued at Berkeley, but in December 1941 the majority of the work shifted to the Met Lab, the Manhattan Project site at the University of Chicago. Seaborg was put in charge of the section devoted to the development of chemical procedures for the extraction and purification of plutonium.

Working with plutonium was challenging. Because it was artificial, it had to be produced. Plutonium-238 has a very long half-life; however, it is produced from neptunium-238, which has a half-life of only 2.1 days. Therefore the quantities that were produced were initially very small, only

---

[1] Most of the plutonium present on Earth is artificially produced, however trace amounts of this transuranium element can be detected in nature; in particular, Darleane Hoffmann discovered traces of plutonium. See Murray and Wade in this volume.

a few micrograms at a time. These small amounts were also impure. A major task of Seaborg's group at the Met Lab was to create pure — or almost pure — plutonium. However, plutonium is allotropic, existing in six different forms, each with its own crystal structure and density. It also has four different oxidation states, which refers to the charge an atom has after a chemical compound undergoes a chemical reaction. Consequently, there were several groups at the Met Lab studying different aspects of plutonium.[2] Everything learned by these groups was applied to the ultimate goal of producing enough pure plutonium for use in an atomic bomb.

Karle at the Met Lab

Karle started work at the Met Lab in January 1944, immediately after completing her PhD at the University of Michigan. Her doctoral research involved electron diffraction by gaseous molecules. In this process, an electron beam is passed through a gas, creating images of the structure of molecules, similar to the way an X-ray creates an image. These images can then be measured and calculations be made to determine the interatomic distances and angles in the gas molecules. In the early 1940s, most of the parts of the apparatus used in this work was made by experimenters rather than ordered from a scientific supply company. "I can see why they selected me," Karle recalled, "although I was working on other problems" [Karle, 2015]. During her doctoral work, she had gained a lot of experience with glass blowing and vacuum lines. This experience was easily transferable to her work at the Met Lab.

Karle was assigned to Group 5, one of the groups in Seaborg's section. Under the leadership of Norman Davidson (1916–2002), this group investigated the preparation and properties of plutonium compounds. In particular, they were using vapour phase procedures to synthesise plutonium halides or oxides. The ultimate goal was to produce pure metallic plutonium, and it was thought that synthesising halides was a step in the production process. Karle's task was to synthesise plutonium chloride, $PuCl_3$, from the plutonium dioxide, PuO_2, that was delivered regularly from the plant at Oak Ridge, Tennessee, that was creating plutonium from

---

[2] See also Rentetzi in this volume.

enriched uranium. The quantities each group were given to work with were extremely small and, as Karle remembered, "[w]e picked out our lumps of material from under the microscope" [Karle, 2005]. For each experiment, she worked on only a few "greasy yellow flakes" [Karle, 2004, p. 94] of impure plutonium.

As the discovery of plutonium had been kept secret, Karle knew nothing about the element before arriving at the Met Lab. Having been told it was similar to uranium, Karle spent her first week in the library learning about the chemistry of that element. She was assigned space in New Chemistry, a hastily erected building across the street from the university's football stadium. Although Seaborg was very concerned about the safety of his staff and their exposure to radiation, the powerful fans in New Chemistry vented directly to the outside, releasing many exhaust gases including plutonium. Karle noted that it must have been "a safe enough disposal for us," as many of the chemists who worked there lived long lives. She quipped, "I always said a little bit of plutonium is good for you" [Karle, 2005].

Before any work could begin, Karle needed to create some of the equipment needed for her experiments. Using the glass blowing skills she had acquired in her doctoral work, Karle created a vacuum apparatus, with glass tubes and stopcocks. As the reactions she was attempting required extremely high temperatures, about 800°C, some typical equipment needed to be made out of atypical materials. Karle made a removable insert of quartz to be the reaction chamber. This was surrounded by a hollowed-out block of copper that would be heated with a torch to act as a furnace. She also made a vessel out of silica, a material with a much higher melting point than glass, to hold the plutonium dioxide.

Each experiment consisted of passing a chlorinated compound in vapour form over the boat containing the plutonium dioxide. After approximately an hour, the copper furnace was allowed to cool and then opened to reveal the crystals that were the result. Karle repeated this experiment many times, trying different compounds until she used carbon tetrachloride. This compound resulted in "brilliant green crystals ... in the place of the grungy yellow oxide" [Karle, 2004, p. 94]. Another attempt, one that was not always successful in creating plutonium chloride, resulted in purple crystals. A version of the experiment that used purified chlorine rather than a chlorinated compound resulted in "a fluffy blue reaction product" [Abraham *et al.*, 1949, p. 741]. Ultimately, Karle was able "to synthesize plutonium chloride in about a dozen different ways" [Karle, 2015]. She also demonstrated that it was a stable compound.

Proof that the results of each experiment were, indeed, plutonium chloride was provided with X-ray diffraction, performed by William H. Zachariasen (1906–1979), who was in charge of X-ray crystallography at the Met Lab. In order to deliver the crystals to Zachariasen, they were transferred into a vial. Karle put it in her pocket and walked across the University of Chicago campus to the building housing the Met Lab's physics team. She "walked around campus freely," as she was only twenty-three and "still wore pigtails," looking like any other co-ed [Karle, 2015]. After performing this walk several times, security discovered she was delivering plutonium samples unaccompanied. Subsequently, she "had two uniformed guards marching with me, probably alerting individuals in view of the campus that there was probably something unusual going on" [Karle, 2004, p. 95]. Security was assigned not just because she was carrying plutonium, a classified substance being studied in a secret war-time project, but because it was plutonium chloride, "the first ever made in this world, that nobody had ever seen before" [Karle, 2005].

***

Plutonium chloride is now referred to as plutonium(III) chloride or plutonium trichloride. Despite all of the work done at the Met Lab, there is still much that is not known about the compound. Research into its structure and synthesis is still being undertaken, particularly at national laboratories such as Argonne and Los Alamos. It has been suggested that plutonium chloride could be used as fuel in molten fuel fast reactors, but this has not happened. However, the synthesis of plutonium chloride from plutonium dioxide remains an important step in the process of purifying plutonium for use in nuclear weapons.

After she left the Met Lab in July 1944, plutonium chloride had no role in Karle's career. She and her husband went to the United States Naval Research Laboratory in 1946 and turned their attention to crystallography. However, she considered the six months she spent in Chicago beneficial, as she "was presented with a number of challenging problems in the form of experimental work that I had to devise in order ... to get the results that I wanted to. So it gave me good practice in attacking quite unknown problems" [Karle, 2015].

Permission to quote from Isabella Karle's interviews is courtesy of the Atomic Heritage Foundation's "Voices of the Manhattan Project" website.

References

Abraham, B. M., Brody, B. B., Davidson, N. R., Hagemann, F., Karle, I., Katz, J. J. and Wolf, M. J. (1949). Preparation and Properties of Plutonium Chlorides and Oxychlorides, in *The Transuranium Elements: Research Papers*, eds. Seaborg, G. T., Katz, J. J. and Manning, W. M. (McGraw-Hill, New York, NY), pp. 740–758.

Howes, R. H. and Herzenberg, C. L. (1999). *Their Day in the Sun: Women of the Manhattan Project* (Temple University Press, Philadelphia, PA).

Karle, I. L. (1973). Crystallographer, *Ann. NY Acad. Sci.*, 208, pp. 11–14.

Karle, I. (2004). My First Professional Assignment, in *Remembering the Manhattan Project: Perspectives on the Making of the Atomic Bomb and its Legacy*, ed. Kelly, C. C. (World Scientific, Singapore), pp. 93–96.

Karle, I. (2005). Isabella Karle's Interview (2005), by Cynthia C. Kelly, Atomic Heritage Foundation, "Voices of the Manhattan Project" (Washington, DC) (https://www.manhattanprojectvoices.org/oral-histories/isabella-karles-interview-2005, last accessed 23 September 2018).

Karle, I. (2015). Isabella Karle's interview (2015), by Alexandra Levy, Atomic Heritage Foundation, "Voices of the Manhattan Project" (Virginia) (https://www.manhattanprojectvoices.org/oral-histories/isabella-karles-interview-2015, last accessed 10 September 2018).

Seaborg, G. T. (1978). *History of Met Lab Section C-I: May 1943 to April 1944*, PUB 112, vol. 2 (Lawrence Berkeley Laboratory, Berkeley, CA).

Select Bibliography

Hargittai, M. (2015). *Women Scientists: Reflections, Challenges, and Breaking Boundaries* (Oxford University Press, New York, NY).

Howes, R. H. and Herzenberg, C. L. (1999). Women in Weapons Development: The Manhattan Project, in *Women and the Use of Military Force*, eds. Howes,

R. H. and Stevenson, M. R. (Lynne Reiner Publishers, Boulder, CO), pp. 95–109.

Julian, M. M. (1986). Isabella L. Karle and a New Mathematical Breakthrough in Crystallography, *J. Chem. Ed.*, 63, pp. 66–67.

Owens, J. (1997). Isabella L. Karle, in *Notable Women in the Physical Sciences: A Biographic Dictionary*, eds. Shearer, B. F. and Shearer, B. S. (Greenwood Press, Westport, CT), pp. 217–222.

Seaborg, G. T. (1994). *The Plutonium Story: The Journals of Professor Glenn T. Seaborg, 1939–1946*, eds. Kathren, R. L., Gough, J. B. and Benefiel, G. T. (Battelle Press, Columbus, OH).

Chien-Shiung Wu and the Nuclear Poison Xenon-135

Ann E. Robinson

Independent Scholar, Cambridge, MA

Chien-Shiung Wu (1912–1997) (Figure 1) is often called "the Chinese Marie Curie" and "the First Lady of Physics" [Chiang, 2014, p. ix]. She was an expert in beta (β-)decay, a process in which a neutron within an atomic nucleus is transformed into a proton or vice versa by the emission of an electron or a positron, respectively. Wu is widely known for the experimental work she conducted that validated the theory of parity violation proposed by her colleagues Tsung-Dao Lee (b. 1926) and Chen-Ning Yang (b. 1922), both of whom won the Nobel Prize in Physics in 1957. Less well-known is her doctoral research on the fission products of uranium. In the course of this research, she identified two radioactive isotopes of the element xenon and established their decay chains. Knowledge of one of these isotopes was crucial to the operation of the nuclear reactors set up in Hanford, Washington, as part of the Manhattan Project.

Figure 1. Chien-Shiung Wu assembling an electro-static generator at Smith College (AIP Emilio Segrè Visual Archive; reproduced by kind permission of AIP Emilio Segrè Visual Archive).

Early Life and Education

Wu was born in Liu He, China, a small town near Shanghai. Her father had trained as an engineer and was headmaster at the Ming De School, one of the earliest schools in China to accept girls. He encouraged Wu's interest in all subjects, especially science and mathematics. After attending Ming De, Wu enrolled in the Normal School programme at the Soochow School for Girls. This programme traditionally led to a career in teaching. It was here that Wu learned about Marie Curie (1867–1934), which only increased her interest in maths and physics. In order to continue her studies in science, she attended the National Central University in Nanjing. Wu completed her bachelor's degree in physics in 1934 and, after teaching for a year at a university in Hangzhou, followed her father's advice to "ignore the obstacles and keep walking ahead" [Benczer-Koller, 2006, p. 277]. She boarded a ship and set sail for the United States to pursue graduate study in physics.

Wu arrived in San Francisco, California, in 1936 and intended to continue on to Ann Arbor, Michigan, where she was to begin graduate study at the University of Michigan. Several factors, however, resulted in her decision to enrol at the University of California at Berkeley instead. First, Wu heard stories about discrimination against women students at Michigan. Second, she met Luke Chia-Liu Yuan (1912–2003), a graduate student in physics at Berkeley who would become her husband in 1942. Third, she visited the Berkeley campus and the Radiation Laboratory, and met Ernest O. Lawrence (1901–1958), who won the Nobel Prize in 1939 for his invention and development of the cyclotron. Lawrence agreed to be her advisor. He had many duties both on and off campus, and Wu worked directly with Emilio Segrè (1905–1989), the Italian physicist who discovered element 43 in 1937 and was engaged in the search for element 85.[1] Segrè was unable to officially be Wu's advisor as he was not a faculty member. By the time she finished her PhD in 1940, "Wu had transformed herself … from an ambitious student to a confident and competent young scientist making contributions at the frontier of her field" [Wang, 2008, p. 365]. One of the contributions she made while at Berkeley related to radioactive isotopes of xenon.

Xenon

Xenon is one of the so-called noble gases. It has an atomic number of 54, and eight stable and more than forty unstable isotopes. Xenon was discovered by William Ramsay (1852–1916) and Morris Travers (1872–1961) in 1898, the same year they discovered two other noble gases, krypton and

---

[1] See also Van Tiggelen on Noddack and on Cauchois, as well as Forstner in this volume.

neon. It is considered to be a trace gas, making up less than one percent of the earth's atmosphere. Xenon and the other noble gases were unique in two ways. First, their discovery was completely unexpected. Dmitri Mendeleev (1834–1907) had boldly predicted the existence of at least sixteen elements, but none of his predictions included an entire family of gases. At first, many were not convinced that xenon and its relatives were elements, despite the fact that they had unique atomic weights and unique spectra, amongst other things. Eventually, an entirely new group of elements, group 0, was added to the end of the periodic table. In today's periodic table, the noble gases comprise group 18.

The second unique thing about the noble gases is that they were considered to be inert. Unlike other elements, xenon and its family members did not react in the presence of other compounds. They were resistant to combining with other elements. This was not because they were gases. Other gaseous elements — hydrogen, oxygen, and nitrogen come immediately to mind — mix well with many other elements. Rather, the noble gases are extremely stable; the outer electron shell in the atom is considered to be "full," which makes them less likely to react. It was not until 1962 that the first xenon compound, xenon hexafluoroplatinate, was created by Neil Bartlett (1932–2008). Since then more than eighty xenon compounds have been created.

Wu's Doctoral Research

After completing the necessary coursework, Wu began the work that would form her two-part doctoral dissertation. Lawrence assigned her an experiment that would become the first part. It was to examine so-called "Bremsstrahlung", the radiation that occurs when a charged particle decelerates. Enrico Fermi (1901–1954), a frequent visitor to the Berkeley campus, suggested she undertake a comparative study of the internal and external radiation caused when electrons are emitted from a nucleus during β-decay.[2] In this case, internal radiation refers to the X-rays that are produced by the deceleration of neutrons as they leave the nucleus, while

---

[2]Electrons or their positive counterparts (positrons) are emitted from the nucleus because neutrons are transformed into protons, electrons and antineutrinos, or because protons are converted into neutrons, positrons and neutrinos.

external radiation refers to X-rays produced when an electron decelerates as it moves through the electromagnetic field of the nucleus. Bremsstrahlung occurs during β-decay. This experiment was Wu's introduction to a field that she "would make her own" [Wang, 2008, p. 365].

The second part of Wu's PhD dissertation involved using two of Lawrence's cyclotrons to investigate the fission products of uranium. At the time, nuclear fission was a new and exciting discovery. During the process of fission, the nucleus of an atom splits into lighter nuclei. Wu and Segrè bombarded uranium with neutrons, using one of Lawrence's cyclotrons. They bombarded their uranium for one hour, waited two hours and then extracted the iodine that resulted from the fission process. Xenon was then extracted from the iodine ten hours later. They found two radioactive isotopes of xenon. One of these isotopes had a half-life of about five days and had been previously observed by Alexander Langsdorf, Jr. (1901–1954). The second isotope was one not previously observed, having a half-life of 9.4 hours. Tantalisingly, they also reported there were "one or more short-life xenons growing out of iodine which we have not yet investigated" [Segrè and Wu, 1940, p. 552]. Details on this short-lived xenon isotope would not be published until after the war.

Segrè "went fishing for 'inspiration'" [Chiang, 2014, p. 89], leaving Wu to continue the work on uranium fission products. While he was away, she was able to identify the two longer-lived radioactive isotopes of xenon. In a letter to the editor of the *Physical Review*, she reported that the isotope with a half-life of 5 days was Xe-133 [Wu, 1940]. The isotope with the 9.4-hour half-life was Xe-135. Wu and Segrè published brief letters, leaving out many of the details of their experiments and their results [Segrè and Wu, 1940; Wu, 1940]. Although fission was an exciting and new research topic, it was also recognised that it could be instrumental in the making of bombs. Wu and Segrè published in 1940, before the United States was actively involved in the war, but they held to an informal ban on revealing too much information about fission.

Xe-135, Nuclear Poison

Wu finished her PhD in 1940 and stayed on at Berkeley as a research fellow for two years. She and her husband then moved to the East Coast. In March 1944, Wu was recruited to work at the Substitute Alloy

Materials (SAM) Laboratories at Columbia University. As part of the Manhattan Project, the SAM Laboratories supported the enrichment of uranium by gaseous diffusion project which was based in Oak Ridge, Tennessee. Wu, along with William W. Havens, Jr. (1920–2004) and James Rainwater (1917–1986), worked on the development of radiation detector instrumentation, as well as studies of neutron cross sections which describe the likelihood that an incidental neutron will interact with a target nucleus. Both of these projects utilised skills Wu had honed at Berkeley, running experiments using cyclotrons and building new equipment.

As important as this work was, Wu's most critical contribution to the Manhattan Project involved her knowledge of Xe-135. The first large-scale nuclear reactor, known as the B Reactor, was built at Hanford, Washington, to produce plutonium for use in atomic bombs. It produced its first chain reaction in September 1944. Things seemed to be going well until the chain reaction suddenly stopped. It restarted again after a few hours. This occurred several times. Fermi theorised that this was being caused by the absorption of neutrons from a fission product of uranium, which was being used to synthesise plutonium. Knowing of Wu's work with Segrè, he contacted her.

Segrè and Wu had produced a manuscript detailing their experiments on the fission products of uranium. They had decided not to publish until after the war, not knowing if their results would have an impact on the production of bombs. In fact, it did. Wu's work had shown that Xe-135, the isotope she had identified with a 9.4-hour half-life, had a large neutron cross section. This gave it an affinity for absorbing a large number of neutrons. By absorbing so many neutrons, Xe-135 was shutting down the chain reaction that synthesised plutonium from uranium. Wu's investigation of this radioactive xenon isotope gave the physicists and engineers the knowledge needed to devise new techniques for controlling the chain reaction.

Today, Xe-135 is known as the most powerful "nuclear poison," the term used for substances with large neutron cross sections. If too much Xe-135 builds up in a reactor, it will shut down and not restart. Nuclear reactors are designed to mitigate this effect, and reactor operators are trained to monitor the absorption of neutrons. A contributing factor to the meltdown of the Chernobyl reactor in 1986 was a failure to account for and manage the build-up of Xe-135 during the steam turbine test that precipitated the accident. As a product of uranium fission, the measurement of Xe-135, as well as several other xenon isotopes, in the atmosphere is used to detect and monitor nuclear explosions.

Small quantities of Xe-133, the other radioactive isotope of xenon that Wu identified in her doctoral research, are sometimes released into the atmosphere from nuclear reactors. The half-life of this isotope is now considered to be 5.2 days and it is relatively harmless. When it is inhaled, it circulates through the body and is exhaled when it returns to the lungs. Xe-133 has several medical applications. It is primarily used in single-photon emission tomography as a radiopharmaceutical imaging agent. As a gas, it is inhaled to image the lungs and to evaluate pulmonary function, and is also used to assess blood flow in the brain.

***

Identifying two radioactive isotopes of xenon and becoming an expert in them was not what Wu expected when she began graduate studies. However, the work she did with Segrè not only solved a major problem during the Manhattan Project, but informed the building and operating of nuclear reactors and contributed to the diagnosing of medical conditions. It also set her on the path to becoming an early expert in nuclear fission and in β-decay. After World War II, Wu carried out experiments that confirmed several important theories and hypotheses in physics. Her expertise as an experimentalist earned her the nickname "Queen of Nuclear Research" and garnered comparisons with Marie Curie. Wu approached all of her work with the same expectation: "You must work very hard at the beginning. It is hard to push the door open and to get inside a subject. But once you understand it, it is *very* interesting" [McGrayne, 1998, p. 264] (emphasis in original).

References

Benczer-Koller, N. (2006). Chien-Shiung Wu, in *Out of the Shadows: Contributions of Twentieth-Century Women to Physics*, eds. Byers, N. and Williams, G. (Cambridge University Press, Cambridge), pp. 272–281.

Chiang, T.-S. (2014). *Madame Wu Chien-Shiung: The First Lady of Physics Research*, trans. Wong, T.-F. (World Scientific, Singapore).

McGrayne, S. B. (1998). *Nobel Prize Women in Science: Their Lives, Struggles, and Momentous Discoveries* (Birch Lane Press, Secaucus, NJ).

Segrè, E. and Wu, C. S. (1940). Some Fission Products of Uranium, *Phys. Rev.*, 57, p. 552.

Wang, Z. (2008). Wu Chien-Shiung, in *New Dictionary of Scientific Biography*, ed. Koertge, N. (Charles Scribner's Sons, Detroit, MI), pp. 363–368.

Wu, C. S. (1940). Identification of Two Radioactive Xenons from Uranium Fission, *Phys. Rev.*, 58, p. 926.

Select Bibliography

Allen, U. (1997). Chien-Shiung Wu, in *Notable Women in the Physical Sciences: A Biographic Dictionary*, eds. Shearer, B. F. and Shearer, B. S. (Greenwood Press, Westport, CT), pp. 423–429.

Benczer-Koller, N. (2009). *Chien-Shiung Wu 1912–1997: A Biographical Memoir* (National Academy of Sciences, Washington, DC) (http://www.nasonline.org/publications/biographical-memoirs/memoir-pdfs/wu-chien-shiung.pdf, last accessed 6 September 2018).

Hargittai, M. (2015). *Women Scientists: Reflections, Challenges, and Breaking Boundaries* (Oxford University Press, New York, NY).

Howes, R. H. and Herzenberg, C. L. (1999). *Their Day in the Sun: Women of the Manhattan Project* (Temple University Press, Philadelphia, PA).

Yost, E. (1959). *Women of Modern Science* (Greenwood Press, Westport, CT).

The Unsung Heroines of the Superheavy Elements

Claire A. Murray* and Jessica A. F. Wade[†]

*Diamond Light Source, Harwell Campus, Didcot, OX11 0DE, UK
[†]Department of Physics, Imperial College London,
London, SW7 2AZ, UK

After the discovery of the last naturally occurring element, francium, in 1939 [Perey, 1939],[1] the pursuit to complete the periodic table became increasingly competitive, experimentally challenging and political. So far, there seems to be no limit on what scientists can create; as long as they can work out the recipe and accelerate beams to high enough energies, they will continue to push the limits of the periodic table. Long gone are the days of the seemingly primitive setup managed by the Curies, who discovered polonium and radium by extracting it from the uranium ore pitchblende [Curie *et al.*, 1898].[2] For elements beyond uranium, i.e. with more than 92 protons ("transuranic elements"), and even heavier ones ("superheavy elements"), more advanced characterisation techniques are

---

[1] See Rayner-Canham and Rayner-Canham on Perey in this volume.
[2] See Roqué on Curie in this volume.

required. Superheavy elements (ca. 100 protons and beyond)[3] have incredibly short lifetimes (microseconds), are extraordinarily difficult to create and have never been detected out of the few laboratories who work on them. Not only are the elements fascinating, but so too are the scientists who discovered them. Their names may not be as familiar as Marie Skłodowska Curie's, but women have played leading roles in every aspect of superheavy discovery; challenging the biases of outdated laboratory members and transforming our understanding of the periodic table.

***

The basic science behind the formation of superheavy elements is not too complicated — combine two lighter nuclei with the combined right number of protons for what you are aiming for. Americium (Am, $Z = 95$, i.e. 95 protons) was combined with calcium (Ca, $Z = 20$) to make moscovium (Mc, $Z = 115$) [Rudolph *et al.*, 2013]. In reality, what is needed is a particle accelerator, an intense beam of ions and a target. As the beam and target (a nucleus) are positively charged, they strongly repel each other, which means that the beam of ions must be accelerated to almost ten per cent the speed of light. At atomic length scales the electrostatic force is not the only interaction at play — the strong nuclear force that keeps the nucleus together is important, too.

Very occasionally, the ions hit the target. When they do, many nuclei break apart immediately, with each nucleus splitting into two fragments and emitting a few neutrons (formally known as nuclear "fission"). To maximise new element success, these interactions must occur for long enough that nuclear fusion can happen and be detected. To do this, scientists start with neutron-rich isotopes — the new elements can shed excess energy via neutron emission. The newly formed elements are not stable, but leave radioactive signatures that guide scientists to isotopes of nearby elements in the periodic table that they have characterised before. For the past century, scientists have raced to fill the gaps on the periodic table — particularly in the second half of the twentieth century. But since the introduction of accelerators, the nature and the means of

---

[3]There is no exact definition for the term "superheavy elements," but it is normally used to describe elements with more than 100 protons [Krämer, 2016].

the quest have changed deeply. The short half-lives, elaborate equipment, larger teams of co-workers and fierce competition mean there are very few scientists who can compete in the quest for new elements. Indeed, this journey has become both expensive and political [Kragh, 2017].

The pursuit of the transuranic elements has been in progress since 1940, when Edwin McMillan (1907–1991) and Philip Abelson (1913–2004) detected neptunium (Np, Z = 93) at the Berkeley Radiation Laboratory [McMillan and Abelson, 1940; Kragh, 2017]. Subsequent experiments by Glenn Seaborg (1912–1999), Arthur Wahl (1917–2006) and Joseph Kennedy (1916–1957) produced plutonium (Pu, Z = 94) [Seaborg *et al.*, 1946], and Seaborg and Albert Ghiorso (1915–2010) would go on to be the most prominent scientists in the identification of the transuranics, until 1960, when the Soviet Union started their own transuranic programmes (see, for example, [Thompson, Ghiorso *et al.*, 1950; Thompson, Street *et al.*, 1950; Ghiorso *et al.*, 1955; 1958; 1970; and Ghiorso, Nitschke *et al.*, 1974]). Laying claim to the discovery of an element accelerated in the 1980s when, akin to the space race, the Soviet Union and USA became increasingly competitive.

Prior to the 1960s, proof for the discovery of an element was sufficient with the provision of an atomic weight and X-ray emission spectrum, but this quickly became irrelevant and impossible for the superheavy elements, as acquiring a suitable spectrum requires far more superheavy material than is possible to generate. X-ray emission can be used indirectly, as the parent superheavy element will decay into daughter nuclides, with distinct X-rays emitted. To address the growing dilemma of how to attribute these discoveries, it became clear that there was a need to define what constitutes the discovery of an element in absolute terms. Hints of the competition between the Soviet Union and American teams are evident in element names. Elements 104 and 105 were particularly controversial, as scientists in California, USA, and Dubna, Russia, attempted to name the elements themselves (kurchatovium, after the father of the Soviet atomic bomb, and nielsbohrium, after Niels Bohr) — elements without formal approval from the International Union for Pure and Applied Chemistry (IUPAC). The naming battle for elements 104 and 105 was only officially resolved in the 1990s, when the Transfermium Working Group, a cluster of concerned scientists chaired by Denys Wilkinson set up in 1985 to address the conflicting claims to elemental discoveries, decided that both parties should share the credit. Element 104 would become "rutherfordium," (Rf) to honour the nuclear physicist Ernest Rutherford (1871–1937), and element 105 would be named "dubnium," (Db) after the Dubna laboratory [IUPAC, 1997]. In the 1980s researchers at the GSI Helmholtz Centre for Heavy Ion Research in Darmstadt, Germany, joined the quest, when they experimentally proved the existence of elements 107–112, and gained the rights to name them.

The Transfermium Working Group issued these formal criteria: "Discovery of a chemical element is the experimental demonstration, beyond reasonable doubt, of the existence of a nuclide with an atomic number Z not identified before, existing for at least 10^{-14} s" [IUPAC/ IUPAP, 1991]. In 1913, the British physicist Henry Moseley (1887–1915) redefined the concept of the atomic number by focussing on the sequential nature of atoms [Moseley, 1913]; eighty years later the new criteria emphasised the importance of the atomic number in the discovery of an element. Following the introduction of the new definition of what an element is, in 2004 a team at Riken Nishina Centre for Accelerator-Based Science in Japan were allowed to name element 113 'nihonium' (Nh) (the Japanese for 'Japan') [Morita *et al.*, 2004] — they had provided what was considered solid evidence for the atomic number of the nuclei produced, whereas the Soviets, who claimed discovery in 2003, had not. Advances in technology mean that further criteria have been developed since then, with requests for a comprehensive and more stringent range of experimental and theoretical data from the scientists who are on a quest for new additions to the periodic table [Hofmann *et al.*, 2018].

Darleane Hoffman: The First Woman Team-Leader in Superheavy Elements

Women have made significant contributions to the field of nuclear chemistry, from Marie Curie (1867–1934) and Lise Meitner (1878–1968) to Irène Joliot-Curie (1897–1956) and Marguerite Perey (1909–1975).[4] It

---

[4]All these women are featured in this volume. See the chapters by Roqué, Jacquemond, and Rayner-Canham and Rayner-Canham, respectively.

was a lucky coincidence that led the young student Darleane Hoffman (b. 1926) to switch her major from applied art to chemistry at Iowa State College [Dagani, 2000]. A mandatory chemistry class delivered by Professor Nellie Naylor (1885–1992) inspired her to stay in the lab, leading to a Bachelor of Science degree in chemistry in 1948. For her following PhD in nuclear and radiochemistry, completed in 1951, she worked with Professor Don Martin, and her first task was to construct a Geiger-Müller counter for detecting and measuring ionising radiation. After completing her PhD she worked at the Oak Ridge National Laboratory for a year, and then she moved to Los Alamos where she would go on to work for 31 years [Dagani, 2000].

This was the early 1950s, and the challenging synthesis routes to the transuranic elements had left many scientists convinced that these elements did not occur in nature. By 1952 scientists had discovered plutonium-244 (with a half-life of 81 million years) in the debris of "Ivy Mike," the first thermonuclear test carried out on the island of Elugelab in the Enewetak Atoll on the Marshall Islands. They proposed that there might still be traces of plutonium on earth. The nuclear weapon tests occurring at Los Alamos in the 1950s unsurprisingly also led to research into finding new heavy elements and isotopes in the "Ivy Mike" debris, and this was the keystone of one of Hoffman's most important discoveries. The fermium-257 (Fm, Z = 100) isotope was the heaviest nuclide that the team could isolate, and it demonstrated curious properties when its spontaneous fission properties were measured. Up to this point, studies of spontaneous or low-energy fission had always resulted in asymmetric fission (i.e., the nucleus split into two fragments of unequal mass).

For more than 20 years Hoffman was involved in multiple projects at Los Alamos, but one of her most significant results came in 1971. After two years of intense work, Hoffman and her team provided definitive evidence of the presence of plutonium-244 (Pu, Z = 94) in 260 kg of precambrian bastnäsite rock formation by using mass spectrometry, a technique which can measure the mass distribution in a sample to very high precision. This was a very challenging experiment — it required 4×10^{-15} g of plutonium for positive detection, a considerable amount relative to naturally-occurring plutonium, and therefore called for the processing of a large volume of the bastnäsite rock [Hoffman *et al.*, 1971]. Hoffman and her team (Figure 1) went on to demonstrate that a significant number of

Figure 1.   Postdoctoral researcher Joseph Weber, staff members Darleane Hoffman and Jerry B. Wilhelmy, and the accelerator operator, in the control room of the Van de Graaf accelerator at the Los Alamos Scientific Laboratory in 1975, at the time of the discovery of the fermium-259 spontaneously fissioning isotope using a target of only a pictogram of fermium-257 and tritium (hydrogen-3) projectiles from the Van de Graaf accelerator (Special Collections and University Archives / Iowa State University Library). The editors wish to thank Darleane Hoffman for providing information about the photograph.

the nuclei of fermium-257 split via symmetric fission, which Seaborg called a "monumental step" [Hoffman and Hoffman, 1974; Dagani, 2000].

   The calibre of Hoffman's work led to her recognition with the prestigious John Simon Guggenheim Memorial Foundation Guggenheim Fellowship in 1978, which she dedicated to investigating the mechanisms of nuclear fission at the Lawrence Berkeley National Lab in association with Seaborg's group. This collaboration would lead to an offer of full professorship in chemistry at University of California (UC), Berkeley in 1984, where she would be appointed leader of the Heavy Element Nuclear and Radiochemistry Group, taking over directly from Seaborg (Figure 2). In 1994, Hoffman would go on to be part of the team who confirmed the 1974 discovery of element 106, which the team named seaborgium (Sg).

Figure 2. The American Chemical Society's Award for Nuclear Chemistry 1983, presented by Glenn T. Seaborg to Darleane C. Hoffman. Hoffman was the first woman ever to receive this award (Special Collections and University Archives / Iowa State University Library).

At UC Berkeley, her skills in handling highly radioactive elements with extremely short half-lives enabled her to carry out very difficult experiments in both the solution and gas phase with lawrencium-260 (Lr, $Z =$ 103, half-life 180 seconds) [Brüchle *et al.*, 1988; Scherer *et al.*, 1988] and dubnium-262 (Db, $Z = 105$, half-life 35 seconds) [Paulus *et al.*, 1999]. One of the team members for the work on dubnium was Dawn Shaughnessy, who is following in Hoffman's footsteps in the fields of nuclear and radiochemistry.

Dawn Shaughnessy: Into the Island of Stability

The heavy element team at the Lawrence Livermore National Laboratory (LLNL) in Berkeley, California are part of the search for superheavy

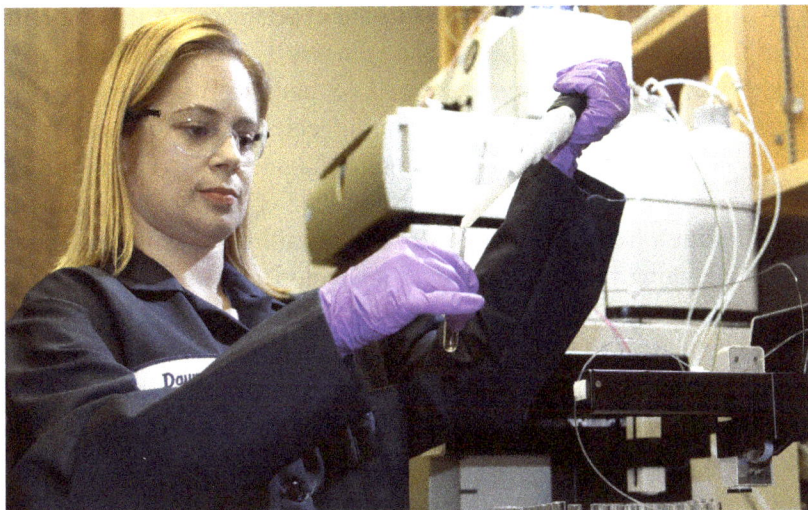

Figure 3. Lawrence Livermore National Laboratory chemist Dawn Shaughnessy. Element 116, livermorium, was so named because it was discovered (by Shaughnessy and her team) at the Livermore lab (Lawrence Livermore National Laboratory photo).

elements. The LLNL team are led by Dawn Shaughnessy (Figure 3) and have collaborated with the Flerov Laboratory of Nuclear Reactions in Dubna to identify six superheavy elements with atomic numbers 113 to 118. Shaughnessy's career in nuclear chemistry began in earnest after completing her doctoral studies with Hoffman, working on einsteinium and the properties of its isotopes in delayed fission [Shaughnessy *et al.*, 2002]. During her postdoctoral research with Heino Nitsche at the Lawrence Berkeley National Laboratory, Shaughnessy learnt how to process plutonium and neptunium, exploring how they could be safely removed from the environment via complexation with manganese minerals [Shaughnessy, Nitsche *et al.*, 2003]. Shaughnessy joined the staff at the LLNL in 2002, and has since held numerous positions, including the simultaneous group leadership of both the Nuclear and Radiochemistry groups in the Nuclear and Chemical Sciences division, and the Joint High Energy Density Science organisation. She has served as Principal Investigator on a number of projects in fields such as plasma nuclear science, nuclear forensics, and heavy element chemistry [LLNL, 2018].

Whilst LLNL have contributed to the experiments to make a number of the superheavy elements, Shaughnessy's team were so involved with element 116 they were allowed to name it livermorium (Lv). Livermorium was made by accelerating calcium-47 (Ca) ions (frequently used in these experiments due to the unusually rich ratio of 28 neutrons to 20 protons) into a californium-248 (Cm) target. Isolating an element with a half-life of some dozens of milliseconds is experimentally challenging, and so far only four isotopes of this element have been produced with mass numbers 290–293 (with 290–292 having half-lives of less than twenty milliseconds, and 293 of 61 milliseconds). Only 35 atoms of livermorium have been observed before it decayed by alpha (α-)decay, losing 2 protons and 2 neutrons to first become element 114, flerovium (Fl), and subsequently element 112, copernicium (Cn). Shaughnessy and her team provided the curium (Cm, Z = 96) target material, which in 2000 made the observation of livermorium possible at the Joint Institute for Nuclear Research.

***

Tennessine (Ts, element 117) was made in April 2010 in a Russian-American collaboration, including the involvement of Shaughnessy, with the production of the $^{293}117$ and $^{294}117$ isotopes from fusion reactions between calcium (^{48}Ca) and berkelium (^{249}Bk) [Oganessian *et al.*, 2010]. The decay chains of tennessine have fascinated chemists worldwide — they show an increased stability for heavier isotopes with Z≥111. This is predicted to be a temporary reversal of the stability decrease in elements heavier than uranium for elements that are approaching the "magic number" — where the nuclides have nucleons arranged in complete shells within the atomic nucleus — which could correspond to the elusive and Seaborg's long sought-after "island of stability" [Oganessian, 2012]. Theoretical predictions suggest that the island of stability could produce elements with sufficient half-lives for applications in industrial processes, which is one of Shaughnessy's many pursuits. The exact location of this island is still unclear — predictions shifted when scientists learned that, due to interactions between protons, superheavy elements were not perfectly spherical atoms. Shaughnessy and other superheavy scientists are not only working to populate the periodic table further, but to better understand the world around us. And while some are dreaming of reaching the island of stability, one achievement can already be safely celebrated: whether it is Meitner, Hoffman or Shaughnessy, the periodic table would not be the same without the research of women scientists.

References

Brüchle, W., Schädel, M., Scherer, U. W., Kratz, J. V., Gregorich, K. E., Lee, D., Nurmia, M., Chasteler, R. M., Hall, H. L., Henderson, R. A. and Hoffman, D. C. (1988). The Hydration Enthalpies of Md^{3+} and Lr^{3+}, *Inorganica Chimica Acta*, 146:2, pp. 267–276.

Curie, P., Curie, M. and Bémont, M. G. (1898). Sur une nouvelle substance fortement radio-active, contenue dans la pechblende, *Comptes Rendus of the Sessions of the Academy of Sciences*, 127, pp. 1215–1217.

Dagani, R. (2000). Priestley Medal Profile, *CENEAR*, 78:13, pp. 31–35.

Ghiorso, A., Nitschke, J. M., Alonso, J. R., Alonso, C. T., Nurmia, M., Seaborg, G. T., Hulet, E. K. and Lougheed, R. W. (1974). Element 106, *Physical Review Letters*, 33:25, pp. 1490–1493.

Ghiorso, A., Nurmia, M., Eskola, K., Harris, J. and Eskola, P. (1970). New Element Hahnium, Atomic Number 105, *Physical Review Letters*, 24:26, pp. 1498–1503.

Ghiorso, A., Sikkeland, T., Walton, J. R. and Seaborg, G. T. (1958). Element No. 102, *Physical Review Letters*, 1:1, pp. 18–21.

Ghiorso, A., Thompson, S. G., Higgins, G. H., Seaborg, G. T., Studier, M. H., Fields, P. R., Fried, S. M., Diamond, H., Mech, J. F., Pyle, G. L., Huizenga, J. R., Hirsch, A., Manning, W. M., Browne, C. I., Smith, H. L. and Spence, R. W. (1955). New Elements Einsteinium and Fermium, Atomic Numbers 99 and 100, *Physical Review*, 99:3, pp. 1048–1049.

Hoffman, D. C. and Hoffman, M. M. (1974). Post-Fission Phenomena, *Annual Review of Nuclear Science*, 24:1, pp. 151–208.

Hoffman, D. C., Lawrence, F. O., Mewherter, J. L. and Rourke, F. M. (1971). Detection of Plutonium-244 in Nature, *Nature*, 234, p. 132.

Hofmann, S., Dmitriev S. N., Fahlander, C., Gates J. M., Roberto J. B. and Sakai, H. (2018). On the Discovery of New Elements (IUPAC/IUPAP Provisional Report), *Pure and Applied Chemistry*, 90:11, p. 1773.

IUPAC (1997). Names and Symbols of Transfermium Elements (IUPAC Recommendations 1997), *Pure and Applied Chemistry*, 69:12, p. 2471.

IUPAC/IUPAP (1991). Criteria that Must be Satisfied for the Discovery of a New Chemical Element to be Recognised, *Pure and Applied Chemistry*, 63:6, pp. 879–886.

Kragh, H. (2017). On the Ontology of Superheavy Elements, *Substantia*, 1:2, pp. 7–17.

Krämer, K. (2016). Explainer: Superheavy Elements. Everything You Ever Wanted to Know About the Last Row of the Periodic Table but Were Afraid to Ask, *ChemistryWorld*, 9 June 2016 (https://www.chemistryworld.com/news/explainer-superheavy-elements/1010345.article, last accessed 6 January 2019).

LLNL (2018). Dawn Shaughnessy, *Lawrence Livermore National Laboratory, Physical and Life Sciences Directory* (https://pls.llnl.gov/people/staff-bios/nacs/shaughnessy-d, last accessed 19 December 2018).

McMillan, E. and Abelson, P. H. (1940). Radioactive Element 93, *Physical Review*, 57:12, pp. 1185–1186.

Morita, K., Morimoto, K., Kaji, D., Akiyama, T., Goto, S., Haba, H., Ideguchi, E., Kanungo, R., Katori, K., Koura, H., Kudo, H., Ohnishi, T., Ozawa, A., Suda, T., Sueki, K., Xu, H., Yamaguchi, T., Yoneda, A., Yoshida, A. and Zhao, Y. (2004). Experiment on the Synthesis of Element 113 in the Reaction ^{209}Bi$(^{70}$Zn,n$)^{278}$113, *Journal of the Physical Society of Japan*, 73:10, pp. 2593–2596.

Moseley, H. G. J. (1913). The high-frequency spectra of the elements, *Philosophical Magazine*, 6:26, pp. 1024–1034.

Oganessian, Y. (2012). Nuclei in the Island of Stability of Superheavy Elements, *Journal of Physics: Conference Series*, 337:1, 012005, pp. 1–6.

Oganessian, Y. T., Abdullin, F. S., Bailey, P. D., Benker, D. E., Bennett, M. E., Dmitriev, S. N., Ezold, J. G., Hamilton, J. H., Henderson, R. A., Itkis, M. G., Lobanov, Y. V., Mezentsev, A. N., Moody, K. J., Nelson, S. L., Polyakov, A. N., Porter, C. E., Ramayya, A. V., Riley, F. D., Roberto, J. B., Ryabinin, M. A., Rykaczewski, K. P., Sagaidak, R. N., Shaughnessy, D. A., Shirokovsky, I. V., Stoyer, M. A., Subbotin, V. G., Sudowe, R., Sukhov, A. M., Tsyganov, Y. S., Utyonkov, V. K., Voinov, A. A., Vostokin, G. K. and Wilk, P. A. (2010). Synthesis of a New Element with Atomic Number $Z = 117$, *Physical Review Letters*, 104:14, 142502, pp. 1–4.

Paulus, W., Kratz, J. V., Strub, E., Brüchle, W., Pershina, V., Schädel, M., Schausten, B., Adams, J. L., Gregorich K. E., Hoffman, D. C., Lane, M. R., Laue, C., Lee, D. M., McGrath, C. A., Shaughnessy, D. K., Strellis, D. A.

and Sylwester, E. R. (1999). Chemical Properties of Element 105 in Aqueous Solution: Extraction of the Fluoride-, Chloride-, and Bromide Complexes of the Group-5 Elements into an Aliphatic Amine, *Radiochimica Acta*, 84, p. 69.

Perey, M. (1939). L'élément 87: AcK, dérivé de l'actinium, *J. Phys. Radium*, 10:10, pp. 435–438.

Rudolph, D., Forsberg, U., Golubev, P., Sarmiento, L. G., Yakushev, A., Andersson, L. L., Di Nitto, A., Düllmann, C. E., Gates, J. M., Gregorich, K. E., Gross, C. J., Heßberger, F. P., Herzberg, R. D., Khuyagbaatar, J., Kratz, J. V., Rykaczewski, K., Schädel, M., Åberg, S., Ackermann, D., Block, M., Brand, H., Carlsson, B. G., Cox, D., Derkx, X., Eberhardt, K., Even, J., Fahlander, C., Gerl, J., Jäger, E., Kindler, B., Krier, J., Kojouharov, I., Kurz, N., Lommel, B., Mistry, A., Mokry, C., Nitsche, H., Omtvedt, J. P., Papadakis, P., Ragnarsson, I., Runke, J., Schaffner, H., Schausten, B., Thörle-Pospiech, P., Torres, T., Traut, T., Trautmann, N., Türler, A., Ward, A., Ward, D. E. and Wiehl, N. (2013). Spectroscopy of Element 115 Decay Chains, *Physical Review Letters*, 111:11, 112502, pp. 1–5.

Scherer, U. W., Kratz, J. V., Schädel, M., Brüchle, W., Gregorich, K. E., Henderson, R. A., Lee, D., Nurmia, M. and Hoffman, D. C. (1988). Lawrencium Chemistry: No Evidence for Oxidation States Lower than 3+ in Aqueous Solution, *Inorganica Chimica Acta*, 146:2, pp. 249–254.

Seaborg, G. T., McMillan, E. M., Kennedy, J. W. and Wahl, A. C. (1946). Radioactive Element 94 from Deuterons on Uranium, *Physical Review*, 69:7–8, pp. 366–367.

Shaughnessy, D. A., Gregorich, K. E., Adams, J. L., Lane, M. R., Laue, C. A., Lee, D. M., McGrath, C. A., Ninov, V., Patin, J. B., Strellis, D. A., Sylwester, E. R., Wilk, P. A. and Hoffman, D. C. (2002). Electron-Capture Delayed Fission Properties of ^{244}Es, *Physical Review C*, 65:2, 024612, pp. 1–7.

Shaughnessy, D. A., Nitsche, H., Booth, C. H., Shuh, D. K., Waychunas, G. A., Wilson, R. E., Gill, H., Cantrell, K. J. and Serne, R. J. (2003). Molecular Interfacial Reactions between Pu(VI) and Manganese Oxide Minerals Manganite and Hausmannite, *Environmental Science & Technology*, 37:15, pp. 3367–3374.

Thompson, S. G., Ghiorso, A. and Seaborg, G. T. (1950). The New Element Berkelium (Atomic Number 97), *Physical Review*, 80:5, pp. 781–789.

Thompson, S. G., Street, K., Ghiorso, A. and Seaborg, G. T. (1950). Element 98, *Physical Review*, 78:3, pp. 298–299.

Select Bibliography

Chapman, K. (2019). *Superheavy. Making and Breaking the Periodic Table* (Bloomsbury Sigma, London).

Hoffman, D. C., Ghiorso, A., Seaborg, G. T. (2000). *The Transuranium People* (Imperial College Press, London).

Kragh, H. (2018). *From Transuranic to Superheavy Elements: A Story of Dispute and Creation* (Springer International Publishing, Cham).

Krämer, K. (2016). Explainer: Superheavy Elements. Everything You Ever Wanted to Know About the Last Row of the Periodic Table but Were Afraid to Ask, *ChemistryWorld*, 9 June 2016 (https://www.chemistryworld.com/news/explainer-superheavy-elements/1010345.article, last accessed 6 January 2019).

Laue, C. (2004). Finding Equality Again, *Science*, 28 May 2004 (http://www.sciencemag.org/careers/2004/05/finding-equality-again, last accessed 6 January 2019).

Schädel, M. and Shaughnessy, D. A., eds. (2014). *The Chemistry of Superheavy Elements* (Springer, Berlin and Heidelberg).

Walter, K. (2010). Collaboration Expands the Periodic Table, One Element at a Time, *Science & Technology Review*, October/November, pp. 16–19 (https://str.llnl.gov/OctNov10/shaughnessy.html, last accessed 19 December 2018).

Part 6

Instrumental Revolution and Interface Between Chemistry and Industry

Sonja Smith-Meyer Hoel — Chemical Engineer in the Ferrosilicon and Aluminium Industry in Norway

Annette Lykknes

Department of Teacher Education, NTNU — Norwegian
University of Science and Technology, 7491 Trondheim, Norway

Sonja Smith-Meyer Hoel (1920–2004) was among Norway's most influential female engineers of her time. She graduated as a chemical engineer at the Norwegian Institute of Technology in 1947, and worked her whole life in chemical industry, at Elkem, one of the twentieth century's most important Norwegian companies. Hoel acquired knowledge and skills related to the production and properties of metals and alloys such as ferrosilicon, steel and aluminium, as well as to mineral wool. For many years, she was the director of the company's patent office, and she was member of several committees and boards on patents in Norway as well as internationally.

***

The beginning of the twentieth century was a remarkable period for industrial Norway. Access to hydroelectric power made it possible for companies like Hydro (formerly, Norsk Hydro) and Elkem (formerly, Elektrokemisk) to establish power-intensive electrochemical and electro-metallurgical processes such as the production of fertilizers, metals and metal alloys. This coincided with an increased emphasis on scientific and technological knowledge in industrial processes in the period which is often referred to as the "second industrial revolution." A new type of engineer, who would develop new processes and establish new industries, was needed. Many Norwegian men went to Germany to a "Technische Hochschule" to train as engineers ("Diplomingenieure"); however, in 1910 Norway established its own Institute of Technology ("Norges tekniske høiskole" (NTH)) in Trondheim [Brandt and Nordal, 2010; Lykknes and Gusland, 2015, pp. 15–16, 140].

Expectations were high. The Institute of Technology was counted upon to train engineers with the scientific as well as the technical-practical competence needed to raise new industries, and thereby gradually to bring the country into prominence. In 1910, this was about building a new, independent nation-state after 400 years of Danish rule and almost 100 years in union with Sweden. After World War II, the engineers were once more the heroes of the nation, expected to rebuild the country out of the devastation of war, with the help of the most recent technical developments. Technical-scientific research was prioritized, which once again raised the expectations towards NTH and its candidates [Lykknes and Gusland, 2015, pp. 298–300].

Sonja Smith-Meyer: Student and Activist

Sonja Smith-Meyer was born in Fredrikstad in the eastern part of Norway in 1920, but in her early childhood the family moved to Kristansand in the south of Norway. Here she matriculated in 1939, and she was amongst the few women who applied and were admitted to the Norwegian Institute of Technology during World War II. She may have been inspired by her father, Otto Smith-Meyer (1887–1941), who had studied to become a construction engineer at Trondheim technical school, and later a "Diplomingenieur" at the "Technische Hoschule" in Dresden (1908–1909). Sonja Smith-Meyer, however, did not initially opt for engineering, as she studied English and German at the University of Oslo directly after her matriculation exam, for which she had specialized in the natural sciences and mathematics. During the war she was part of the student resistance movement and was one of many who were arrested at the University in 1943.[1] In the same year she took up her chemistry studies at NTH in Trondheim. In November 1943, the university in Oslo closed down when students were suspected of setting fire to the assembly hall [Fure, 2007, pp. 263–282]. The NTH in Trondheim remained open, but in 1944 students were urged by the resistance movement to leave the Institute. Sonja went back to her hometown, Kristiansand, where she was co-editor of the illegal newspaper *Nordmannen* ("The Norwegian") [Alming *et al.*, 1995, p. 161; Bugge, 2004].

In 1947 Smith-Meyer became the twenty-fourth woman to complete her degree in chemical engineering at NTH. The first woman had been admitted as early as in 1910, but left within two years. A total of fifteen women were enrolled before World War II. In the 1950s the number of women students increased rapidly; in fact, the number of those admitted during this decade matched that of those admitted in the first thirty years of the Institute's history. Chemistry and architecture were the most popular subjects for women at NTH, so most female engineers were chemical engineers. We know little about Smith-Meyer's experiences as a woman in her class, but accounts of those who preceded her include examples of both harassment and acceptance [Lykknes and Gusland, 2015, pp. 103–105]. Smith-Meyer seems to have been an independent woman, unafraid of mingling with her male colleagues in the student newspaper *Sugepumpen* ("The Suction Pump"), initiating practical jokes, and eventually becoming the newspaper's editor [Lykknes and Gusland, 2015, pp. 252–253].

Chemical Engineer at Elkem

After a short engagement at the Institute, Smith-Meyer was employed as a chemical engineer at Elkem in 1947. Established in 1904 as *Det Norske Aktieselskab for Elektrokemisk Industri* ("The Norwegian Electrochemical Industry Corporation") at the initiative of the industrialist Sam Eyde (1866–1940), Elkem aspired from early on to be a large enterprise. At its peak in 1918, before the post-World War I crisis, Elkem's equity corresponded to 0.5 per cent of the country's gross national product, testifying

---

[1] She was arrested on 3 March and released on 16 March [Ottosen and Knudsen, 2004].

to its importance for the small and emerging nation state [Sogner, 2003, p. 9]. In 1919, Elkem encompassed fifty companies, with one hundred and fifty staff members at the main office in Oslo [Snl, 2018a]. It was from Elkem that Hydro branched out in 1905 as the company specialized in fertilizers, and it was Elkem which patented the much-used Söderberg Self-Baking Electrode[2] in the industrial smelting of carbide, ferro-alloy and pig iron in 1917; this would later also be utilized for the production of aluminium by electrolysis. More than 300 patents were linked to the Söderberg electrode [Snl, 2018b]. In fact, Elkem built a business around the sale of the right to use the Söderberg principle [Sogner, 2008]. In the beginning, Smith-Meyer worked at Elkem's research station "Fiskaa verk" in her hometown, Kristiansand.

During World War II, just before Smith-Meyer came to Fiskaa, research on the production of ferrosilicon (an iron-silicon alloy) was important. The production of mineral wool was among the emerging projects. Ferrosilicon, which contains a sufficient amount of silicon to reduce the amount of oxygen in steel is cheaper to produce than pure silicon, and is used to produce steel with certain properties. Mineral wool (rock wool) was developed as a by-product from slag (waste from some smelting processes) and used for insulation [Sogner, 2003, pp. 101, 107–108]. In 1940 the research station had been expanded to include research on the production of aluminium as well.

Notably, Elkem became instrumental to the Germans' plans to make Norway the aluminium capital of Europe. Aluminium was used in military aircraft, and was considered the political metal of World War II. Germany was the world's largest manufacturer of aluminium, and aimed to expand the production to Norway to produce the same amount there. The metal was made by electrolysis, and many aircraft were produced using the Söderberg electrode. Thus, Norwegian hydroelectric power was important to fulfil the occupants' grand-scale goal, and the expansion of the power plants was part of this plan [Sogner, 2003, pp. 101–114].

The massive German investments opened new possibilities for growth. In Norway, after the war, i.e. at a time when modernization was

---

[2]The Söderberg Self-Baking Electrode was named after its inventor, the Swedish-born (Carl) Wilhelm Söderberg (1876–1955).

equal to industrialization, aluminium production became a priority [Sandvik, 2008]. In fact, two state-owned aluminium plants were established in 1947 and 1954 respectively (and an ironworks as well) — testimony to the role given to electrochemical industry for the rebuilding of the country. The climate was non-competitive. Elkem, too, collaborated closely with the government in the post-war years, and was granted access to cheap hydroelectric power. An aluminium production plant was established in Mosjøen in the north of Norway in 1958, and has been described as the largest isolated investment in Elkem's history [Sogner, 2003, p. 141].

Immediately after the war, however, when aware of the loss of income due to the expiration of the patents for the Söderberg electrode, Elkem started producing smelting furnaces for other companies. They also took up work in the Skorovas pyrite mines in Trøndelag country, started the production of mineral wool, and continued to prioritize the smelting of ferrosilicon at the Fiskaa plant. Indeed, the company became one of Europe's largest producers of ferro-alloys. As a chemical engineer at Fiskaa, Smith-Meyer is likely to have been highly involved with the production of ferrosilicon. Figure 1 shows Sonja as the engineer involved in the production of mineral wool.

Director of the Patent Office

In 1949, after a few years at the Fiskaa plant, Smith-Meyer started to work with the company's patents at the Oslo office. As she explained in an interview many years later, when she arrived at the patent office she hardly knew what a patent was [ES, 1980]. Over the years she learnt how the patent system worked, but it was overwhelming, and Smith-Meyer (from 1952 Smith-Meyer Hoel) put down her ability to master it to her stubbornness. When she was interviewed in 1980, patents filed from Elkem counted more than 550, and this was an enormous amount of paper to handle and supervise (Figure 2). In fact, the establishment of the trademark and the protection of designs were also part of Smith-Meyer Hoel's job at the patent office. At the office, the team working with Smith-Meyer Hoel was further responsible for following up on competitors' activities in relevant areas. In 1965 she became the director of the patent office. Fifteen

Figure 1.   Sonja Smith-Meyer producing mineral wool at Fiskaa research station in 1959 (Elkem, reproduced by kind permission of Elkem).

years later, when Smith-Meyer Hoel was 60 years old, five people were working full time at the office.

Normally, the time from filing for a patent at the patent office to the patent being granted was between two and five years. Smith-Meyer Hoel and her colleagues would help the engineer identify the novelty in the invention and highlight this in the application, supervise the writing and initiate procedures. According to an engineer working closely with her on patents, Smith-Meyer-Hoel carried authority, and she definitely knew the patent law very well. Furthermore, because of her technical training she understood the processes related to the patent. Thanks to her competence, in many cases Smith-Meyer Hoel and her team realized that a patent application could be filed for more than one invention [ES, 1980]. Patents filed from Elkem's patent office during Smith-Meyer Hoel's time include those for aluminium, raw or processed (1957); metal-alloy addition in steel (1968); ferro-alloy addition in ductile iron (1972); rockwool for insulation (1974); and amorphous silicate dust (1982) [Norwegian Industrial Property Office]. They demonstrate her wide range of insight

Figure 2. Sonja Smith-Meyer Hoel, 1980. Hoel worked at Elkem's patent office from 1949, and became its director in 1965. Despite the apparent chaos on her desk, according to one of her former colleagues, she always knew where to find the right paper (from ES har egen patentavdeling, *Elkem-Spigerverkets bedriftsavis*, June 1980, p. 6, reproduced by kind permission of Elkem).

into metal elements and alloys such as aluminium, iron, steel, ferrosilicon and silicon, as well as mineral wool.

Over the years, Smith-Meyer Hoel became involved in both nation-wide and international patent work. In particular, she made an effort to increase the knowledge about patent systems in industries, and to establish a system for compensating patent holders employed by companies [NTB, 1995]. From 1974 to 1989 she was president of the Norwegian

Association for Industrial Patent Engineers ("Norsk forening for industriens patentingeniører"). She also served as board member and vice president of the European Federation of Agents of Industry in Industrial Property, and held a number of other offices nationwide and internationally. In 1996 she published the book *Immatrielle rettigheter: Patenter, mønstre, varemerker* ("Immaterial Rights: Patents, Designs, Trademarks") [Smith-Meyer, 1996], and she was frequently invited to give lectures about patents at her *alma mater* as well as at the Norwegian School of Economics (NHH) in Bergen [NTB, 1995; Bugge, 2004].

***

Sonja Smith Meyer Hoel has been described as one of the earliest influential women in the Norwegian industry [Bugge, 2004]. As a student of chemical engineering at the Norwegian Institute of Technology in the 1940s, she was one of a very small number of women who studied to become engineers, but she was not among the first to do this; twenty-three women had completed their degrees in chemical engineering before her. However, few of these women — especially those graduating before the war — actually stayed in the chemical industry [Lykknes and Gusland, 2015, pp. 105, 254]. Some took on other jobs, some left employment because they got married. In 1950 in Norway, only 4–5 per cent of married women were registered as working [Melby, 2005, p. 306]. This was a period when the "ideology of marriage" with its implications for wifely subordination was particularly strong [Abir-Am and Outram, 1987]. Smith-Meyer Hoel married and had one son (who also became a chemical engineer), but remained active in the company at which she was first employed. At Elkem, she was gradually given more responsibility, and her competence on patent law was noticed across Norway and also abroad. One of the engineers who worked with her while filing a patent described her as a highly respected woman at Elkem, very competent in her area of expertise, both in the technical and in the legal realm, and as someone with integrity who was never afraid to stand up for her beliefs and values. Through her work at Elkem, both at the Fiskaa research station and at the patent office, she acquired comprehensive insights into the chemistry and technology of the metal elements aluminium, iron and silicon, as well as into products such as steel and mineral wool.

Acknowledgements

The author wishes to thank Smith-Meyer Hoel's son Jan-Otto Smith-Meyer Hoel and her former colleagues Vegard Olderheim and Magnar Storset for information on her life and work, the professor of metallurgy Leiv Kolbeinsen for guidance on metallurgical processes, the historians Jorunn Sem Fure, Knut Sogner and Bjørn L. Basberg for their valuable advice, Brigitte Van Tiggelen for her comments on an earlier draft of this article, and Anke Timmermann for her careful editorial work on it.

References

Abir-Am, P. G. and Outram, D. (1987). Introduction, in *Uneasy Careers and Intimate Lives: Women in Science, 1789–1969*, eds. Abir-Am, P. G. and Outram, D. (Rutgers University Press, New Brunswick, NJ), pp. 1–16.

Alming, K. *et al.* [book committee] (1995). *Holdningskamp og motstandsvilje* (Tapir, Trondheim).

Brandt, T. and Nordal. O. (2010). *Turbulens og tankekraft: Historien om NTNU* (Pax, Oslo).

Bugge, N. E. (2004). Nekrolog: Sonja Smith-Meyer Hoel, *Adresseavisen*, 27 January 2004, via https://web.retriever-info.com/services/archive.html, last accessed 19 December 2018.

ES (1980). ES har egen patentavdeling, *Elkem-Spigerverkets bedriftsavis*, June 1980, pp. 4–6.

Fure, J. S. (2007). *Universitetet i kamp 1940–1945* (Vidarforlaget, Oslo).

Lykknes, A. and Gusland, J. Z. (2015). *Akademi og industri: Kjemiutdanning og -forskning ved NTNU gjennom 100 år* (Fagbokforlaget, Trondheim).

Melby, K. (2005). Del IV. Husmortid. 1900–1950, in *Med kjønnsperspektiv på norsk historie*, eds. Blom, I. and Sogner, S. (Cappelen akademisk, Oslo), pp. 255–331.

NTB (1995). Sonja Smith-Meyer Hoel 75 år 27. Februar, *NTB tekst*, 3 February 1995.

Norwegian Industrial Property Office. Patent register for Norway (https://dbsearch2.patentstyret.no/Search.aspx?from=tlist, last accessed 15 December 2019).

Ottosen, K. and Knudsen, A., eds. (2004). *Nordmenn i fangenskap 1940–1945* (Universitetsforlaget, Oslo).

Sandvik, P. T. (2008). European, Global or Norwegian? The Norwegian Aluminium Companies, 1946–2005, in *The European Enterprise*, ed. Schröter, H. G. (Springer, Berlin), pp. 241–252.

Smith-Meyer, S. (1996). *Immatrielle rettigheter: Patenter, mønstre, varemerker* (Norges eksportråd, Oslo).

Snl (2018a). Elkem, in *Store norske leksikon* (https://snl.no/Elkem, last accessed on 22 December 2018).

Snl (2018b). Wilhelm Søderberg, in *Store norske leksikon* (https://snl.no/ Wilhelm_S%C3%B8derberg, last accessed on 22 December 2018).

Sogner, K. (2003). *Skaperkraft: Elkem gjennom 100 år* (Messel forlag, Oslo).

Sogner, K. (2008). Changing Transnational Affections: Orkla, Elkem and Norwegian Big Business, 1960–2004, in *The European Enterprise*, ed. Schröter, H. G. (Springer, Berlin), pp. 253–267.

Select Bibliography

Bix, A. S. (2013). *Girls Coming to Tech! A History of American Engineering Education for Women* (MIT Press, Cambridge, MA).

Lykknes, A. and Gusland, J. Z. (2015). *Akademi og industri: Kjemiutdanning og -forskning ved NTNU gjennom 100 år* (Fagbokforlaget, Trondheim).

Sogner, K. (2003). *Skaperkraft: Elkem gjennom 100 år* (Messel forlag, Oslo).

Tobies, R. and Vogt, A., eds. (2014). *Women in Industrial Research* (Franz Steiner, Stuttgart).

Toshiko Mayeda and the Isotopes of Oxygen

Matthew Shindell

Space History Department, Smithsonian National Air and
Space Museum, Washington, D.C., USA

Humans have learned a lot about the history of the Earth and other planets from the study of stable isotopes in nature. This isotope geochemistry emerged mainly in the years after World War II, as chemists and geologists began using physical instruments like mass spectrometers to analyse soil, rocks, and meteorites. Toshiko "Tosh" Mayeda (1923–2004) (Figure 1), a Japanese-American woman interned by her own government during the war, was one of the most talented and experienced mass spectrometer operators during this period, and helped to establish methods of using the isotopes of oxygen to study historic ocean temperatures and the history of the solar system. Although she never rose above the rank of laboratory assistant, she collaborated with two very influential chemists who became known as the founders of the new discipline. Moreover, her position allowed her effectively to run the laboratory in which she worked.

***

Figure 1. Toshiko "Tosh" Mayeda (The University of Chicago, reproduced by kind permission of the University of Chicago News Office).

Toshiko Kuki was born on 7 February 1923 in Tacoma, Washington to Matsusaburo Kuki (1887–1952) and Haruko (Okada) Kuki (1900–1927). The family was separated by illness when Haruko contracted tuberculosis. With the children in tow, Haruko left her husband in the USA and returned to her native Japan to be cared for by her family. Tragically, the children lost their mother shortly after arriving in Japan, leaving them to be raised by extended family in Osaka. When she turned 18, the family sent Toshiko back to the USA to find her father and bring him back to Japan. She found her father, who by that point was earning a comfortable living as a dentist, and the two spent the next several months preparing to leave. Their plans soon changed, however. On 7 December 1941 the Japanese attacked the US Naval base at Pearl Harbor, leading the US to enter World War II. On 19 February 1942, US President Franklin D. Roosevelt (1882–1945) signed Executive Order 9066, authorising the forced relocation and incarceration of all people of Japanese ancestry living on the West Coast [Robinson, 2001, p. 108]. Toshiko and her father were sent to Tule Lake internment camp in California, where they were interned until the end of the war.

During her time at the camp, Kuki met her future husband, Harry Keith Mayeda (1911–2003). After the war, the two married and left the West Coast, resettling in Chicago, Illinois. She decided to continue her education, enrolling in Wilbur Wright College. She gravitated toward chemistry, and received a bachelors in chemistry in 1950. As explored by Margaret Rossiter in her book *Women Scientists in America: 1940–1972*, the postwar period is often considered to be a golden age in American science, and yet women had an especially difficult time making careers for themselves [Rossiter, 1998]. Non-white women, like Mayeda, faced even greater obstacles. Despite the enormous growth in funding, available jobs, and technological advances following World War II, women did not experience a surge in career opportunities. Record numbers of women worked as scientists during the 1950s and 1960s, and yet these women were practically invisible in their fields, relegated to the role of research and lab assistants or unrecognised for their published work if at the PhD level.

Mayeda's story reflects the difficulties faced by women scientists during this period. After completing her bachelors degree, she became a laboratory technician for the Nobel Prize-winning physical chemist, Harold C. Urey (1893–1981), then at the University of Chicago. Urey had migrated to Chicago from Columbia University after the war, and had transformed his research program into a study of isotopes in nature. During these years, Urey's laboratory developed new methods for applying mass spectrometry to questions in earth and planetary science. They developed a so-called "oxygen thermometer," which used the relative amounts of the stable isotopes of oxygen trapped in marine fossils to determine temperature records for the pre-historic oceans (records that today are used in climate change models). Light and heavy oxygen (oxygen-16 and -18, respectively) both exist in nature. When marine shellfish grow their shells, the amounts of light and heavy oxygen that make it into the shell is dependent on the temperature of the surrounding water. Urey and his collaborators set out to determine the precise relationship between water temperature and oxygen isotope ratios, and to develop reliable tools and methods to measure these ratios with great precision. This work required building and operating very sensitive mass spectrometers that were able to analyse isotope ratios in fossil shells and other specimens, and developing new laboratory methods for treating and analysing specimens.

Mayeda began her career in Urey's lab washing glassware. Urey soon realised that she had a strong grasp of chemical methods and an aptitude for mass spectrometer work. She became a crucial member of Urey's research team, and an expert at maintaining and operating the complicated instruments in the lab. There was almost as much art as science to using the "home-built" instruments the work required (commercially-built mass spectrometers of the type Urey needed did not exist in the 1950s). It was as important to have an able technician like Mayeda, who was familiar with the instruments and their quirks, as it was to have the instruments themselves.

Although Mayeda never rose above the rank of laboratory assistant, she did gain a certain amount of respect and autonomy in the lab that many women scientists at the time were unable to obtain. Much of this had to do with the encouragement of Harold Urey. Urey had a progressive attitude toward women in science, and helped to promote the careers of women. He championed the career of the German-born American physicist Maria Goeppert-Mayer (1906–1972), who was awarded one fourth of the Nobel Prize in Physics in 1963 for her discovery concerning the nuclear shell structure of the atomic nucleus. Urey also did not believe that marriage should be the end of a woman's career in science. Mildred Cohn (1913–2009), a biochemist and graduate student of Urey's at Columbia University in the late 1930's, said that Urey believed she could have a career whether she was married or not. Urey even offered Cohn a loan from some of his Nobel Prize money so that she could be spared long hours of work outside the lab and instead pursue her education. Cohn remembered how this was different from other male professors at the time, who would not try to help their female graduate students get jobs [Shindell, 2019]. When Mayeda arrived in Urey's lab in 1950, he already had an established history of providing research assistants with career opportunities.

After Urey left Chicago in 1958, the cosmo-chemist Robert Clayton (1930–2017) took over the laboratory, including the mass spectrometers [Humayun and O'Neil, 2003]. Mayeda stayed on as Clayton's assistant, and ultimately became one of his closest collaborators. She continued to perform the sample analysis in the laboratory. It is fair to say that, although she never received a PhD, Mayeda's laboratory work and her collaboration with Urey and Clayton provided her with the equivalent of a PhD education in physical chemistry and mass spectrometry (despite Urey's and Clayton's reliance upon Mayeda, the two men may not have encouraged her to pursue further formal education). She and Clayton published extensively on oxygen isotope ratios in the solar system, studied meteorites, and even analysed lunar soil samples returned by the National Aeronautics and Space Administration's Apollo Program [Onuma *et al.*, 1970; Clayton *et al.* 1973]. Mayeda was an integral part of the lab's success — meteorites and samples were sent to her and Clayton from all over the world for analysis. The lab's mass spectrometer is now on display in the University of Chicago's Henry Hinds Laboratory for Geophysical Sciences. In Clayton's 2017 obituary, the *UChicago News* wrote that "[f]or a quarter of a century, oxygen from virtually every new type of meteorite passed through that spectrometer as Clayton and Mayeda mapped out the isotope ratios for every class of meteorites" [Lerner, 2018; Clayton, 2007]. Together they developed the Clayton-Mayeda model of oxygen isotope abundances in the solar system. The ratios they measured helped them to begin opening up new questions about the history of the solar system and the formation of meteorites — many of which have yet to be answered. For her contributions to cosmochemistry, the Geochemical Society of Japan awarded Mayeda their Society Merit Prize in 2002 [Koppes, 2004].

Mayeda continued working with Clayton until her death in 2004; however, cancer slowed her down in her later years. Although her career does not seem to have advanced in step with her accomplishments, her work did not go unrecognised. Clayton and his colleagues acknowledged that much of their success was due to her abilities as his "indomitable" research assistant [Wood, 1982]. In addition to her research, Mayeda mentored the graduate students and postdoctoral fellows in the lab and became a sort of mother figure to them. Clayton noted that Mayeda not only helped students improve their research techniques, but she also took a personal interest in them. She "contributed to their personal development in a direct but friendly way" and was not shy about improving their attitudes and habits. In fact, she was a "great favorite" with the graduate students and many called her "mom" [Koppes, 2004].

References

Clayton, R. N., Hurd, J. M. and Mayeda, T. K. (1973). Oxygen Isotopic Compositions of Apollo 15, 16, and 17 Samples, and their Bearing on Lunar Origin and Petrogenesis, in *Proceedings of the Fourth Lunar Science Conference* (Supplement 4, *Geochemica et Cosmochemica Acta*) vol. 2, pp. 1535–1542.

Clayton, R. N. (2007). Isotopes: From Earth to the Solar System, *Annual Review of Earth and Planetary Sciences*, 35:1, pp. 1–19.

Humayun, M. and O'Neil, J. R. (2003). A Special Issue Dedicated to Robert N. Clayton, *Geochimica et Cosmochimica Acta*, 67:17, pp. 3097–3099.

Koppes, S. (2004). Toshiko K. Mayeda, Chemist, 1923–2004, The University of Chicago News Office, 18 February 2004 (http://www-news.uchicago.edu/releases/04/040218.mayeda.shtml, last accessed 5 February 2019).

Lerner, L. (2018). Robert N. Clayton, "One of the Giants" of Cosmochemistry, 1930–2017, *University of Chicago News*, 18 January 2018 (https://news.uchicago.edu/story/robert-n-clayton-one-giants-cosmochemistry-1930-2017, last accessed 05 February 2019).

Onuma, N., Clayton, R. N. and Mayeda, T. K. (1970). Apollo 11 Rocks: Oxygen Isotope Fractionation Between Minerals, and an Estimate of the Temperature, *Geochimica et Cosmochimica Acta Supplement*, 1, p. 1429.

Robinson, G. (2001). *By Order of the President: FDR and the Internment of Japanese Americans* (Harvard University Press, Cambridge, MA).

Rossiter, M. W. (1998). *Women Scientists in America: Before Affirmative Action, 1940–1972*, 1st ed. (Johns Hopkins University Press, Baltimore, MD, London).

Shindell, M. (2019). *The Life and Science of Harold C. Urey* (University Of Chicago Press, Chicago, IL).

Wood, J. A. (1982). Robert N. Clayton Received the Leonard Medal of the Meteoritical Society, *Meteoritics*, 17, pp. 171–176.

Select Bibliography

Brush, S. G. (1996). *A History of Modern Planetary Physics: Fruitful Encounters* (Cambridge University Press, New York, NY).

Clayton, R. N. (2007). Isotopes: From Earth to the Solar System, *Annual Review of Earth and Planetary Sciences*, 35:1, pp. 1–19.

MacPherson, G. J. (2008). *Oxygen in the Solar System*, Reviews in Mineralogy and Geochemistry, vol. 68 (Mineralogical Society of America, Chantilly, VA).

Rayner-Canham, M. F. and Rayner-Canham, G. (2001). *Women in Chemistry: Their Changing Roles from Alchemical Times to the Mid-Twentieth Century* (Chemical Heritage Foundation, Philadelphia, PA).

Rossiter, M. W. (1998). *Women Scientists in America: Before Affirmative Action, 1940–1972* (Johns Hopkins University Press, Baltimore, MD, London).

Simpson, C. C. (2002). *An Absent Presence: Japanese Americans in Postwar American Culture, 1945–1960* (Duke University Press, Durham, NC).

Mary Almond, Iron in Rocks and Continental Drift

Sally Horrocks*, Tom Lean[†] and Paul Merchant[†]

*School of History, Politics and International Relations,
University of Leicester, UK
[†]National Life Stories, The British Library, UK

Mary Almond's (b. 1928) work with iron occurred in the period 1951–1954, at the University of Manchester and Imperial College, London. It was the iron in rocks that concerned her — specifically iron minerals in sedimentary rocks that had aligned themselves with the Earth's magnetic field at the time they were formed, hundreds of millions of years ago. Working as part of a small team assembled by physicist Patrick Blackett (1897–1974), Almond used an instrument called an astatic magnetometer to measure the direction and dip of iron minerals in samples of sandstones collected from nine sites across Britain. The aim was to test the theory of continental drift, which was far from accepted in the early 1950s. If these rock samples contained iron minerals that pointed in directions other than the current poles, and dipped at an angle consistent with formation in latitudes other than 60-ish degrees north, this would suggest that they had been formed in a "Britain" at a different place on Earth. Almond took the majority of the measurements and discovered that Britain did seem to have, in her characteristically understated words, "moved a bit" [Merchant, 2011, Track 1, [1:22:13–1:23:00]].

The co-authored paper reporting on the measurements concluded that "the most likely explanation of the observed horizontal direction of magnetization of the sediments studied is that the whole of the land mass which now constitutes England has rotated clockwise through 34° relative to the earth's geographical axis," and that it "seems probable that the magnetic dip at the time of magnetization was appreciably less than today ... [which] would imply an increase in magnetic latitude, since magnetization" [Clegg *et al.*, 1954a, pp. 596–597]. The paper contained at once the "first fully described oblique paleomagnetic directions from Britain" and "the first palaeomagnetic support for continental drift" [Frankel, 2012, p. 104]. The revelations of plate tectonics would follow in the 1960s.

Though Almond went on to work and co-author publications on other samples [Almond *et al.*, 1956], she did not use success with iron in rocks as a foundation for her future academic career. By the time the rest of the team published a key review paper in 1960 [Blackett *et al.*, 1960], Almond had left work in palaeomagnetism to have children. She then retrained as a computer scientist and lectured in that subject at several universities. Her story serves as an example of how women's contributions to successful collaborative teams in the mid-twentieth century did not always provide them with the same career opportunities available to male colleagues.

Education

Mary Almond was born in Manchester in 1928. Her father, a commercial traveller with a wide range of interests including photography, sold paper bags for his father's firm. Her mother, who had started a Latin degree but dropped out to look after her brothers, was keen for her daughters to have the education denied to her. Both Mary and her younger sister, Joyce, attended Withington Girls School, which encouraged its pupils to excel in the physical sciences and mathematics. At the University of Manchester, they both successfully completed undergraduate degrees and were awarded PhDs.

As an undergraduate Almond was taught by two of Manchester's most distinguished physicists, Bernard Lovell (1913–2012) and Blackett. Each

Figure 1.   Research team at Jodrell Bank Observatory in the 1950s. Bernard Lovell centre, Mary Almond left. Courtesy of Mary Almond. Mary was interviewed for An Oral History of British Science at the British Library in 2011.

had decisive influences on her career. Lovell invited his students to contribute to the construction of facilities for radio astronomy at Jodrell Bank, and Almond and her friend Marjorie found themselves living in a caravan among "these old army trailers in a sea of mud," spending days sandpapering "rust off an old army search light that they were going to attach an aerial to" [Merchant, 2011, Track 1, [41:15–42:00]] (Figure 1). Following her degree, Almond worked on a PhD on the orbits of meteor showers, under Lovell's supervision.

Recording Iron in Rocks

Almond's PhD was successful, and she published two papers from the research [Almond *et al.*, 1951; 1952]. Rather than opting to continue in radio astronomy she applied for and accepted a job with Ferranti, a

Figure 2. Mary Almond, 1950s. Courtesy of Mary Almond. Mary was interviewed for An Oral History of British Science at the British Library in 2011.

Manchester based electronics firm. At this point, Blackett intervened, offering her the research post in the small palaeomagnetism group that he was assembling, with funds from the UK's Department of Scientific and Industrial Research. Almond (Figure 2) joined the group in 1952 along with a geologist, Peter Stubbs. By this time the "operational head" of the group, the physicist John Clegg (1913–1987), had already constructed a new astatic magnetometer — based on Blackett's, but more suitable for the measurement of magnetised iron in rock samples — in the basement of Manchester University's Physical Laboratories [Frankel, 2012, p. 60]. When the group moved, with Blackett, to Imperial College in 1953, this magnetometer was moved to a special non-magnetic hut "in the sort of ... back garden of ... Imperial College" [Merchant, 2011, Track 1, [1:21:47–1:22:01]].

Almond assisted in the collection of the samples of sandstones from sites across Britain, but her forte was the use of the magnetometer. Although others also carried out measurements, she was seen as the most efficient operator of this sensitive and complex apparatus. In Clegg's view, "she could measure the magnetism of rock samples twice as fast ... and more accurately" than himself [Nye, 1999, p. 86 n. 85]. Stubbs cut a core from each sample and then a slice from each core. It was these slices of rock that Almond measured with the magnetometer. She recalls the process as follows:

> Interviewer: You walk into the hut, you're, holding a disc [of rock] what do you do next in order to say this is the magnetisation of this disc?
>
> Almond: Put it inside the magnetometer, raised it up to the little magnets there which made the — a light beam on this metre rule thing and see how far that swung.
>
> Interviewer: And you — how did you record your — you know, the number?
>
> Almond: Wrote it down I think [both laugh] ... I took thousands of measurements to be honest [laughs], I mean my life was spent in the hut measuring the deflection of this spot [Merchant, 2011, Track 1, [1:27:29–1:28:40]].

Asked whether she was involved in "presenting" the outcomes of this work, Almond replies, "I wasn't — just did all the donkey work measuring the samples" [Merchant, 2011, Track 1, [1:32:04–1:32:44]]. Her comments on analysis are equally brief and matter of fact: "I would measure the direction of magnetisation in these discs and we found that England had moved a bit" [Merchant, 2011, Track 1, [1:22:13–1:23:00]]. But this short description belies the significant role she played in analysing the results. She is specifically credited by Blackett with finding, "a marked correlation of μ with colour, the darker as opposed to redder rocks having the large susceptibility and the low coercive forces" [Blackett, 1956, p. 52]. The credit is valued by Almond, and makes sense to her in relation to a specific memory:

> I remember telling Blackett about it because I'd been measuring this susceptibility of twenty or thirty of these samples and just for fun I'd arranged them right across my desk in order of susceptibility, going from very pale in colour to very dark [laughs] and when Blackett — I mean he came to see me almost every day and I said, "Look at these, they're in order of susceptibility," [laughs] so he — obviously he'd remembered

my comment and put it in his book [Merchant, 2011, Track 1, [1:31:10–1:31:51]].

Almond was central to this research and the co-author of the resulting paper — the first ever paper to conclude, based on the evidence of iron minerals in rocks, that Britain had moved north and rotated relative to the Earth's geographic axis. This was a crucial step in the confirmation of continental drift [Clegg *et al.*, 1954a; Nye, 1999; Frankel, 2012]. Almond contributed to further research on Jurassic tuffs and dolerites from Tasmania [Almond *et al.*, 1956] before leaving to start a family.

Marriage, Family and a New Career

Whilst she was working on iron in rocks, Almond married Jim Pickering, Blackett's laboratory assistant. In 1954, two years after she joined Blackett's group, she left Imperial College to be a full-time mother. Pickering accepted a job with the oil company Shell, and the family moved to a Shell housing estate near to the firm's refinery at Stanlow, later relocating to another company estate near Shell Haven in Essex. This was not a positive experience for Almond, who felt isolated in the company of women with whom she had few common interests, and who she believed were "stewing in their own juices" [Merchant, 2011, Track 2, [3:02–3:45]]. When her son John was two, Almond opted not to join her husband on an overseas posting and instead got a job teaching science at a local school, with childcare provided by "unmarried mothers" who advertised in *The Lady* magazine for accommodation in return for providing domestic help.

After around six months' teaching, Almond developed an interest in computer programming and started applying for jobs in this area, indicating a willingness to accept an entry level post despite her academic qualifications. Eventually she was offered employment in the Electrical Engineering Department of Queen Mary College, writing programmes for calculations required by members of the department, which had previously been carried out manually by female calculators using calculating machines. Soon afterwards, by now a single mother, she took up a post as lecturer in computer science at Queen Mary College, London where she co-authored a book on programming [Collins and Almond, 1966]. She speaks in her interview of the "difficulty" of combining, at this time, a career in science with childcare:

> Almond: I worked late at night on preparing my lectures and spent what time I could with the children...
>
> Interviewer: And presumably you worked — you gave lectures at times of the day when they were at school, you were able to do that?
>
> Almond: Yes, yes. Yes, Professor Ferraro was very considerate ... I used to take them in with me if I was desperate, usually at half-term [week-long school holidays in the UK] 'cause you know universities don't have a half-term.
>
> Interviewer: What did your children do while you were lecturing in the —
>
> Almond: Came and sat in the back of the class [laughs] [Merchant, 2011, Track 2, [24:11–25:30]].

In 1966, in order to be nearer to the support of family and friends, she returned to the University of Manchester, to its new Department of Computer Science, which was led by Tom Kilburn (1921–2001). Here she concentrated on undergraduate teaching and departmental administration, and also worked for the Open University in order to fund the cost of sending her son, who was dyslexic, to a specialised school (Figure 3).

Asked to reflect on the impact of her family on her career, Almond responded, "well I never became a professor." Probed further on whether this might have been a realistic possibility, she expressed doubt:

> Interviewer: Would it have — could it have been possible to become a professor?
>
> Almond: I doubt it.
>
> Interviewer: Why not?
>
> Almond: Not good enough...
>
> Interviewer: What makes you — what gives you that impression, what makes you say that about yourself?

Figure 3. Mary Almond at her desk at the University of Manchester, in the 1970s. Photo credit: Courtesy of Mary Almond. Mary was interviewed for An Oral History of British Science at the British Library in 2011.

Almond: The things I've done in life.

Interviewer: At what points in your career have you felt that, then, that you reached the limit of your ability, that you couldn't go further but you might have wanted to?

Almond: No, never really, I just [pause] I didn't get involved with the research ... busy doing everything else [Merchant, 2011, Track 2, [57:38–58:46]].

***

Characterising her work with iron, Almond said that she "just did all the donkey work measuring the samples." This echoes a more general self-assessment of herself as a rather secondary and second-rate scientist, capable of taking crucial measurements and writing important computer programmes in a way that, in her view, did not amount to being "involved with the research." But there is plenty of evidence that this self-assessment is overly modest. During her career Almond researched and published in three very different areas of science: radio astronomy, palaeomagnetism and computer programming. In the case of palaeomagnetism, she was responsible for contributing, through attention to iron in rocks, to a revolution in the earth sciences. When she turned to computer programming she rapidly achieved a very high level of competence in a new field whilst also raising two children as a single mother at a time when childcare provision was extremely limited. To most people these would count as impressive achievements, and not the career of someone who was, as she says, "not good enough."

It is possible that this modesty arises from having worked in close proximity with three men — Lovell, Blackett and Kilburn — who rank among the most distinguished of mid-twentieth century British scientists [Lovell, 1975; Wilkes and Kahn, 2003; Davies, Graham-Smith, and Lyne, 2016]. In comparison to their achievements, the vast majority of scientists would probably see their own careers as rather modest. Almond's story suggests that it might be useful to recalibrate how success in science is measured and regarded, according more recognition to those who made significant contributions to team projects and supported the work of others, and more respect to those who did so in the face of significant personal challenges.

References

Almond, M., Davies, J. G. and Lovell, A. C. B. (1951). The Velocity Distribution of Sporadic Meteors, part I, *Mon. Not. Roy. Astron. Soc.*, 111, pp. 585–608.

Almond, M., Davies, J. G. and Lovell, A. C. B. (1952). The Velocity Distribution of Sporadic Meteors, part II, *Mon. Not. Roy. Astron. Soc.*, 112, pp. 21–38.

Almond, M., Clegg, J. A. and Jaeger, J. C. (1956). Remanent Magnetism of Some Dolerites, Basalts and Volcanic Tuffs from Tasmania, *Phil. Mag.*, 1, pp. 771–782.

Blackett, P. M. S. (1956). *Lectures on Rock Magnetism* (The Weizmann Science Press of Israel, Jerusalem).

Blackett, P. M. S., Clegg, J. A., and Stubbs, P. H. S. (1960). An Analysis of Rock Magnetic Data, *Proc. R. Soc. Lond. A.*, 256, pp. 291–322.

Clegg, J. A., Almond, M. and Stubbs, P. H. S. (1954a). The Remanent Magnetization of Some Sedimentary Rocks in Britain, *Phil. Mag.*, 45, pp. 583–598.

Clegg, J. A., Almond, M. and Stubbs, P. H. S. (1954b). Some Recent Studies of the Pre-History of the Earth's Magnetic Field, *J. Geomagn. Geoelectr.*, 6, pp. 194–199.

Collins, J. S. and Almond, M. (1966). *Principles of ALGOL 60 Programming* (Harrap, London).

Davies, R. D., Graham–Smith, F. and Lyne, A. G. (2016). Sir Alfred Charles Bernard Lovell OBE, 31 August 1913–6 August 2012, *Biogr. Mems Fell. R. Soc.*, 62, pp. 323–344.

Frankel, H. R. (2012). *The Continental Drift Controversy: vol. 2* (Cambridge University Press, Cambridge).

Lovell, B. (1975). Patrick Maynard Stuart Blackett, Baron Blackett, of Chelsea,18 November 1897–13 July 1974, *Biogr. Mems Fell. R. Soc.*, 21, pp. 1–115.

Merchant, P. (2011). Interview with Mary Almond, British Library ref. C1379/38.

Nye, M. J. (1999). Temptations of Theory, Strategies of Evidence: PMS Blackett and the Earth's Magnetism, 1947–52, *Br. J. Hist. Sci.*, 32, pp. 69–92.

Wilkes, M. and Kahn, H. J. (2003). Tom Kilburn CBE FREng, 11 August 1921–17 January 2001, *Biogr. Mems Fell. R. Soc.*, 49, pp. 283–297.

Select Bibliography

Frankel, H. R. (2012). *The Continental Drift Controversy: vol. 2* (Cambridge University Press, Cambridge).

Nye, M. J. (1999). Temptations of Theory, Strategies of Evidence: PMS Blackett and the Earth's Magnetism, 1947–52, *Br. J. Hist. Sci.*, 32, pp. 69–92.

Barbara Bowen and the Ozone Hole

Sally Horrocks*, Tom Lean[†] and Paul Merchant[†]

*School of History, Politics and International Relations,
University of Leicester, UK
[†]National Life Stories, The British Library, UK

Barbara Bowen's (b. 1932) work with atmospheric ozone (O_3), an "allotrope" of oxygen consisting of three oxygen atoms, occurred in the period 1976–1980 at the headquarters of the British Antarctic Survey (BAS) in Cambridge, England. Here she was responsible for digitising the geophysical and atmospheric data collected at the two British bases in Antarctica, including measurements of total atmospheric ozone recorded by instruments called Dobson spectrophotometers. BAS had been recording the amount of ozone in the Antarctic atmosphere — along with many other variables — since 1956, because the ozone layer, about 50 km above sea level, was of interest to scientists concerned with improving understanding of the circulation of the atmosphere [Sullivan, 1961, pp. 239–241; Dobson, 1968, pp. 399–401].

Shortly before Bowen started work at BAS, papers discussing the possibility of depletion of the ozone layer by pollutants, including industrially produced chlorofluorocarbons (CFCs), had appeared in science journals and attracted some popular comment. Nevertheless, ozone remained a rather low-profile gas until 1985 (after Bowen had moved on to other roles in BAS), when the scientists Bowen had worked with — Joseph Farman (1930–2013), Brian Gardiner and Jonathan Shanklin — published a now-famous paper in the influential journal *Nature*, reporting "large losses of total ozone in Antarctica" [Farman *et al.*, 1985]. Suddenly this form of oxygen, a very small constituent of the atmosphere (800 times less abundant than carbon dioxide, for example), was big news. Public concern over a "hole" in the ozone layer — known to shield the earth's surface from harmful ultraviolet light — led to the signing of the Montreal Protocol on Substances that Deplete the Ozone Layer and the phasing out of CFCs [Pyle and Harris, 2013]. Bowen's work on O_3 had contributed to finding the "ozone hole" over Antarctica, which had been missed by a satellite from the National Aeronautics and Space Administration (NASA).

Education and Early Career

Barbara Bowen was born in Lancaster, north-west England in March 1932, as the elder of two daughters. Her mother had been a teacher before her marriage, and her father, who had a PhD in chemistry, was a manager for Lansil, a rayon producer. In her final years at school Bowen specialised in science and geography, and after a further year at technical college improving her skills in mathematics, she studied "general science" at King's College, Durham, the predecessor of the University of Newcastle-Upon-Tyne. Though she passed her main subjects — botany and bacteriology — she was less successful in her subsidiary subject of chemistry, and was not awarded a degree. Nevertheless, Bowen was able to find a job that made use of her scientific expertise, as an abstract writer at the Department of Scientific and Industrial Research Water Pollution Research Laboratory in Stevenage. She explains that this suited her wish to work in science but not in a laboratory:

> Well we were then wearing tights and I ruined so many pairs and plus the odd skirt, because you had lab coats but ... they got holes in them and then ... if you got something else on — another chemical on it, it went through that hole and onto your clothes [laughs] ... I'd just decided that

I didn't want to do lab work but I wanted to be, obviously as I'd trained as a scientist, I wanted to do something scientifically inclined but not work in a lab [Merchant, 2010b, Track 4, [48:40–49:44]].

Whilst working in Stevenage Bowen met her future husband, whom she married in 1958. Bowen left paid employment when her two daughters were born, but later, now living near Cambridge, returned to work part-time for Cambridge University Press as an editor and indexer. This work was chosen because it fitted around her caring commitments while also contributing to the family finances at a time when she felt her marriage was becoming strained. The job was, however, poorly paid for the amount of time required, causing Bowen to look for alternative employment. When she learned that the BAS sought part-time scientific assistants to digitise data she applied successfully for this job.

Digitising Ozone Data at BAS

Bowen was appointed in 1976 as one of two technicians digitising measurements of ozone — along with other atmospheric and geophysical measurements — recorded at two Antarctic bases by the staff of BAS' Physics Group, led by Joseph Farman. Turning graphical and hand-written records into digital records, stored on a computer, would allow them to be processed more quickly, and this was important because, at least in the case of the ozone records, "a backlog was building up" [Shanklin, 2010, p. 34]. Her equipment included a digitising table and a hand-held cursor, rather like a computer mouse (Figure 1). Bowen would attach each data chart to the table, perform a zero calibration, identify the chart with a number, and then follow whatever plot or curve was on the chart, pressing a button at intervals. Each press of the button was picked up by the mesh of fine wires under the surface of the table, and stored on the computer as a data point. Bowen remembers that her shifts followed the repetitive pattern of "press, move it, press ... press ... press ... press ... press ... press... press ... press ... press ... press ... press, about that sort of interval" [Merchant, 2010b, Track 5, [42:05–42:17]].

Bowen vividly remembers digitising a series of charts (not capturing ozone measurements in this case) in which the values involved

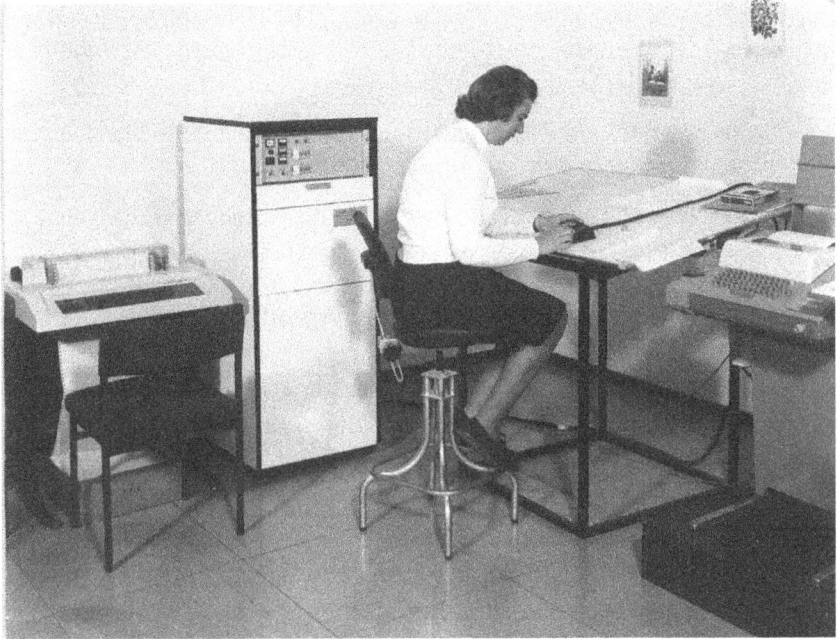

Figure 1.   Digitisation tables at the British Antarctic Survey, 1978. Courtesy of Barbara Bowen. Barbara was interviewed for An Oral History of British Science at the British Library in 2010. Note, this shows Barbara Bowen's colleague Dawn (surname not certain) at work not Barbara herself.

were unusually high — "off the charts" — due to an explosion at chemical works in Flixborough, England; but in general she did this work with little sense of how the data was being used or of its potential significance:

> I understood what I had got to do so — but I didn't understand sort of the reason for it until I came to the charts, which went off the chart, it was such an extreme reaction in my magnetic charts that I was working on, that as I said was the reflection of the Flixborough disaster being bounced off the ionosphere and back down again [Merchant, 2010b, Track 5, [33:41–34:05]].

These recollections suggest that this was a moment when the relationship between her work and what was happening out in the world became clear. She says that her work on ozone records was completed with no such clarity:

> Interviewer: What did Joe [Farman] tell you about what you were doing, what the importance of it was, what it involved?
>
> Bowen: Not a lot really [laughs], he did — I didn't know at all, just that what I had to do was measuring these — to measure these charts and enter the numbers in, digitising these numbers into the computer, he didn't say until it sort of came out and, you know, he was broadcasting and getting his paper written I suppose that is what it was all about really and so, you know, he didn't give any indication as far as I recollect as to [laughs] what it was all for [Merchant, 2010b, Track 5, [47:37–48:18]].

We might say, then, that Bowen (and her colleague Dawn) were extensions of the digitising equipment itself rather than active constructors of new knowledge. However, it would be a mistake to simply contrast Bowen's disinterested view of ozone with that of the "real scientists" around her. Ozone had been recorded in Antarctica since 1956 as one variable among many, partly because, as Farman said, it lent itself to being measured:

> It does happen to be a gas with a highly peculiar spectrum and it's one of the few constituents you can measure in such an easy way with such precision. I mean there are lots of other things you'd love to be able to measure but they all turn out to be much more difficult in essence because it's the easiest measurement you can make of something in the whole thickness of the atmosphere [Merchant, 2010a, Track 4, [20:54–21:22]].

Even though there was some concern about the potential effects of CFCs on ozone at the time of Bowen's work, any depletion was expected to occur in the tropics, not over Antarctica, so that "no one was searching for long-term patterns in springtime data for the Antarctic" [Shanklin, 2010, p. 34]. And right up until the publication of the key *Nature* paper, Farman says that other scientists, including very senior scientists, expressed strong doubts about the value of long terms measurements of ozone, and other variables, in Antarctica [Merchant, 2017, pp. 35–36]. Farman persisted, even though for a long time he himself might not have been able to state clearly, in Bowen's words, "what it was all for."

Inclusion and Recognition

Barbara Bowen, then, was not necessarily different from the scientists around her by not appreciating the value of paying attention to ozone. Nor, could she have felt, at this time, that the only line of inclusion/exclusion at BAS concerned the difference between scientists and technical staff; it was not until 1983 that the first female BAS scientist, Janet Thomson (b. 1942), was permitted to work in Antarctica [Merchant, 2016]. But more significantly perhaps, Bowen did not seek or value inclusion in a world of atmospheric science, whether focused on ozone or not. She regarded her digitisation work as "a bit boring really" [Merchant, 2010b, Track 1, [30:34–34:09]] (in fact she uses the word "boring" four times to describe the work during the interview), and this led her to move to a different role within BAS, so that by the time the *Nature* paper was published she was working as an editorial assistant (enabling her to make better use of her previous experience), and she later transferred to a role in the BAS archives.

What she did value — at the time of her work on ozone — was a job that accommodated her family commitments. Bowen stresses that Farman in particular, and BAS in general, included her in the work of the team by allowing her choose to work in school term times, during the school day. The job also provided a reasonable salary, made use of her scientific education and involved congenial coffee-room relations with other staff, including scientists.

Bowen's account does not suggest that she was motivated by the expectation that she might contribute to important new discoveries, or that she would be recognised for doing so. It is only with hindsight that she was able to reflect on the significance of the work she did as part of BAS' Physics Group and express satisfaction about what they had achieved: "it was very important work and — and still is I suppose and so to be commended for it, so I appreciate that mine was a very lowly cog in that — in his [Farman's] big machine" [Merchant, 2010b, Track 5, [26:00–26:16]]. Nor did she express any rancour that her contribution to this project had remained little evident in the public domain. Instead she drew attention to an occasion later in her career, when she felt a senior colleague had taken credit for an innovation in the BAS archive that she had initiated:

I can't remember what it was now but at any rate it was fairly fundamental to the — the way we archived these old documents and stuff and — but it was not — it was taken up by her but I wasn't acknowledged that it was my idea ... so I was a bit — a bit miffed to say the least, but there we are, that is the scientific reality I think [Merchant, 2010b, Track 1, [29:43–30:34]].

***

Bowen received no credit for her role in the discovery of the ozone hole, even though her digitisation work almost certainly contributed to that discovery. But then, neither did the many, many men who took the measurements in Antarctica that she digitised. Or those who built the wooden huts at BAS bases from which the ozone spectrometer poked out (Figure 2). What should be the "cut-off point" for inclusion in scientific papers (in the list of authors or acknowledgements) and for inclusion in the journalistic

Figure 2. Dobson spectrometer measuring ozone concentration in stratosphere over Antarctica for the first time, mid 1950s. Courtesy of Stanley Evans. Stanley was interviewed for An Oral History of British Science at the British Library in 2011.

and popular accounts of Roan [1989] and Booth [1994]. Though clearly, in one sense, central to the discovery of the ozone hole, Bowen does not herself wish to claim any credit. She is much more concerned about an incident in which her innovative idea on an archive cataloguing system went unrecognised.

References

Booth, N. (1994). *How Soon Is Now: The Truth About the Ozone Hole* (Simon and Schuster, London).

Dobson, G. M. B. (1968). Forty Years' Research on Atmospheric Ozone at Oxford: A History, *Appl. Opt.*, 7:3, pp. 387–405.

Farman, J., Gardiner, B. and Shanklin, J. (1985). Large Losses of Total Ozone in Antarctica Reveal Seasonal ClO_x/NO_x Interaction, *Nature*, 315, pp. 207–210.

Merchant, P. (2010a). Interview with Joseph Farman, British Library ref. C1379/07.

Merchant, P. (2010b). Interview with Barbara Bowen, British Library ref. C1379/18.

Merchant, P. (2016). Janet Thompson: "A Woman's Place is in Antarctica", *Dangerous Women's Project* (http://dangerouswomenproject.org/2016/12/16/janet-thomson/, last accessed 9 January 2019).

Merchant, P. (2017). "He Didn't Go Round the Conference Circuit Talking About It": Oral Histories of Joseph Farman and the Ozone Hole, in *The Silences of Science: Gaps and Pauses in the Communication of Science*, eds. Mellor, F. and Webster, S. (Routledge, London), pp. 31–47.

Pyle, J. and Harris, N. (2013). Joe Farman (1930–2013): Discoverer of the Ozone Hole, *Nature*, 498, p. 435.

Roan, S. (1989). *The Ozone Crisis: The 15-Year Evolution of a Sudden Global Emergency* (Wiley, Chichester).

Shanklin, J. (2010). Reflections on the Ozone Hole, *Nature*, 465, pp. 34–35.

Sullivan, W. (1961). *Assault on the Unknown: The International Geophysical Year* (Hodder and Stoughton, London).

Select Bibliography

Booth, N. (1994). *How Soon Is Now: The Truth About the Ozone Hole* (Simon and Schuster, London).

Merchant, P. (2017). "He Didn't Go Round the Conference Circuit Talking About It": Oral Histories of Joseph Farman and the Ozone Hole, in *The Silences of Science: Gaps and Pauses in the Communication of Science*, eds. Mellor, F. and Webster, S. (Routledge, London), pp. 31–47.

Roan, S. (1989). *The Ozone Crisis: The 15-Year Evolution of a Sudden Global Emergency* (Wiley, Chichester).

Reatha Clark King, Starting Fires with Fluorine

Mary Mark Ockerbloom

Science History Institute, 315 Chestnut Street,
Philadelphia, PA 19119, USA

In the early 1960s, the Reverend Martin Luther King, Jr. (1929–1968) spoke out for civil rights, and President John F. Kennedy (1917–1963) committed Americans to landing on the moon. Both inspired Reatha Clark King (b. 1938), the first African American woman scientist to work at the National Bureau of Standards in Washington, DC. Between 1963 and 1968, King tested chemical compounds being proposed for use in rocket propellants. The most dangerous were gaseous fluorine compounds, among them oxygen difluoride and chlorine trifluoride. They were toxic, corrosive, and exploded on contact with other substances, including those that were generally unreactive like steel and glass. To ensure that NASA had reliable data, King needed to design flame calorimetry equipment capable of controlling and measuring the activity of fluorine and other dangerous gases.

441

Heat of Formation

In thermochemistry, under standard conditions, the energy stored in a set amount of a substance when it is formed (its "heat of formation" or "enthalpy of formation") and the energy that substance can release when burned (its "heat of combustion" or "enthalpy of combustion") will be equal.[1] By Hess's Law, the enthalpy change in a complex chemical reaction must equal the sum of the enthalpy changes for the individual reaction steps it involves. In thermochemistry, if you can describe the reactions that formed something precisely enough, you can predict the energy that it is capable of producing [Atkins *et al.*, 2018, pp. 42–58].

The conditions that form a person are far more complex, and their effects less predictable. Reatha Belle Clark was born on 11 April 1938, in Pavo, Georgia. Her father Willie Clark was a "smart man" but illiterate [Jenkins, 1996, p. 343]. Her mother, Ola Mae Watts Campbell, had a third-grade education. The second of three sisters, Clark went to live with her grandmother in Coolidge, Georgia around fourth grade, so that her grandmother would not be alone [King, 1999, pp. 13–14]. When Clark was about 10, her father left the family. All three sisters lived with their grandmother while their mother went north to work as a maid. Mother and daughters reunited in Moultrie, Georgia, where Clark entered eighth grade [Brown, 2005, pp. 2–3]. The girls helped support their family by picking cotton and tobacco. At age 12, Clark got up at 4 a.m. and picked 200 pounds of cotton in a day, to earn about $6. Sun burn, dirt and tobacco stains marked her skin. It was common to miss school in the spring and fall to do agricultural work [King, 1999, p. 16].

---

[1]Enthalpy and heat are different ways of conceptualizing the physical functioning of a thermodynamic system. When heat is transferred between a thermodynamic system and its surroundings during a chemical reaction, the temperature change depends on the amount of heat involved and the heat capacity of the system. Measures of heat alone are therefore insufficient to describe the state of a thermodynamic system. Enthalpy is a state function that fully defines the state of the system. It can be obtained by measuring the change in the heat transferred between a system and its surroundings at constant pressure. The change in enthalpy of the system during a chemical reaction is equal to the change in its internal energy plus the change in the product of the pressure times the volume of the system. Under standard conditions for a fixed amount of a substance, the terms "heat" and "enthalpy" are sometimes used interchangeably [Atkins et al., 2018, pp. 42–58].

Clark first attended school at age 4, in a segregated one-room schoolhouse at Mt. Zion Baptist Church in Pavo, Georgia (grades 1-7 were taught in one room). Clark's intelligence and determination were recognized [Brown, 2005, pp. 2–3] and she was encouraged by her teacher Florence Frazier and other church members [Brown, 2012, p. 116]. The community's pride in her achievements made Clark aware of the importance of continuing to succeed [Brown, 2005, pp. 3–4]. In 1954, the same year that the Supreme Court of the United States ruled in *Brown v. Board of Education* that segregated schools were illegal,[2] Clark graduated from Moultrie High School for Negro Youth as valedictorian of her class [Brown, 2012, p. 116]. In an interview in 2005, Clark recalled:

> Did I have encouragement from other adults outside of my family? Yes, and very clearly the people in our church because we, sort of, reflected their pride. It always seemed like we were carrying a beacon, a torch for them and they were depending on us to help bring blacks into an era of better opportunities [Brown, 2005, p. 4].

Charlton R. Hamilton, Dean of Students for the historically black Clark College in Atlanta, Georgia, visited Moultrie as a recruiter. He offered Clark a scholarship and part-time work in the registrar's office. At the time, young black women's professional options were to be hairdressers, teachers, or nurses. Clark enrolled in home economics with the expectation of becoming a teacher [King, 1999, pp. 17–20]. One of her required classes was chemistry. As Clark describes it, she fell in love with the chemistry laboratory [Brown, 2005, pp. 6–7]. Her teacher, Alfred Spriggs, encouraged her to major in chemistry and become a research chemist. Clark earned a Bachelor of Science in chemistry with a minor in mathematics from Clark College in 1958 [Brown, 2012, p. 116].

---

[2] *Brown v. Board of Education of Topeka, 347 U.S. 483 (1954)*, was a landmark decision in which the United States Supreme Court declared that state laws establishing separate public schools for black and white students were unconstitutional, because the schools were "inherently unequal," and so violated the Equal Protection Clause of the Fourteenth Amendment of the United States Constitution. Although the ruling was a major victory for the Civil Rights Movement, the decision did not outline a process for ending racial segregation in schools [Kluger, 1975, pp. 710–711].

Clark applied for a Woodrow Wilson Fellowship for financial aid for graduate school, listing the University of Chicago in Illinois as her first choice [Brown, 2005, pp. 8–9]. Because the state university, which she was qualified to attend, was still segregated, the State of Georgia was required to give her a grant to study elsewhere. She earned her Master of Science and PhD degrees in physical chemistry from the University of Chicago in 1960 and 1963. In Chicago she studied thermodynamics, thermochemistry and metallurgy, getting a solid grounding in both materials science and research methods [Brown, 2012, pp. 116–118]. For her PhD thesis, she used calorimetry (measurement of the transfer of energy as heat)[3] to examine thermochemical properties of alloys including intermetallic compounds [King and Kleppa, 1964, p. 87].

While in Chicago, Clark also met and married N. Justice King. This influenced her selection of possible places to work when she applied for jobs. It also influenced some potential employers [Brown, 2012, pp. 118–119].

> [In one case] it became clear that my gender would work against me because the recruiter asked me outright if I was — well, this was before I got married. He asked me if I was planning to get married and if I got married, would I have children. So those questions were very open and blatant and the stereotyping was very clear... Race was an issue, but gender by then had clearly emerged as a second issue, which was just as much of a barrier [Brown, 2005, pp. 19–20].

The Bureau of Standards

In 1963, Reatha Clark King joined the calorimetry laboratory in the heat division of the National Bureau of Standards in Washington, DC [Brown, 2005, pp. 21–23]. Given that there were several qualified candidates, the director George Thomson Armstrong (1916–1982) chose to set a new standard by hiring the first African American woman scientist to work at the Bureau [King, 1999, pp. 24–25; Goldberg, 1982, p. 4]. King later described him as "a very enlightened person" about race [Brown, 2005, p. 21].

---

[3] Measurements of the transfer of energy as heat in calorimetry are associated with changes of state due to chemical reactions, physical changes, or phase transitions [Atkins *et al.*, 2018, pp. 42, 121].

King worked with Armstrong on materials science and high-temperature chemistry. An early project examined aluminium carbide (Al_4C_3) using bomb calorimetry [King and Armstrong, 1964; 1965]. In bomb calorimetry the amount of heat transferred between a thermodynamic system and its surroundings during a chemical reaction is measured. Reactants are placed in an explosive-proof container or "bomb," a closed system whose contents are held at a constant volume. The bomb is submerged in a liquid which can absorb heat. A combustion reaction is set off inside the bomb, and heat transfer to the surrounding liquid is measured [Atkins *et al.*, 2018, pp. 42–50]. The goal is to measure the energy involved without losing any products of combustion [Armstrong and King, 1979, pp. 335–336, 356].

Identifying the intermediate steps in a complex chemical reaction and establishing the standard reaction enthalpies for each step makes it possible to more precisely calculate values for the overall reaction [Atkins *et al.*, 2018, pp. 52–53]. When substances do not combust completely the calculations become harder, as any remaining masses must be taken into consideration. Failing to do so increases the possible error of the calculations and can compromise the experiment [King and Armstrong, 1964, pp. 661–667; 1965, pp. 26–27].

King and Armstrong corrected for several possible sources of error in previous studies. They determined heat of combustion by igniting aluminium carbide in an atmosphere of highly-purified oxygen to minimize combustible impurities. They also determined that two different forms of aluminia were produced. In addition to a "massive lump" in the combustion zone, aluminia condensed on the walls inside the calorimetry bomb as a fine powder. Taking account of both forms of aluminia was essential to determining an accurate heat of combustion for aluminium carbide and calculating its heat of formation. The Bureau's *Technical News Bulletin* for February 1965 included a summary of the aluminium carbide work and a photograph of King (who was pregnant at the time) on its cover (see Figure 1) [King and Armstrong, 1965; Brown, 2005, p. 25].

The Dangers of Fluorine

As well as reviewing and maintaining standards of measurement, the National Bureau of Standards carried out contract research projects for

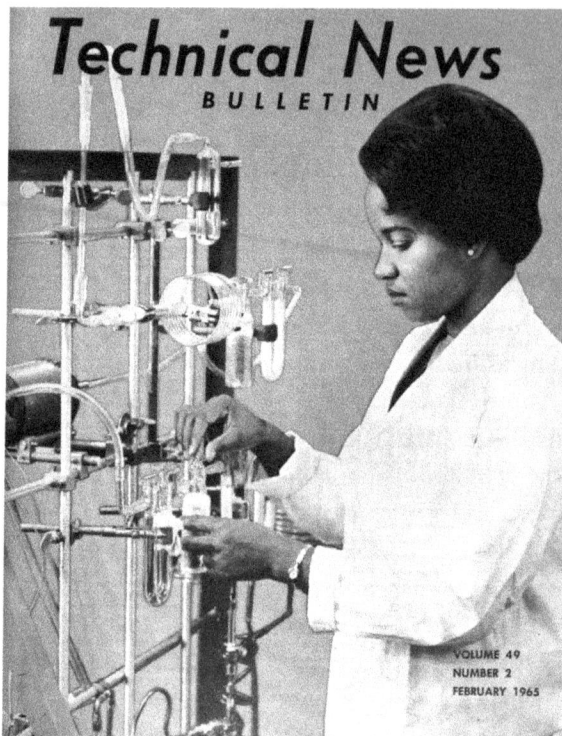

Figure 1.   Reatha Clark King, cover of the *Technical News Bulletin*, 49:2 (National Bureau of Standards, Washington, DC, 1965) (Federal government publication, public domain).

other agencies. One of those agencies was the National Aeronautics and Space Administration (NASA). King was assigned to lead a project to determine the heats of formation of gaseous fluorine compounds such as oxygen difluoride (OF_2) for NASA. Two previous teams had tried to study fluorine without success [Jenkins, 1996, p. 358].

Fluorine and related compounds were of interest to NASA for their possible use in propellants for liquid chemical rockets. At least 100 chemicals were tested for possible use in propellants. Thousands of possible bipropellants, which combine a fuel (e.g. kerosene, alcohol, hydrazine or liquid hydrogen) with an oxidizer (nitric acid, nitrogen tetroxide, liquid oxygen or liquid fluorine), were proposed. As one of the most highly reactive elements in the periodic system, with theoretical predictions of high performance, fluorine was an element of particular interest. High reactivity could offer advantages when trading off weight and energy to lift a rocket into space. Fluorine was also hypergolic: it could ignite other substances upon exposure. That suggested possibilities for fluorine as an oxidizer, capable of repeated ignition without an external stimulus [Sutton, 2005, pp. 25–26, 34–45].

The qualities that made fluorine interesting also made it extremely difficult to control. Fluorine reacts suddenly and explosively with most substances, including many that are generally considered unreactive, such as glass, steel and asbestos. Liquid fluorine and its possible products of combustion, including hydrogen fluoride and hydrofluoric acid, are toxic and corrosive as well as explosive. While many fuels are liquids at ordinary temperatures, hydrogen, oxygen and fluorine exist as liquids only at very low temperatures [Landau and Rosen, 1947; Tylenda, 2003, pp. 187–193]. Working with fluorine was extremely dangerous, required meticulous care, and posed significant technical problems [Clark, 1972, pp. 73–75, 110–112; Sutton, 2005, pp. 26–27].

Fluorine Flame Calorimetry

Early work on propellants was hampered by inadequate theory and a lack of good data. In the 1960s and 1970s the situation improved due to better methods of thermochemical analysis, the use of computers and painstaking laboratory work on the thermodynamic properties of chemicals, their reactions and their products. Data about heats of combustion and formation, enthalpies[1] and other properties helped to refine theoretical models. Engineers used the information to eliminate poor candidates, reduce testing, and estimate theoretical performance values to compare to their results [Sutton, 2005, pp. 40–41].

The combustion of gaseous substances is often measured using a type of calorimetry known as flame or flow calorimetry. In flame calorimetry, the combustion process is held at constant pressure, and the change in enthalpy is determined. As in bomb calorimetry, all possible reaction products and sources of error must be considered [Domalski, 1971, p. 234; Armstrong and King, 1979, pp. 333–336, 356–358]. Because of fluorine's reactivity, standard laboratory equipment could not be used to calculate heats of formation for liquid fluorine, oxygen difluoride and other fluorine-oxygen mixtures. Over the next three years, King carefully reviewed previous work, examined existing equipment and techniques, and developed an experimental apparatus and procedures that could be used to safely contain and test fluorine and other dangerous gases [Jenkins, 1996, pp. 258–360].

> The work was challenging because it required the original designs of flow calorimetric equipment that would resist corrosion by these materials, applications of computers, and the learning of new computer languages. I used every bit of material science that I had learned [King, 1999, p. 25].

The calorimeter's design and use are described in detail in King's and Armstrong's 1968 article "Constant Pressure Flame Calorimetry With Fluorine II: The Heat of Formation of Oxygen Difluoride." The layout of the gas flow system resembles previous flame calorimetric systems. The general method involves reacting the oxidizer in a flowing atmosphere of excess hydrogen, and then forming an aqueous solution of the products. Major differences include the solution of hydrogen fluoride in water, and the design of the burner, a two-chambered reaction vessel (see Figure 2). The primary solution vessel is made of nickel-plated copper with a Teflon liner. The rest of the reaction vessel is composed mostly of Monel, a nickel-copper alloy with high tensile strength and resistance to corrosion, with silver solder at permanent joints. The reaction was initiated by an electric spark, using a high voltage electrode made of nickel and insulated by Teflon from a Monel sheath [King and Armstrong, 1967, p. 144; 1968, pp. 114–116].

King's previous work on metallic alloys helped her to identify materials capable of containing fluorine and related compounds. Her use of a coiled tube heat exchanger enabled hot liquids to cool without exploding [Yount, 2008, p. 156]. Armstrong considered the work "remarkable for its completeness, and for the thoroughness with which all aspects of the problem was investigated" [Jenkins, 1996, p. 360]. R. P. Hudson, the head of the heat division, gave King a Meritorious Publication Award [Jenkins, 1996, p. 367; Brown, 2005, p. 25].

King used the redesigned fluorine flame calorimeter to determine the heats of formation of gaseous fluorine compounds including oxygen dif-

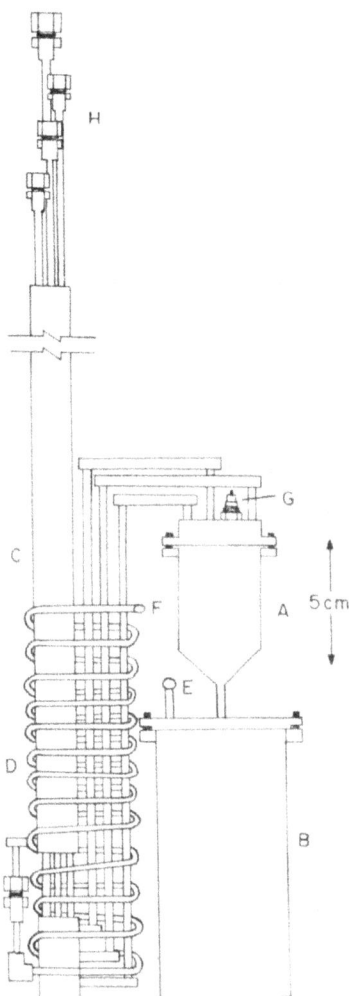

Figure 2. King, R. C. and Armstrong, G. T. (1968). Constant Pressure Flame Calorimetry with Fluorine II: The Heat of Formation of Oxygen Difluoride, *Journal of Research of the National Bureau of Standards (A. Physics and Chemistry)*, 72A:2 (National Bureau of Standards, Washington, DC, 1968). Burner for fluorine flame calorimetry, p. 115 (Federal government publication, public domain).[4]

---

[4]In the combustion chamber (A) oxidizer and hydrogen are mixed and ignited. Reaction products move with effluent hydrogen to the primary solution vessel (B). There gas disperses through water, removing hydrogen fluoride. Monel tubes carry gases from the exterior of the calorimeter to the burner, through the heat exchanger (C). Gases exit

luoride, fluorine and oxygen mixtures with hydrogen, and fluorine-containing oxidizers [King and Armstrong, 1967; Armstrong and King, 1969]. She also determined the heat of formation of chlorine trifluoride (ClF_3) [King and Armstrong, 1970], a gas so dangerous that any trace of oil or grease — even a fingerprint — left inside a reaction vessel could cause an explosion [Clark, 1972, pp. 73–75, 110–112]. King described ClF_3 as a "vigorous fluorinating agent" [Jenkins, 1996, p. 359].

NASA eventually decided that liquid fluorine was too dangerous to use as a fuel [Sutton, 2005, pp. 44–46], but fluorinated oxidizers are still discussed for their possible use in propellants [Valluri *et al.*, 2019]. The equipment and methods that King devised for fluorine flame calorimetry continue to be used for the study of gaseous compounds involving hydrogen, oxygen and fluorine. The apparatus and procedures that King developed to control and measure the combustion of dangerous gases in the lab were later adapted for use in rockets by NASA [Yount, 2008, p. 156].

A Torch for Others

In 1968, King left the Bureau of Standards to move with her family to Long Island [Brown, 2005, p. 23]. She joined the newly established City University of New York as a faculty member in science. She went on to a distinguished career in academics and philanthropy at City University (1968–1977), as president of the Metropolitan State University in Minneapolis and St. Paul (1977–1988) and as vice president of General Mills Corporation and executive director of General Mills Foundation (1988–2002). [Yount, 2008, pp. 156–157; King, 1999, p. 26–36]. In moving into educational and philanthropic roles, King felt that she had an opportunity to build on the legacies of those who had inspired her.

> One objective was to further the purposes of the organizations, and my other objective was to help pave the way to opportunity for other blacks

---

through the cooling coil (D) into the outer tube of the heat exchanger where they circulate among the tubes of entering gases before leaving the calorimeter to the secondary solution vessel at outlet (E). Effluent gases leave the secondary solution vessel at (F) and pass through a helix of Monel tubing before entering the heat exchanger [King and Armstrong, 1968, p. 115].

and women, as well as for myself. I was very conscious of this latter responsibility, so much so that I expected myself to be able to deal with bad treatment because of gender and race as it happened, and not to let it sidetrack me from my major objectives. ... The "first blacks" or "first women" or the first of any group that has a history of discrimination against it, carries a torch for others — those who sacrificed so we could be there, and those of us who will come later [King, 1999, p. 12].

References

Armstrong, G. T. and King, R. C. (1969). The Heats of Formation of Some Fluorine-Containing Oxidizers, *Proc. Conf. Interagency Chemical Rocket Propulsion Group, Thermochemistry Working Group, Cleveland OH, April 9–11, 1969, 7th Meeting, Bulletin 1* (Chemical Propulsion Information Agency, Johns Hopkins University, Applied Physics Laboratory, Silver Spring, MD), pp. 19–40.

Armstrong, G. T. and King, R. C. (1979). Chapter 15: Fluorine Flame Calorimetry, in *Combustion Calorimetry*, eds. Sunner, S. and Mansson, M. (Pergamon, Oxford), pp. 333–359.

Atkins, P., de Paula, J. and Keeler, J. (2018). *Atkin's Physical Chemistry* (Oxford University Press, Oxford).

Brown, J. E. (2005). *Reatha Clark King: Transcript of an Interview Conducted by Jeannette E. Brown at Minneapolis, Minnesota on 1 May 2005* (Chemical Heritage Foundation, Philadelphia, PA).

Brown, J. E. (2012). *African American Women Chemists* (Oxford University Press, New York, NY).

Clark, John D. (1972). *Ignition! An Informal History of Liquid Rocket Propellants* (Rutgers University Press, New Brunswick, NJ).

Domalski, E. S. (1971). Evaluating Experimental Data on Heats of Combustion, *Journal of Chemical Documentation*, 11:4, pp. 234–238.

Goldberg, R. N. (1982). In Memoriam: George T. Armstrong, *CODATA Newsletter*, 23, pp. 4–5.

Jenkins, E. S. (1996). Reatha Clark King: Chemistry of the Halogen Compounds (1938–), in *To Fathom More: African American Scientists and Inventors*, ed. Jenkins, E. S. (University Press of America, Lanham, MD), pp. 342–368.

King, R. C. (1999). Succeeding Against the Odds in Higher Education: Advancing Society by Overcoming Obstacles Due to Race and Gender, in *Grass Roots and Glass Ceilings: African American Administrators in Predominantly White Colleges and Universities*, ed. Harvey, W. B. (State University of New York Press, Albany, NY), pp. 1–38.

King, R. C. and Armstrong, G. T. (1964). Heat of Combustion and Heat of Formation of Aluminum Carbide, *Journal of Research of the National Bureau of Standards (A. Physics and Chemistry)*, 68A:6, pp. 661–668.

King, R. C. and Armstrong, G. T. (1965). Heat of Formation of Aluminum Carbide, *Technical News Bulletin*, 49:2 (National Bureau of Standards, Washington, DC, 1965), pp. 26–27.

King, R. C. and Armstrong, G. T. (1967). Chapter 8: Fluorine Flame Calorimetry, II: The Heats of Reaction of Oxygen Difluoride, Fluorine and Oxygen, with Hydrogen. The Heat of Formation of Oxygen Difluoride, in *National Bureau of Standards NBS Report 9500: Preliminary Report on the Thermodynamic Properties of Selected Light-Element and Some Related Compounds. U.S. Air Force Order No. OAR ISSA 65–8* (National Bureau of Standards, Washington, DC), pp. 140–224.

King, R. C. and Armstrong, G. T. (1968). Constant Pressure Flame Calorimetry With Fluorine II: The Heat of Formation of Oxygen Difluoride, *Journal of Research of the National Bureau of Standards (A. Physics and Chemistry)*, 72A:2 (National Bureau of Standards, Washington, DC, 1968), pp. 113–131.

King, R. C. and Armstrong, G. T. (1970). Fluorine Flame Calorimetry, III: The Heat of Formation of Chlorine Trifluoride at 298.15 K, *Journal of Research of the National Bureau of Standards (A. Physics and Chemistry)*, 74A:6, pp. 769–779.

King, R. C. and Kleppa, O. J. (1964). A Thermochemical Study of Some Selected Laves Phases, *Acta Metallurgica*, 12:1, pp. 87–97.

Kluger, R. (1975). *Simple Justice: The History of Brown v. Board of Education and Black America's Struggle for Equality* (Knopf, New York, NY).

Landau, R. and Rosen, R. (1947). Industrial Handling of Fluorine, *Industrial & Engineering Chemistry*, 39:3, pp. 281–286.

Sutton, G. P. (2005). *History of Liquid Propellant Rocket Engines* (AIAA, Reston, VA).

Tylenda, C. A. (2003). *Toxicological Profile for Fluorides, Hydrogen Fluoride, and Fluorine* (U.S. Department for Health and Human Services, Atlanta, GA).

Valluri, S. K., Schoenitz, M. and Dreizin, E. (2019). Fluorine-Containing Oxidizers for Metal Fuels in Energetic Formulations, *Defence Technology*, 15:1, pp. 1–22.

Yount, L. (2008). *A to Z of Women in Science and Math* (Facts On File, Inc., New York, NY).

Select Bibliography

Brown, J. E. (2012). *African American Women Chemists* (Oxford University Press, New York, NY).

Clark, J. D. (1972). *Ignition! An Informal History of Liquid Rocket Propellants* (Rutgers University Press, New Brunswick, NJ).

King, R. C. and Armstrong, G. T. (1968). Constant Pressure Flame Calorimetry with Fluorine II: The Heat of Formation of Oxygen Difluoride, *Journal of Research of the National Bureau of Standards (A. Physics and Chemistry)*, 72A:2, pp. 113–131.

King, R. C. (1999). Succeeding Against the Odds in Higher Education: Advancing Society by Overcoming Obstacles Due to Race and Gender, in *Grass Roots and Glass Ceilings: African American Administrators in Predominantly White Colleges and Universities*, ed. Harvey, W. B. (State University of New York Press, Albany, NY), pp. 1–38.

Sutton, G. P. (2005). *History of Liquid Propellant Rocket Engines* (AIAA, Reston, VA).

Yount, L. (2008). *A to Z of Women in Science and Math* (Facts On File, Inc., New York, NY).

Part 7

Social Activism, Sisters in Arms

Cupcakes and Chemical Composition: Ida Freund's Legacy

Donald L. Opitz

School of Continuing and Professional Studies,
DePaul University, Chicago, USA

The periodic system of elements formulated by Dmitri Mendeleev (1834–1907) is widely recognized within the public sphere today largely thanks to the impact of innovative school teaching. One of the most durable object lessons has been the didactic use of cupcakes to depict the elements, a method invented in 1908 by Ida Freund (1863–1914), the first woman to be appointed to a university lectureship in chemistry in the United Kingdom. Cupcakes notwithstanding, Freund made a significant impact in chemical education, and she ardently fought for women's professional status in chemistry on par with men's. During her career, she advanced the collegiate science education of women in the UK at a time when women's educational opportunities in science were relatively limited. Beyond her lifetime, Freund's books remained influential for many decades, for their value both in teaching and in promoting the history of chemistry. Her excellent exposition of Mendeleev's achievement reached wide audiences as a result of the popularity of her writing.

This chapter will emphasise these themes in Freund's life and work, showing how her accomplishments as a pioneering woman chemist in the UK contributed to the dissemination of Mendeleev's ideas, particularly among women entering chemistry. Her staunch defence of women's abilities amid popular scepticism advanced women's science education along lines that preserved the integrity of science curriculum alongside the rise of "domestic science" courses in the early twentieth century. Ironically, Freund's renown today stems from her confectionary object lesson, a periodic table of cupcakes, now popularly employed in chemical education contexts worldwide.

Biographical Sketch

Born to Austrian-Jewish parents, Fedor and Sophie Freund, Ida (Figure 1) was educated at state schools in Vienna, including a training college for teachers. At a young age, she had a cycling accident that resulted in the amputation of one of her legs, a disability she overcame with the help of an artificial leg and a tricycle she propelled with her arms. Also while she was young, her mother died prematurely, and her maternal grandparents raised her to adulthood. After her grandmother died, her maternal uncle, the violinist Ludwig Strauss (1892–1953), brought her to the UK, where she became a naturalised citizen. Despite Freund's own protests, her uncle enrolled her at one of Cambridge's new colleges for women, Girton College, in 1882, to study natural science [Hill and Dronsfield, 2004; Ogilvie, 2004; Rubinstein, 2011]. This also afforded her the opportunity to study in Cambridge's Cavendish Laboratory, famous as a site for experimental research in physics and just recently opened to women students by the director, John William Strutt, third Baron Rayleigh (1842–1919) [Gould, 1997]. In 1886 she sat for Cambridge's natural sciences tripos examination — the final examination taken by students at the end of their programmes of study. She placed in the first class in both parts of the examination, with the second part focussing on chemistry.

Upon completing her studies, Freund secured a one-year lectureship at the Cambridge Training College for Women. In 1887 she accepted a

Figure 1. Ida Freund, portrait by Window & Grove, date unknown (reproduced by permission of the Principal and Fellows, Newnham College, Cambridge).

position as a demonstrator in chemistry at Newnham College, Cambridge. In 1893, she was promoted to lecturer in chemistry, a position she held until she retired in 1912. She also served on the Newnham College Council, first as an associate and then as a full member. As a result of complications from an operation, she died in 1914 [Gardner, 1914, p. 37; White, 1973, p. 7].

Despite her heavy teaching responsibilities, Freund carried out original research in physical chemistry. She published her study on the effects of temperature on the volume of salts neutralised in solution at different concentrations in the leading German periodical *Zeitschrift für physikalis-*

che Chemie [Freund, 1909]. While this research was underway, she presented an essay on "double salts" to the Cambridge University Chemistry Club, afterwards published in *Science Progress in the Twentieth Century* [Freund, 1907]. She also won Cambridge's Gamble Prize in 1903 for her essay on the early history of atomic theory, a subject she treated more fully in her landmark text, *The Study of Chemical Composition: An Account of its Method and Historical Development*, published in the following year [Rayner-Canham and Rayner-Canham, 2008, p. 228; Freund, 1904]. She distinguished herself as the sole woman chemist who contributed to the 1889 edition of Watts' *Dictionary of Chemistry*, in authoring the essay on relative densities [Muir and Morley, 1889, pp. 370–377].[1] Reviews noted her singular contribution in gendered terms; the noted chemist Henry Roscoe [1889, p. 641] remarked, "the presence of articles from the pen of a lady — Miss Ida Freund, of Newnham — indicates that scientific research and exposition are no longer to be confined to the hands and heads of the so-called stronger sex."

Freund and Women's Science Education

Roscoe's attention to Freund's "sex" positioned her as an anomalous woman chemist of the time. More generally, women enjoyed few opportunities to conduct scientific research beyond informally collaborating with established men of science, mostly male family members and friends, which was not unusual in Victorian Cambridge [Macleod and Moseley, 1978]. Freund's appointment at Newnham College, founded during a broader expansion of women's access to higher education, provided her with facilities for conducting research while she taught chemistry to generations of women students. The laboratories at Girton and Newnham, modest in terms of space and instrumentation, nevertheless enabled women science students at Cambridge to gain practical experience before their admittance to the University Chemical Laboratory, permitted only after passing the first part of the natural sciences tripos examination. The chemical laboratory Freund directed, Newnham's "Old Laboratory," closed upon her retirement in 1912 (Figure 2). It was reopened in the

---

[1] See also Rae on Four Women Chemists in this volume.

Figure 2. Ida Freund (wearing hat) in the "Old Laboratory," with three students (identities unknown), c. 1910 (reproduced by permission of the Principal and Fellows, Newnham College, Cambridge).

1990s as a site for public concerts [McWilliams Tullberg, 1998; Freund, 1920, p. v; Richmond, 1997; Ogilvie, 2004].

However, educators debated over the suitability of standard science courses for girls and women, with many arguing that "domestic science" (in different contexts also called "household science" and "home economics") was more gender appropriate. At an education conference in 1911, Freund vehemently opposed the idea that domestic science could serve as an effective vehicle for educating girls in science. She conveyed her position in poignant terms in letters and essays printed in a range of periodicals, particularly *Education*, *The School World*, *The Englishwoman*, and *Common Cause* (a suffragist magazine). A typical expression of her view strongly demarcated science teaching from housekeeping concerns: "[A] large number of experienced women science-teachers, among whom not a few can claim first-hand experience of household work and household management … deny the possibility that 'science can be directly and adequately taught in the kitchen'" [Faithfull *et al.*, 1912, p. 453].[2] Freund consistently defended a basis for women's science education equal to men's, and she advocated for improving what she found to be substandard curricula in girls' schools, through strategies that included her own summer courses for teachers at Newnham. Begun in 1897, these laboratory-based courses led teachers in the construction of equipment and the design of experiments that could then be applied in their schools [Blakestad, 1994, pp. 151–177; Dyhouse, 1977; Rayner-Canham and Rayner-Canham, 2011; Freund, 1911].

Beyond these education debates, she also pressed the case for equity in the profession more generally, for instance when she collaborated with her former student Ida Smedley Maclean (1877–1944), and her colleague Martha Annie Whiteley (1866–1956), to pressure the Chemical Society of London to admit qualified women as fellows. Although their petition, signed by 19 women chemists, initially met resistance, ultimately their campaign succeeded. The society elected Maclean as its first woman fellow in 1920 — unfortunately not in time for Freund to witness before her death [Creese, 1991, p. 283; Creese, 1997; Mason, 1991; Rayner-Canham and Rayner-Canham, 2008, pp. 72–79].

Influence through Teaching and Textbooks

In relation to the chemical elements and the periodic system, Freund's contributions are those of an influential educator and expositor. She opened the field of chemistry to generations of women students, several of whom achieved noted careers. A prime example is Maclean, who for many years held a research position at the prestigious Lister Institute in London, where she became an expert on the synthesis of animal fat and its metabolism. Another success story is that of Mary Beatrice Thomas (1873–1954), who held teaching positions at Royal Holloway College and Girton College, Cambridge, serving for many years as Girton's Director of Studies in Natural Sciences [White, 1973, p. 127]. As a tribute to her teacher, Thomas co-edited, with Cambridge mineralogist Arthur Hutchinson (1866–1937), Freund's second book for posthumous publication, which

---

[2] See also Lundgren in this volume.

Freund had left only half-finished [Freund, 1920]. Her text, *The Experimental Basis of Chemistry: Suggestions for a Series of Experiments Illustrative of the Fundamental Principles of Chemistry*, provided ample lessons for use by other chemistry teachers, emphasising empirical, practical work to promote students' grasp of chemical principles. Also after Freund's death, Thomas founded the Ida Freund Memorial Fund, endowed to provide scholarships for women science teachers who wished to bolster their training through summer courses in the spirit of Freund's own earlier efforts [White, 1973, p. 127; Rayner-Canham and Rayner-Canham, 2008, pp. 226–229].

Through her teaching and widely circulated texts, Freund impressed upon students the important interplay between theory and experiment. She illustrated this for the periodicity of elements, precisely at a moment when chemists were rapidly discovering new elements. As she wrote in *The Study of Chemical Compositions*: "I endeavoured to keep sharp and clear the boundary between facts and hypotheses," particularly for subjects based on "the atomic and molecular theory," which was then undergoing much refinement as a result of new discoveries [Freund, 1904, p. v]. She devoted a full chapter to "Mendeleeff [*sic*] and the Periodic Law," providing a clear exposition of the historical background to Mendeleev's crowning achievement. She judged Mendeleev's genius superior to other contenders, especially Julius Lothar Meyer (1830–1895), a critic of Mendeleev's periodic law. She summarized her assessment in the following passage, which also conveys the high calibre of her own critical acumen:

> Comparison of Mendeeff's and Lothar Meyer's treatment of the inductive part of the subject shows that whilst the latter pays greater attention to the consideration of the physical properties, the former devotes himself more to a comparative study of the chemical relations … . But it is especially in the deductive application of the system, that we find the Russian scientist much in advance of the German; the scope of the phenomena encompassed, the definiteness and lucidity of the reasons adduced for the conclusions arrived at, the number and importance of the predictions made together with the marvellous way in which these have been verified, have combined to make this part of Mendeleeff's work one of the greatest scientific achievements of the century, one of the most striking confirmations of the modern method [Freund, 1904, p. 474].

In print for many decades, her book was read by generations of chemistry students worldwide. Historians of chemistry have recognized its value as a pioneering history of the field, providing an early manifesto for chemistry as a modern science [Bensaude-Vincent, 1996, p. 2; Siegfried, 2002, p. 1].

Periodic Table of Cupcakes

Ironically, however, the lesson most remembered of Freund's teaching is her preparation of a periodic table made of confectionary goods and cupcakes — products typically associated with the domestic realm. Despite this apparent breach of her principled separation of science education from the kitchen, the lesson illustrated Freund's application of a standard science pedagogy, one emphasising tactile learning with familiar objects [Keene, 2014]. More importantly, her lessons generated a positive spirit among her students in the days leading up to the tripos examinations, an outcome conveyed in the reminiscence of Hilda Wilson (1886–1970), who sat for the examination in 1908, with the focus being Mendeleev's periodic system:

> Every year just before the Tripos she would summon her Chemistry students to do some special study. It was of course a hoax. In 1907 she urged them to go to the lab, to study again the lives of certain Chemists. They found large boxes of lovely chocolates ... with a different life-history and picture of some famous Chemist in each. In my year [1908] we were requested to go and make a further study of the 'Periodic Table of the Elements'. We found a very large board with the Table set out. The divisions across and down were made with Edinburgh Rock [a confection], numbers were made of chocolate, and the elements were iced cakes each showing its name and atomic weight in icing. The nonvalent atoms were round, univalent had a protruding corner, bivalent two, trivalent triangular, and so on [Phillips, 1988, p. 72].

Freund's lecture accompanying the object lesson pointed out how this version of Mendeleev's table contained more information than usual and would be worthy of her students' further contemplation:

> Your attention is drawn to the desirability — in fact the necessity — of perfecting your knowledge of the Periodic System of Classification of

the Elements. Whether considered from the point of view of theoretical or descriptive and classificatory Chemistry, Mendeleef's [*sic*] system demands extensive and detailed knowledge ... . This table, whilst in the main following the usual lines, tries to bring out, by means of a tentative symbolism, more facts than it is usual to try and convey. Whether however it is of a kind that would lend itself to extended use as an adjunct to the study of Chemistry must be considered doubtful [Phillips, 1988, pp. 72–73].

Despite her scepticism about the "extended use" of her lesson beyond this instance at Cambridge in 1908, her idea did in fact achieve wide popularity, and it remains a favourite object lesson in chemistry today. Images of modern-day cupcake periodic tables are readily discoverable on the Internet, with the occasions ranging from school chemistry classes, to birthday parties of chemistry nerds, to chemical society events, such as the Royal Society of Chemistry's launch of its Visual Elements Periodic Table in 2012 [Salisbury, 2013; Walker 2012]. Further popularising the invention, American baker Rosanna Pansino offered a recipe for the "Periodic Table of Cupcakes" in *The Nerdy Nummies Cookbook* [Pansino, 2015, pp. 52–57]. Associations of the periodic system with the domestic and familiar thus remained durable in the teaching and popularisation of chemistry to today.

References

Bensaude-Vincent, B. and Stengers, I. (1996). *A History of Chemistry* (Harvard University Press, Cambridge, MA).

Blakestad, N. (1994). King's College of Household & Social Science and the Household Science Movement in English Higher Education *c.* 1908–1939, unpublished PhD thesis (University of Oxford, Oxford).

Creese, M. R. S. (1991). British Women of the Nineteenth and Early Twentieth Centuries Who Contributed to Research in the Chemical Sciences, *The British Journal for the History of Science*, 24, pp. 275–305.

Creese, M. R. S. (1997). Martha Anna Whitely (1866–1956): Chemist and Editor, *Bulletin for the History of Chemistry*, 20, pp. 42–45.

Dyhouse, C. (1977). Good Wives and Little Mothers: Social Anxieties and the Schoolgirl's Curriculum, 1890–1920, *Oxford Review of Education*, 3, pp. 21–35.

Faithfull, L. M., *et al.* (1912). Science in Girls' Schools, *The School World*, 14, pp. 452–465.

Freund, I. (1904). *The Study of Chemical Composition: An Account of its Method and Historical Development* (Cambridge University Press, Cambridge).

Freund, I. (1907). Double Salts, *Science Progress in the Twentieth Century: A Quarterly Journal of Scientific Thought*, 2, pp. 135–159.

Freund, I. (1909). Der Einfluss der Temperatur auf die Volumenänderung bei der Neutralisation für verschiedene Salze bei verschiedenen Konzentrationen, *Zeitschrift für physikalische Chemie*, 66, pp. 555–613.

Freund, I. (1911). Domestic Science — a Protest, *The Englishwoman*, 10, pp. 147–163, 279–296.

Freund, I. (1920). *The Experimental Basis of Chemistry: Suggestions for a Series of Experiments Illustrative of the Fundamental Principles of Chemistry* (Cambridge University Press, Cambridge).

Gardner, A. (1914). Miss Ida Freund, *The Cambridge Review*, 35, pp. 470–471.

Gould, P. (1997). Women and the Culture of University Physics in Late Nineteenth-Century Cambridge, *British Journal for the History of Science*, 30, pp. 127–149.

Hill, M. and Dronsfield, A. (2004). Ida Freund: Pioneer of Women's Education in Chemistry, *Education in Chemistry*, 41, pp. 136–137.

Keene, M. (2014). Familiar Science in Nineteenth-Century Britain, *History of Science*, 52, pp. 53–71.

Macleod, R. and Moseley, R. (1978). Fathers and Daughters: Reflections on Women, Science, and Victorian Cambridge, *History of Education*, 8, pp. 321–333.

Mason, J. (1991). A Forty Years' War, *Chemistry in Britain*, 27, pp. 233–238.

McWilliams Tullberg, R. (1998). *Women at Cambridge* (Cambridge University Press, Cambridge).

Muir, M. M. P. and Morley, H. F. (1889). *Watts' Dictionary of Chemistry* (Longmans, Green, and Co., London).

Pansino, R. (2015). *The Nerdy Nummies Cookbook: Sweet Treats for the Geek in All of Us* (Simon and Schuster, New York, NY).

Phillips, A., ed. (1988). *A Newnham Anthology* (Newnham College, Cambridge).

Rayner-Canham, M. and Rayner-Canham, G. (2008). *Chemistry Was Their Life: Pioneering British Women Chemists, 1880–1949* (Imperial College Press, London).

Rayner-Canham, M. and Rayner-Canham, G. (2011). The Rise and Fall of Domestic Chemistry in Higher Education in England During the Early 20th Century, *Bulletin for the History of Chemistry*, 36, pp. 35–42.

Richmond, M. L. (1997). "A Lab of One's Own": The Balfour Biological Laboratory for Women at Cambridge University, 1884–1914, *Isis*, 88, pp. 422–455.

Roscoe, H. E. (1889). Watts' Dictionary of Chemistry, *Nature*, 40, pp. 640–641.

Rubinstein, W. D. (2011). Freund, Ida, *The Palgrave Dictionary of Anglo-Jewish History*, eds. Rubinstein, W. D., Jolles, M. and Rubinstein, H. L. (Palgrave Macmillan, Basingstoke), p. 301.

Salisbury, H. (2013). Ida Freund, *175 Faces of Chemistry: Celebrating Diversity in Science* (http://www.rsc.org/diversity/175-faces/all-faces/ida-freund/, last accessed 23 November 2018).

Siegfried, R. (2002). *From Elements to Atoms: A History of Chemical Composition* (American Philosophical Society, Philadelphia, PA).

Walker, R. (2012, January 1). Visual Elements Periodic Table, *Education in Chemistry* (https://eic.rsc.org/news/visual-elements-periodic-table/2020951. article, last accessed 1 December 2018).

White, A. B. (1973). *Newnham College Register*, 2nd ed. (Newnham College, Cambridge).

Select Bibliography

Creese, M. R. S. (1991). British Women of the Nineteenth and Early Twentieth Centuries who Contributed to Research in the Chemical Sciences, *British Journal for the History of Science*, 24, pp. 275–305.

Kaji, M., Kragh, H. and Palló, G. (2015). *Early Responses to the Periodic System* (Oxford University Press, Oxford).

Rayner-Canham, M. and Rayner-Canham, G. (2008). *Chemistry Was Their Life: Pioneering British Women Chemists, 1880–1949* (Imperial College Press, London).

Alice Hamilton, Confronting the Dangers of Lead

Mary Mark Ockerbloom

Science History Institute, Philadelphia, PA 19119, USA

In 1908, when Alice Hamilton (1869–1970) (Figure 1) wrote her first article on industrial poisons, industrial medicine was largely non-existent in the United States — along with safety standards, labour laws, social insurance and workers' compensation. Lead and other suspected toxic substances were widely used in American industry, and their potential dangers were widely ignored. The accepted American narrative was that American industry was progressive: that American factories were larger, better and more productive; and that American workers were healthier, better paid and better fed than those in other countries. Hamilton questioned such beliefs and empirically tested them. Her work led to widespread change in the United States.

Becoming Alice Hamilton

Hamilton's career was formed by two major influences. The first influence was her medical training. In 1893, with limited career options, Hamilton earned a medical degree from the University of Michigan. The

Figure 1. Alice Hamilton (Acc. 90–105 — Science Service, Records, 1920s–1970s, Smithsonian Institution Archives; public domain).

University of Michigan, in the mid-western United States, was founded as a public research university in 1817. It began to accept women as undergraduates in 1870. Michigan was much less conservative than older universities in the eastern United States like Harvard (established 1636) and Yale (1701). Hamilton found less overt sexism at the University of Michigan than in eastern schools [Musil, 2014, pp. 71–80]. She learned clinical and research skills from "a remarkable group of men" [Hamilton, 1943, p. 39] including the pharmacologist John Abel (1857–1938), the physiologist William Howell (1860–1945), the bacteriologist Fred Novy (1864–1957), the biochemist Victor Vaughan (1851–1929) and the medical doctor George Dock (1860–1951). When she joined the Woman's Medical School of Northwestern University in Chicago in 1897 as a professor of pathology, her research interests were clearly formed: bacteriology, pathology and public health [Hamilton, 1943, pp. 39–53].

The second major influence on Hamilton's career was her residence at Hull House. Hull House was a leader in the settlement movement, which sought to improve urban conditions by reaching across economic and social barriers. As part of the settlement movement, people who were wealthy or well educated chose to live in settlement houses in poor, often diverse neighbourhoods with the intention of using their knowledge and resources for the community's welfare. Settlement houses like Hull House were an important force for change during the Progressive Era. Jane Addams (1860–1935), who co-founded Hull House in Chicago's nineteenth ward in 1889, described its goals as residence, research and reform. Living in a poor area supported "close cooperation with neighborhood people, scientific studies of the causes of poverty and dependence, communication of the facts to the public, and persistent pressure for reforms" [Wade, 1967, p. 414]. Accepted as a resident at Hull House in 1897, Hamilton learned to deal with the diverse population in the ward and with wealthy influential visitors to Hull House [Hamilton, 1943, pp. 54–64; Stebner, 1997, pp. 128–131].

Residence

By contributing to Hull House activities as a doctor and "woman-of-all-work," [Sicherman, 1984, p. 118] Hamilton learned about the daily lives of African Americans and immigrants in her neighbourhood. Some were workers in the dangerous trades, occupations in which high rates of sickness or mortality were suspected of relating to working conditions [Hamilton, 1943, p. 115]. Thomas Oliver's comprehensive work *Dangerous Trades* (1902) introduced Hamilton to British and European research on the subject [Sicherman, 1984, p. 153].

Of all the dangerous substances used in the dangerous trades, lead was one of the earliest elements to be identified as toxic. As far back in time as ancient Rome, the symptoms of acute lead toxicity were described among the "diseases of slaves" [Hamilton, 1943, p. 122]. With the industrial revolution, new manufacturing processes increased the number of people exposed to such hazardous substances. Workers were readily available, and those who became ill or died were easily replaced [Hamilton, 1943, p. 151].

Based on what she was seeing on the Chicago streets, and learning from international sources, Hamilton recognised that working conditions in America (Figure 2) were far behind the standards achieved in

Figure 2. Alice Hamilton, *Lead Poisoning in the Smelting and Refining of Lead* (Washington, 1914), plate 2: Ore Hearth or Scotch Hearth, p. 22 (Government Printing Office, public domain).

England and Germany. An American factory with 142 men might have twenty-five cases of lead poisoning in a year. A German factory with 150 men had two, and an English factory with 182 men had none [Sicherman, 1984, p. 167].

As an example, Hamilton describes the ore hearth in Figure 2:

[The ore hearth] is protected by a double hood, the flaring outer hood coming over the work plate, but not over the lead well, to the right, nor the car for gray slag, to the left. The photograph was taken purposely during an interval when there was no slag on the work plate, so that fumes would not obscure the view [Hamilton, 1914, p. 22].

Research

In 1907 the Illinois General Assembly authorised the creation of the Illinois State Commission on Occupational Diseases, reporting to Governor Charles Deneen (1863–1940) [Webster, 1911, pp. 5–6]. The chairman of the commission, Charles Henderson (1843–1915), was familiar with Hamilton's work. On his recommendation, Hamilton was appointed to the commission in 1908, with Henderson and seven other men [Musil, 2014, pp. 76–77]. Charged to report on occupational illness, they focused on the narrower and more easily defined scope of industrial poisons [Hamilton, 1943, p. 119]. In 1910 Hamilton resigned as commissioner to take a new position of chief medical examiner for the Commission's Survey of Occupational Diseases [Moye, 1986, p. 24]. As chief examiner, she managed a team of twenty-three official "assistants, doctors, medical students, and social workers" and many more contributors [Hamilton, 1943, p. 119; Sicherman, 1984, pp. 156–158]. She made the investigation of lead her personal responsibility. Lacking diagnostic tests to measure the presence of lower levels of lead in the body, Hamilton used a clinical sign of chronic lead poisoning as a standard indicator. She looked for the presence of blueish-black lines along the edge of the workers' gums, at the base of the teeth [Hamilton, 1943, pp. 138–139]. She reported over 500 confirmed cases using this diagnostic indicator as a standard, and described industry conditions as similar to those in France a century earlier [Kovarik, 2005, p. 385].

The Illinois Commission's Survey of Occupational Diseases, led by Hamilton, was the first of its kind in the United States [Sicherman, 1984, p. 4]. The Commission's report became a model for further work and legislation at both state and federal levels. In 1911 the federal commissioner of labour, Charles Neill (1865–1942), asked Hamilton to investigate lead use on a national level. The result was the first of many reports that Hamilton would produce as a federal special investigator during her career. Several of her reports focused specifically on risks to women workers, including *Women in the Lead Industries* (1919) [Hamilton, 1919; Sicherman, 1984, p. 159].

Hamilton used similar tactics in state and federal investigations. Doing what Alexander Langmuir (1910–1993) would later call "shoe-leather epidemiology" [Schaffner and LaForce, 1996, p. S17], Hamilton painstakingly gathered information through face-to-face interviews in homes, union halls, bars, hospitals, dispensaries and factories. Each worker or factory visited offered opportunities to collect gossip about others. Hamilton had to identify the processes and chemicals used in different industries, understand their effects on workers, discover the locations of plants, trace the sick or dead who had worked there, and interview them and their families. Without any legal right of entry, she had to talk her way into each plant to see conditions, alert for dust, fumes, illness and misdirection [Sicherman, 1984, pp. 166–169]. At one plant in the mid-west, a worker's wife told Hamilton that ceilings had been torn out and sick men sent home before her visit [Moye, 1986, p. 26]. She carefully recorded and tabulated her findings, giving details of medically diagnosed cases and correlating illnesses with factories and processes [Sicherman, 1984, p. 158; Ginsberg, 2002, p. 2].

Careful examination of hospital records could suggest possible symptoms of lead poisoning or reveal the plant where a patient had worked. A Carnegie Steel Mill in Pittsburgh, Pennsylvania had so many injuries that the local emergency room at West Penn Hospital had a rubber stamp made to record its name in their records [Hamilton, 1943, pp. 132–133]. Hamilton might find the location of a factory by following workers who showed signs of illness, or looking for smoke stacks and pollution. In one case, Hamilton found a workman with signs of neurological damage through a hospital. A Polish immigrant, the man had worked in a sanitary-ware factory enamelling bathtubs. Hamilton was unaware that such work involved lead, and the factory processes she was initially shown did not use it. Hamilton revisited the sick man and determined that he had worked in a different, more hazardous location.

> I found it and discovered that enameling means sprinkling a finely ground enamel over a red-hot tub where it melts and flows over the surface. I learned that the air is thick with enamel dust and that this may be rich in red oxide of lead. A specimen of it which I secured from a workman, who said he often took some home to his wife for scouring pans and knives, proved to contain as much as 20 per cent soluble lead. ... Thus I nailed down the fact that sanitary-ware enameling was a dangerous lead trade in the United States [Hamilton, 1943, p. 121].

Reform

Hamilton was paid piecemeal for her government reports [Moye, 1986, p. 25], but her self-appointed job was to reform working conditions. She gave the companies she visited complete reports of her findings, with extensive recommendations for improvement. She had the rare ability of connecting with someone and explaining what they were doing badly, without alienating them. She approached people with the expectation that they would act in good faith, although her letters show that she was not always as optimistic as she chose to appear. She used publicity as a potent weapon for change [Sicherman, 1984, pp. 232, 313].

Hamilton's impact was dramatic. Manufacturers and industrialists implemented her recommendations, doctors were trained and hired, insurance companies demanded safer conditions, and workers and their families were better treated [Sicherman, 1984, pp. 153–159, 359]. Hamilton credited some of her effectiveness to Julia Lathrop of Hull House:

> Julia Lathrop never roused one to a fighting pitch, but then fighting was not her method. … She taught me a much-needed lesson, that harmony and peaceful relations with one's adversary were not in themselves of value, only if they went with a steady pushing of what one was trying to achieve. So often, when I have succeeded in breaking down the hostility of an employer and in establishing a friendly relation with him, I have been tempted to let it go at that, to depart without risking unpleasantness. Then I have remembered Julia Lathrop and have forced myself to say the unpleasant things which had to be said [Hamilton, 1943, p. 63].

In 1919, Hamilton's expertise was recognized when she was the first woman to be hired as a professor at Harvard, in the new field of industrial medicine. In 1920, with funding from the American Institute of Lead Manufacturers, Hamilton studied the metabolism of lead in the human body [Moye, 1986, p. 26]. She showed that lead accumulated in bones and tissue, where it tended to remain rather than being metabolised and excreted. The poison's cumulative nature made it particularly insidious. People exposed to low levels of lead might be unaware of it and show few symptoms, until continued exposure reached a level at which severe illness or death occurred. Hamilton's findings made it difficult for lead manufacturers to avoid paying compensation claims and civil damage suits to their workers [Sicherman, 1984, pp. 238–240].

The mechanisms by which lead acted were not yet understood. Nonetheless, Hamilton clearly saw the threat that lead posed to both occupational safety and environmental health. When leaded gasoline was introduced in the 1920s, she spoke out strongly against its use, raising concerns that would later be shown to be well-founded. "You may control conditions within a factory, but how are you going to control the whole country?" [Rosner and Markowitz, 1985, p. 349]. She would later write:

> I am not one of those who believe that the use of this leaded gasoline can ever be made safe. No lead industry has ever, even under the strictest control, lost all its dangers. Where there is lead some case of lead poisoning sooner or later develops, even under the strictest supervision [Rosner and Markowitz, 1985, p. 349].

In 1925 Hamilton published *Industrial Poisons in the United States*, a landmark book codifying knowledge in her field. Almost half of the book discusses lead and its effects on human health. The book also includes chapters on the effects of the elements arsenic, mercury, copper, zinc, manganese, antimony, cadmium, selenium, tellurium, vanadium and phosphorus, and of common compounds of these elements, as well as of more complex chemical compounds. Hamilton later published *Industrial Toxicology* (1934) and an autobiography, *Exploring the Dangerous Trades* (1943) [Sicherman, 1984, p. 240; Hamilton, 1925, pp. ix–x: 1934; 1943].

***

Hamilton is recognised as a pioneer of occupational epidemiology and industrial hygiene. Part detective, part scientist and part investigative journalist, she was an exceptional communicator. Her work led to new legislation, new insurance requirements, modifications to industrial processes, and the creation of an entire field of scientific research and medical practice in the United States [Sicherman, 1984, pp. 1–5, 180–181]. Hamilton said of her work:

> It was pioneering, exploration of an unknown field. No young doctor nowadays can hope for work as exciting and rewarding.

> Everything I discovered was new and most of it was really valuable [Hamilton, 1943, p. 121].

Hamilton addressed issues that we still struggle with: the need for sound scientific evidence as a basis for decision making, the dangers of conflict of interest, the role of government in establishing and enforcing safety standards and regulating industry, and the importance and difficulty of determining and acting in support of the public good. These issues are relevant today as we face problems of climate change, scientific credibility, industrial responsibility, and the balance between business and government.

References

Ginsberg, J. (2002). *Alice Hamilton and the Development of Occupational Medicine. A National Historic Chemical Landmark* (American Chemical Society, Washington, DC).

Hamilton, A. (1914). *Lead Poisoning in the Smelting and Refining of Lead: Bulletin of the United States Bureau of Labor Statistics, No. 141* (Government Printing Office, Washington, DC).

Hamilton, A. (1919). *Women in the Lead Industries: Bulletin of the United States Bureau of Labor Statistics, No. 253* (Government Printing Office, Washington, DC).

Hamilton, A. (1925). *Industrial Poisons in the United States* (Macmillan, New York, NY).

Hamilton, A. (1934). *Industrial Toxicology* (Harper, New York, NY).

Hamilton, A. (1943). *Exploring the Dangerous Trades: The Autobiography of Alice Hamilton, M.D.* (Little, Brown, Boston, MA).

Kovarik, W. (2005). Ethyl-Leaded Gasoline: How a Classic Occupational Disease Became an International Public Health Disaster, *International Journal of Occupational and Environmental Health*, 11, pp. 384–397.

Moye, W. T. (1986). BLS and Alice Hamilton: Pioneers in Industrial Health, *Monthly Labor Review*, 109:6, pp. 24–27.

Musil, R. K. (2014). Don't Harm the People: Ellen Swallow Richards, Dr. Alice Hamilton, and Their Heirs Take On Polluting Industries, in Musil, R. K., *Rachel Carson and Her Sisters: Extraordinary Women Who Have Shaped America's Environment* (Rutgers University Press, New Brunswick, NJ), pp. 53–88.

Rosner, D. and Markowitz, G. (1985). A 'Gift of God'? The Public Health Controversy over Leaded Gasoline during the 1920s, *American Journal of Public Health*, 75:4, pp. 344–352.

Schaffner, W. and LaForce, F. M. (1996). Training Field Epidemiologists: Alexander D. Langmuir and the Epidemic Intelligence Service, *Am. J. Epidemiol*, 144:8 (suppl.), pp. S16–S22.

Sicherman, B. (1984). *Alice Hamilton: A Life in Letters* (Harvard University Press, Cambridge, MA).

Stebner, E. J. (1997). *The Women of Hull-House: A Study in Spirituality, Vocation, and Friendship* (SUNY Press, Albany, NY).

Wade, L. C. (1967). The Heritage from Chicago's Early Settlement Houses, *Journal of the Illinois State Historical Society (1908–1984)*, 60:4, pp. 411–441.

Webster, G. W., ed. (1911). *Report of Commission on Occupational Diseases to His Excellency Governor Charles S. Deneen* (Warner Printing Company, Chicago, IL).

Select Bibliography

Ginsberg, J. (2002). *Alice Hamilton and the Development of Occupational Medicine: A National Historic Chemical Landmark* (American Chemical Society, Washington, DC).

Hamilton, A. (1925). *Industrial Poisons in the United States* (Macmillan, New York, NY).

Hamilton, A. (1934). *Industrial Toxicology* (Harper, New York, NY).

Hamilton, A. (1943). *Exploring the Dangerous Trades: The Autobiography of Alice Hamilton, M.D.* (Little, Brown, Boston, MA).

Oliver, T. (1902). *Dangerous Trades: The historical, Social, and Legal Aspects of Industrial Occupations as Affecting Health, by a Number of Experts* (E.P. Dutton, New York, NY and J. Murray, London).

Sicherman, B. (1984). *Alice Hamilton: A Life in Letters* (Harvard University Press, Cambridge, MA).

Stebner, E. J. (1997). *The Women of Hull-House: A Study in Spirituality, Vocation, and Friendship* (SUNY Press, Albany, New York, NY).

Wade, L. C. (1967). The Heritage from Chicago's Early Settlement Houses, *Journal of the Illinois State Historical Society (1908–1984)*, 60:4, pp. 411–441.

Gertrud Johanna Woker (1878–1968) and Tetraethyl Lead

Gisela Boeck

Institute of Chemistry, University of Rostock,
D-18051 Rostock, Germany

The element lead with the symbol Pb (from Latin "plumbum"), with a relative atomic mass of 207,2 is counted among the heavy metals. Lead's atomic number is 82, and it can be found in Group 14 and the sixth period of the periodic system. Three stable isotopes exist: ^{206}Pb, ^{207}Pb and ^{208}Pb. Compared with other metals lead has a low melting point (327,43°C). This blue-white metal can be deformed easily. In nature it can be found as lead compounds and in ores [Hollemann, 2007, pp. 1002–1041].

Lead was already used in antiquity. The earliest discoveries go back to 6500 B.C. In ancient Rome lead was used in the construction of drinking water pipes, lead roofs, leaded windows, etc. The sweet lead acetate $(Pb(CH_3COO)_2)$ was added to wine [Rumpf, 1973].

In the first century B.C., the Roman writer Vitruvius already showed some concern about the toxicity of lead [Vitruvius, 1964, Liber octavus, VI, 10–11]. Nevertheless, lead was used without any restrictions for millennia. Today, breathing in lead dust is known to be hazardous. Also, several lead compounds are extremely toxic. Lead accumulates in the body, especially in the bones, and it has a long retention time. The toxicity is based on a disturbance of the haemoglobin synthesis.

Thanks to our strict modern regulations concerning chemicals in the work place and beyond, and the replacement of old lead pipes with safe alternatives, acute and chronic lead poisoning are now rarely diagnosed by German doctors. The remaining common sources of lead poisoning are lead-based paints in old houses, car batteries, cosmetics, or toys, which are then subjected to uncontrolled recycling procedures [Busse *et al.*, 2008].

Tetraethyl Lead

The lead concentration in the air worldwide increased greatly between 1920 and 1980. The reason for this was the addition of tetraethyl lead ((C_2H_5)$_4$Pb) to petrol as an antiknock agent. From 1923 onwards, manufacturers including the Refiners Oil Company and the Standard Oil Company used it systematically.

Tetraethyl lead is an oily liquid which is not soluble in water. If it is added to petrol during the combustion in the car engine metallic lead (Pb) and lead (II) oxide (PbO) are produced and accumulate. These substances have the potential of damaging the motor. To avoid potential damage, organic bromine and chlorine compounds were added to the petrol. Pb and PbO react with them to form lead (II) chloride ($PbCl_2$) and lead (II) bromide ($PbBr_2$). Due to the high temperature in the motor, both substances are gases. They are emitted into environment and increase the concentration of lead within it [Breu, 2002]. One of the earliest scientists who, as early as in 1932, warned of this antiknock agent's dangers was the Swiss chemist Gertrud Woker [Woker, 1932, pp. 72, 207, 238–240].

Gertrud Woker: Female Chemist, Fighter for Peace and Disarmament, and Women's Rights

The Swiss chemist Gertrud Johanna Woker was born on 16 December 1878 in Bern. After completing her primary and secondary education she passed her final exams at the (boy-only) gymnasium as an external candidate in 1898. From 1898 until 1900 she attended courses to become a teacher. In 1900, Woker began to study chemistry and biology at the University of Bern [Vogt, 2011].

Woker investigated the class of substances called flavones, which are present in nature, e.g., as water-soluble plant pigments. In her dissertation she wrote about the synthesis of 3,4-dioxyflavones [Woker, 1903]. She continued her studies as a visiting student at the University of Berlin between 1903 and 1905. There she enrolled in classes given by the physical chemist Jacobus Henricus van't Hoff (1852–1911) and the pharmacist Hermann Thoms (1859–1931). After returning to Bern she worked as a teacher at a gymnasium, while continuing her scientific work at the same time. In October 1906 she applied for the "Habilitation" in theoretical and physiological-biological chemistry. This was the prerequisite for a professorship.

In 1907 Woker received the "venia legendi," the admission to teach at the university as a "Privatdozent." But the Faculty of Philosophy at the University of Bern rejected her request for the "venia legendi" in physical chemistry, and it was awarded for history of chemistry and physics instead. The Faculty professors explained that the subjects do not play a role in the process, and that Woker was allowed to teach in modern fields. However, a misogynistic sentiment may be detected [von Leitner, 1998, p. 43].

For her inaugural lecture Woker chose a modern topic and described her research program, which was dedicated to problems of catalysis in connection with biochemical problems [Woker, 1907]. From 1911 to 1951 Woker was the head of the physico-chemical biology laboratory at the University in Bern. She would not be named professor here until 1933.

In the 1910s Woker published several papers on enzymes. She was able to demonstrate that these substances accelerate the reaction time of biochemical processes. Her research added to the understanding the catalytic processes in living bodies. In 1910 she was asked to write a book on catalysis — one of the interesting scientific topics at the turn of the century [Ostwald, 2017, pp. 283–293] — in the series *The Chemical Analysis*. Indeed, she published four books on catalysis between 1910 and 1931 [Woker, 1910; 1915; 1924a; 1931]. In the first volume she provided a comprehensive overview over the history of catalysis. Her work clearly shows that she collected a large amount of information on catalytic problems. Her publication discussed the results, sometimes in comparison

Figure 1.   Gertrud Woker (1878–1968) at work, without date (Gosteli-Foundation, Switzerland, photo collection (n.k.) Gertrud Woker; reproduced by kind permission of the Gosteli-Foundation).

with her own findings. The book was appreciated by many scientists. Wilhelm Ostwald (1853–1932), who received the 1909 Nobel Prize in Chemistry for his work on catalysis, recommended it in a letter as a comprehensive collection of facts and materials [von Leitner, 1998, p. 60].

In later years Woker (Figure 1) investigated alkaloids and their biogenesis, and published her results in the 1950s [Woker, 1953; 1956]. Throughout her scientific life she faced the hostility of many of her male colleagues, as has been mentioned above in connection with the "Habilitation." For a long time she did not receive the regular salary of a docent, her requests for an extraordinary professorship were denied, and the work in her laboratory was interfered with [von Leitner 1998, pp. 123–132]. The affronts multiplied around 1915, when she started to advocate against gas warfare, for peace and disarmament, and also for women's rights [von Leitner, 1998, for example pp. 208–209 and 281–302].

From 1915 and for the rest of her life, Woker was a member of the Women´s International League for Peace and Freedom. It was probably because of her membership in this League that she was asked to contribute an autobiographical piece to the first volume of the compilation *Leading Women of Europe*, edited by the journalist Elga Kern (1888–1957) [Woker, 1928].

But Woker was not only a chemist working in the lab, publishing scientific articles and books and speaking at meetings of the Women's International League for Peace and Freedom. She also wrote poems and tales. Gertrud Woker died on 13 September 1968 in Préfargier [von Leitner, 1998; Vogt, 2011; Chiquet, 2017].

Gertrud Woker: The Tetraethyl Lead and the Chemical Weapons

Woker's work on the human body's reactions to chemical substances was an important prerequisite to the evaluation of the terrible consequences of chemical and nuclear poisons, including the effects of tetraethyl lead. Consequently, in her articles and books she was often warning against scientific research for military purposes, and more generally against the militarisation of science, see e.g. [Woker, 1924b and Woker 1925].

In her arguments against the construction of weapons Woker often did not use her own scientific results, but her knowledge of the relevant literature. Her ability to evaluate these results allowed her to outline the consequences of toxic compound use.

Woker knew of the work of Heinrich Zangger (1874–1957), the former director of the Forensic Institute at the University in Zurich. He inferred that tetraethyl lead is 100 times more toxic than the well-known poison strychnine [Zangger, 1925]. Woker discussed his work in her book *Der kommende Gift- und Brandkrieg und seine Auswirkungen gegenüber der Zivilbevölkerung* ("The Coming War of Poison and Fire and its Effects on the Civilian Population") [Woker, 1932, pp. 238–240], and suggested a substitution of tetraethyl lead in petrol with other non-toxic compounds. From the present perspective, with the ongoing debates about the car industry in mind, her statement still rings true:

> For this peaceful purpose enough cheap and harmless additives exist, so there is no benefit to producing a highly toxic, metallorganic

antiknock agent, which is dangerous for the workers who produce it and for the inhabitants of the towns in which the cars exhaust it.[1] [Woker, 1932, p. 239]

Today, tetraethyl lead is no longer used in car petrol as an antiknock agent. As a result the lead concentration in the environment has decreased. Despite the known risks, however, the aviation industry still uses gasoline containing tetraethyl lead.

***

Gertrud Woker was one of the earliest female chemists of the twentieth century who were able to build an academic career and to do scientific research in their own laboratory. Furthermore, she was involved in the Women's Movement and in peace and disarmament activism. Her warnings against the construction and the use of chemical weapons were issued after World War I, at a time when their danger was quite familiar to the public. As a scientist she also felt obligated to discuss the problems of other toxic substances and to present scientific arguments against them based on her profound knowledge of a scientist.

References

Breu, M. (2002). *Bleibenzin — eine schwere Geschichte: Die Geschichte der Benzinverbleiung aus Sicht der Politik, des Rechts, der Wirtschaft und der Ökologie* (Ökom, Munich).

Busse, F. P., Fiedler, G. M., Leichtle, A., Hentschel H. and Stumvoll, M. (2008). Lead Poisoning Due to Adulterated Marijuana in Leipzig, *Deutsches Ärzteblatt*, 105: 44, pp. 757–762.

Chiquet, F. (2017). Gertrud Woker: Geschichte einer Berner Heldin. Eine dokumentarische Videoinstallation (http://fabianchiquet.net/projects/all-0/gertrud-woker-1/, last accessed 1 February 2019).

Hollemann, A. F. (2007). *Lehrbuch der anorganischen Chemie*, 102nd edition (Walter de Gruyter, Berlin and New York, NY).

Hug-Rüegger, H. (1982). Ehre, wem Ehre gebührt! [letter to the editor], *Neue Zürcher Zeitung*, 26 January 1982, p. 47.

Ostwald, W. (2017). *The Autobiography*, eds. Jack, R. S. and Scholz, F. (Springer, Cham).

---

[1] All translations from the German are the author's unless otherwise indicated.

Rumpf, K. (1973). Blei Teil A 1: Geschichtliches, in *Gmelins Handbuch der anorganischen Chemie,* ed. Gmelin-Institut für anorganische Chemie und Grenzgebiete (Springer, Berlin and Heidelberg).

Vitruvius [Pollio, M. V.] (1964). Vitruvii *De architectura libri decem,* ed. Fensterbusch, C. (Akademie-Verlag, Berlin).

Vogt, A. B. (2011). Gertrud Johanna Woker (1878-1968), in *European Women in Chemistry,* eds. Apotheker, J. and Sarkadi, L. S. (Wiley, Weinheim), pp. 65–67.

von Leitner, G. (1998). *Wollen wir unsere Hände in Unschuld waschen? Gertrud Woker (1878–1968), Chemikerin & Internationale Frauenliga 1915–1968* (Weidler, Berlin).

Woker, G. (1903). *Synthesen des 3,4-Dioxyflavons* (Jensen, Bern).

Woker, G. (1907). *Probleme der katalytischen Forschung* (Veit & Comp., Leipzig).

Woker, G. (1910). *Katalyse,* vol. 1 (Enke, Stuttgart).

Woker, G. (1915). *Katalyse,* vol. 2 (Enke, Stuttgart).

Woker, G. (1924a). *Katalyse,* vol. 3 (Enke, Stuttgart).

Woker, G. (1924b). *Wissenschaft und wissenschaftlicher Krieg* (Schweizerische Zentralstelle für Friedensarbeit, Zurich).

Woker, G. (1925). *Der kommende Giftgaskrieg* (Oldenburg, Leipzig).

Woker, G. (1928). Dr. Gertrud Woker, in *Führende Frauen Europas,* ed. Kern, E. (Ernst Reinhardt, Munich), pp. 138–169.

Woker, G. (1931). *Katalyse,* vol. 4 (Enke, Stuttgart).

Woker, G. (1932). *Der kommende Gift- und Brandkrieg und seine Auswirkungen gegenüber der Zivilbevölkerung,* 6th–9th ed. (Oldenburg, Leipzig).

Woker, G. (1953). *Die Chemie der natürlichen Alkaloide,* vol. 1 (Enke, Stuttgart).

Woker, G. (1956). *Die Chemie der natürlichen Alkaloide,* vol. 2 (Enke, Stuttgart).

Zangger, H. (1925): Eine gefährliche Verbesserung des Automobilbenzins, *Schweizerische medizinische Wochenschrift,* 6, pp. 26–29.

Select Bibliography

Breu, M. (2002). *Bleibenzin — eine schwere Geschichte: die Geschichte der Benzinverbleiung aus Sicht der Politik, des Rechts, der Wirtschaft und der Ökologie* (Ökom, Munich).

Chiquet, F. (2017). Gertrud Woker: Geschichte einer Berner Heldin. Eine dokumentarische Videoinstallation (http://fabianchiquet.net/projects/all-0/gertrud-woker-1/, last accessed 1 February 2019).

Hollemann, A. F. (2007). *Lehrbuch der anorganischen Chemie,* 102nd edition (Walter de Gruyter, Berlin and New York, NY).

Muscionico, D. and Sprecher, M. (2011). *Starke Schweizer Frauen: 24 Porträts* (Limmat, Zurich).

Regula, L. (2018). Woker, Gertrud, in *Historisches Lexikon der Schweiz* (HLS) http://www.hls-dhs-dss.ch/textes/d/D9411.php, last accessed 1 February 2019).

Rumpf, K. (1973). Blei Teil A 1: Geschichtliches, in *Gmelins Handbuch der anorganischen Chemie,* ed. Gmelin-Institut für anorganische Chemie und Grenzgebiete (Springer, Berlin and Heidelberg).

Vogt, A. B. (2011). Gertrud Johanna Woker (1878–1968), in *European Women in Chemistry*, eds. Apotheker, J. and Sarkadi, L. S. (Wiley, Weinheim), pp. 65–67.

von Leitner, G. (1998). *Wollen wir unsere Hände in Unschuld waschen? Gertrud Woker (1878–1968), Chemikerin & Internationale Frauenliga 1915–1968* (Weidler, Berlin).

Woker, G. (1928). Dr. Gertrud Woker, in *Führende Frauen Europas,* ed. Kern, E. (Ernst Reinhardt, Munich), pp. 138–169.

Woker, G. (1932). *Der kommende Gift- und Brandkrieg und seine Auswirkungen gegenüber der Zivilbevölkerung*, 6th–9th ed. (Oldenburg, Leipzig).

Zangger, H. (1925): Eine gefährliche Verbesserung des Automobilbenzins, *Schweizerische medizinische Wochenschrift*, 6, pp. 26–29.

Authors' Biographies

Gisela Boeck is an associate professor at the Institute of Chemistry of the University of Rostock. She first graduated in chemistry and then, for her doctorate, specialised in quantum chemistry. Today she teaches general, inorganic and organic chemistry to medical students. She is a board member for the "History of Chemistry" division of the German Chemical Society and represents the Society in the Working Party on the History of Chemistry (EuChemS). Her research interests concern pedagogy, didactics, and the history of chemistry. Boeck's publications include "The Periodic System and its Influence on Research and Education in Germany Between 1870 and 1910" (2015); "Else Hirschberg (1892–1942): The Rediscovery of the Private and Professional Life of the First Female Chemistry Graduate at Rostock University in a Digitised World," (with T. Peppel, 2018); and "Julius Adolph Stöckhardt (1809–1886): Res naturales qua de causa perscrutandae, qua methodo docendae et tractandae, quomodo naturae convenienter disponendae" (with S. H. Michael, 2018).

Patrice Bret is a senior researcher at the Centre Alexandre Koyré and the general secretary of the Comité Lavoisier at the Académie des Sciences in Paris. He has been editing the correspondence of Antoine Lavoisier (as well as that of Guyton de Morveau) since 1993. Bret's early publications investigated the links between science and the military from the late eighteenth to the early twentieth centuries in France, especially the military laboratories from Lavoisier to Le Châtelier. His more recent research includes work on scientific translation (*La traduction au siècle des Lumières: Enjeux et pratiques scientifiques, intellectuels et politiques, 1660–1840* (co-edited with Jean-Luc Chappey, 2017–2018); (co-authored) chapters in *Histoire des traductions en langue française*, vols. on the XVIIth–XXth centuries (2012–2019), and on women in science (*Madame d'Arconville (1720–1805): Une femme de lettres et de sciences au siècle des Lumières* (co-edited with Brigitte Van Tiggelen, 2011); *Femmes de sciences de l'Antiquité au XIXe siècle: Réalités et représentations* (co-edited with Adeline Gargam, 2014); articles in *Dictionnaire des femmes des Lumières* (2015). He is also author of articles and editor of special issues on Guyton de Morveau (*Annales historiques de la Révolution française*, 2016; *SABIX*, 2017) and on industrial chemical heritage (*Patrimoine industriel*, 2016). Bret is currently writing a biography of Mme Picardet.

Joanne A. Charbonneau is a retired professor and previously taught medieval to modern literature, humanities, and women's studies courses at several universities in the US. She has co-authored papers with Richard E. Rice on Ellen Swallow Richards for the 3rd British–North American History of Science Meeting in Edinburgh (1996) and for History of Chemistry symposia at American Chemical Society meetings in New Orleans (1996) and Boston (1998). At the 5th British–North American Joint Meeting of the British Society for the History of Science, Canadian Society for the History and Philosophy of Science and History of Science Society, "Circulating Knowledge," in Halifax (2004), they presented a paper on 19th- and early 20th-century North American women who furthered their study of physics and chemistry in Europe.

Peter Childs has a first degree and a D.Phil. in Chemistry from the University of Oxford. He taught chemistry at university level from 1970 in Uganda, the UK and Ireland, and retired in 2009. He still works closely together with chemistry teachers in Ireland, and founded the magazine *Chemistry in Action!* in 1980, and the annual ChemEd-Ireland conferences in 1982. As well as doing research in chemical education he also has an interest in the history of chemistry, including the language of chemistry. Childs is currently writing a series of articles on pioneers of science education for the journal of the Irish Science Teachers Association. He has written on Kathleen Lonsdale (including in *Lab Coats and Lace*, 2009), on the history of the Irish chemical industry, and on the seaweed industry. He is a former President of the Irish Science Teachers Association and of the Institute of Chemistry of Ireland, and former chairman of the EuCheMS (European Association for Chemical and Molecular Sciences, today: European Chemical Society, EuChemS) Division of Chemical Education. From 2010 to 2019 he has been involved with three EU funded projects to improve the teaching of science.

Kristina Espmark received her PhD in History of Science and Ideas in 2012, at Umeå University. Her dissertation was on Astrid Cleve von Euler, the first woman in Sweden to receive a PhD in science (1898), and focussed on Cleve's research and its development over time. More recently, as part of the multi-disciplinary forest research program Future Forests (a collaboration between the Swedish University of Agricultural Sciences, Umeå University and the Forestry Research Institute of Sweden), Espmark has studied debates about continuous cover forestry in Swedish daily press between 1994 and 2013, aiming for a deeper under-standing of different stakeholders' views on silviculture.

Christian Forstner is a historian of science at the Goethe University in Frankfurt am Main. He graduated in physics, and for his PhD in history of science focussed on the history of quantum mechanics and its interpre-tations during the Cold War. Since 2011 he has been the chairman of the Division for the History of Physics of the German Physical Society. Forstner has authored and edited books on physics and the Cold War, and physics and the military. His latest book *Kernphysik, Forschungsreaktoren und Atomenergie: Transnationale Wissensströme und das Scheitern einer Innovation in Österreich* is currently in press, as is a volume of articles he has edited, titled *Biographies in the History of Physics: Actors, Objects, and Institutions*.

Bretislav Friedrich is the head of the research group "Interactions of Molecules In and With Fields" at the Fritz Haber Institute of the Max Planck Society in Berlin, and an honorary professor at the Technical

University Berlin. After the completion of his PhD in chemical physics he was an Alexander von Humboldt Fellow at the Max Planck Institute in Göttingen and a Senior Research Fellow and Lecturer at Harvard University. Alongside his research in molecular physics (directional states of molecules, quantum computing, analytic solvability, and the structure of solvated molecular complexes) he maintains an abiding interest in the history of science (the emergence of quantum mechanics, the rise of physical and theoretical chemistry, scientific biography) and is engaged in efforts to eliminate chemical and other weapons of mass destruction. Friedrich has (co-)authored and co-(edited) numerous publications, including *Cold Molecules: Theory, Experiment, Applications* (co-edited with Roman Krems and William C. Stwalley, 2009) and *One Hundred Years of Chemical Warfare: Research, Deployment, Consequences* (co-edited with Dieter Hoffmann, Jürgen Renn, Florian Schmaltz and Martin Wolf, 2017). He serves on several editorial boards, among them that of the journal *Frontiers in Chemistry and Chemical Physics* (as specialty chief editor) and as member of the advisory board of the *Springer Briefs in the History of Science and Technology* book series.

Dieter Hoffmann has been a research fellow at the Max Planck Institute for the History of Science since 1995, and was an adjunct professor at the Humboldt University in Berlin from 1995 until his retirement in 2014. He received his diploma in physics from the Humboldt University in Berlin in 1972, and then became a historian of science, finishing his PhD in 1976 and completing his "Habilitation" in 1989. From 1976 to 1990 he was a Research Fellow at the GDR Academy of Sciences in (East)Berlin. In 2002 he was elected a member of the International Academy of the History of Science, and in 2010 of the National Academy of Sciences, Leopoldina; in the same year he was also awarded the DPG "Ehrennadel", an honour awarded by the German Physical Society. Hoffmann's research is on the history of science and physics in the nineteenth and twentieth centuries, in particular scientific biographies and institutional histories, by focussing on Berlin as a leading centre of science and technology. He further works on science in totalitarian regimes, in particular during the Third Reich and in the GDR. His publications include *One Hundred Years of Chemical Warfare: Research, Deployment, Consequences* (co-edited with Bretislav

Friedrich, Jürgen Renn, Florian Schmaltz and Martin Wolf, 2017); *The German Physical Society in the Third Reich. Physicists between Autonomy and Accomodation* (co-edited with Mark Walker, 2011*); One Hundred Years at the Intersection of Chemistry and Physics. The Fritz Haber Institute of the Max Planck Society 1911–2011* (co-authored with Bretislav Friedrich, Jeremiah James and Thomas Steinhauser, 2011); *Max Planck: Annalen Papers* (2008); and *Science under Socialism. East Germany in Comparative Perspective* (co-edited with K. Macrakis (1999).

Sally Horrocks is an associate professor in contemporary British history at the University of Leicester. Since 2011 she has been the senior academic advisor to National Life Stories at the British Library, working on "An Oral History of British Science" and "An Oral History of the Electricity Supply Industry in the UK." Horrocks' research concentrates on the history of women in science and engineering, industrial research, and science and food manufacturing. Published articles include "Doing it for Britain: Science and Service in Oral History with Government Scientists" (With Thomas Lean, 2016); "World War II, Postwar Reconstruction and British Women Chemists" (2011) and "'A Promising Pioneer Profession? Women in Industrial Chemistry in Inter-War Britain" (2000). From 2010 to 2012 she was President of the British Society for the History of Science.

John Hudson spent most of his career teaching chemistry in Cambridge at Anglia Ruskin University. In 1992 he published *The History of Chemistry*, designed as a course text to accompany a module he devised and taught on the BSc chemistry course. His main research interest has been analytical chemistry in nineteenth-century Britain, and in 2012 with Colin Russell he co-authored *Early Railway Chemistry and its Legacy,* which dealt with the activities of chemists employed by the railway companies prior to 1923. Hudson was the Secretary of the Society for the History of Alchemy and Chemistry for nine years, and until recently was Chair of the Historical Group of the Royal Society of Chemistry. He now enjoys retirement in the English Lake District and continues to research aspects of the history of chemistry.

Louis-Pascal Jacquemond is an "inspecteur d'Académie honoraire". He holds a PhD in history and is a "professeur agregé" specialized in the contemporary history of women and science. He is a lecturer at the Institut d'Etudes Politiques de Paris (Sciences Po). He is a member of the Board of Mnemosyne (Association for the Development of Women's and Gender History). He contributed to *La place des femmes dans l'Histoire: Une histoire* (2010), a textbook for teachers that promotes the history of women and gender. He has also contributed to the *Dictionnaire des féministes, France XVIIIème–XXIème siècle*, edited by Christine Bard and Sylvie Chaperon, and to *L'Europe des Femmes XVIIIème-XXIème siècle*, edited by Fabrice Virgili and Julie Le Gac (both 2017). His further publications include *Irène Joliot-Curie: Biographie* (2014) and *L'espoir brisé: 1936, les femmes et le Front populaire* (2016).

Jeffrey Allan Johnson is an emeritus professor of history at Villanova University. He was (2017–2019) a Visiting Scholar in the Research Group on the History of the Max Planck Society at the Max Planck Institute for History of Science, Berlin, studying postwar biochemistry in the Max Planck Society. In 2011–2017 he presided over the Commission on the History of Modern Chemistry (IUHPST). He received his doctorate in modern European history from Princeton University in 1980; his dissertation became *The Kaiser's Chemists: Science and Modernization in Imperial Germany* (1990). Subsequent studies of the history of modern chemistry and its interactions with the chemical industry and the military include *German Industry and Global Enterprise. BASF: The History of a Company* (co-authored with Werner Abelshauser *et al.*, 2004) and *Frontline and Factory: Comparative Perspectives on the Chemical Industry at War, 1914–1924* (co-edited and co-authored with Roy M. MacLeod, 2006). The first of several works on women chemists was a two-part article in *NTM*, "German Women in Chemistry, 1895–1945" (1998). Johnson's next project will be a history of artificial life in fiction and in fact.

Keiko Kawashima is a professor at the Nagoya Institute of Technology. She is the secretary of the Japanese Society for Eighteenth-Century

Studies, and a member of several history of science societies in Japan as well as the Gender History Association of Japan. Her early publications were devoted to the relation between gender and science, especially in eighteenth-century France. She is author of the monograph *Émilie du Châtelet et Marie-Anne Lavoisier* (2013) and of an article of Mme Lavoisier in *Dictionnaire des femmes des Lumières* (2015). Recently, she has also worked on the female scientists around Marie Curie, in particular, the first Japanese female physicist, Toshiko Yuasa, who was disciple of Frédéric Joliot-Curie ("Nobuo Yamada and Toshiko Yuasa: Two Japanese Scientists and the Curie Family" (2017)). Kawashima received the Nao Aoyama Prize (2006) and the Yamazaki Prize (2010) for her studies on gender and science. Her original manga, available on her homepage, are part of her endeavour to make the history of science better known.

Alexander Kraft is a chemist and works as a consultant in the field of electrochromics and smart windows. Between 1998 and 2017 he was active at Gesimat GmbH in Berlin, a company that developed smart switchable electrochromic glazing. He was the co-founder of Gesimat in 1998, and was mainly responsible for research and development. He had previously developed electrochemical water treatment technologies and devices. Kraft finished his PhD on semiconductor electrochemistry at the Humboldt University in Berlin in 1994. He began working in the field of the history of chemistry in 2007, with a focus on the history of Prussian blue, and the history of chemistry in Berlin and in seventeenth- and eighteenth-century Germany. His publications include *Chemie in Berlin. Geschichte, Spuren, Persönlichkeiten*, a book on the history of chemistry in Berlin (2012), as well as articles on Prussian blue, and on the chymists Jacob Roesser (Johann de Monte Raphaim), Dorothea Juliana Wallich, Johann Heinrich Pott, Caspar Neumann and Andreas Sigismund Marggraf.

Thomas Lean is a historian of technology and computing. Since 2009 he has worked for National Life Stories at the British Library, on "An Oral History of British Science" and "An Oral History of the Electricity Supply Industry in the UK," recording and analysing extended "life story" interviews with scientists and engineers in a range of fields. He is the author of *Electronic Dreams* (2016).

Anders Lundgren is a professor emeritus in History of Ideas and Science at Uppsala University. His work focusses on the history of chemistry, the chemical revolution in Sweden, and the relationship between mining and chemistry during the eighteenth century, the significance of textbooks in chemistry, and on pharmaceuticals in Sweden during the twentieth century. He has recently published a monograph on knowledge and the chemical industry in Sweden during the nineteenth century (*Kunskap och kemisk indusri i Sverige under 1800-talet*). His current interests concern the role of smell and taste in chemistry, and the life of a little-known nineteenth-century chemist, Lars Fredrik Svanberg.

Annette Lykknes is a professor of chemistry education and historian of chemistry at NTNU-Norwegian University of Science and Technology. Her research interests include the history of twentieth-century chemistry, the history of education in chemistry, women in science, collaborative couples in the sciences, cultural studies of chemistry and teaching practices in chemistry. For her PhD (2005) she researched the Norwegian radiochemist Ellen Gleditsch. Lykknes has been the vice-chair of the Working Party on History of Chemistry (EuCheMS) since 2011. She is the co-editor (with Donald L. Opitz and Brigitte Van Tiggelen) of *For Better or For Worse: Collaborative Couples in the Sciences* (2012) and co-author (with Joakim Ziegler Gusland) of *Akademi og industri: Kjemiutdanning og -forskning ved NTNU gjennom 100 år*, an anniversary volume on the 100 years of chemistry teaching and research at NTNU (2015).

Paul Merchant is an oral historian and researcher at National Life Stories at the British Library with a background in cultural geography. Since 2009 he has worked on "An Oral History of British Science" and the interdisciplinary project "Science and Religion: Exploring the Spectrum." He has articles drawing on his interviews in *Notes and Records of the Royal Society*, *Oral History*, *Forum d'histoire orale* and *Oral History*.

Claire Murray is an Irish scientist who works at the UK's national synchrotron. Alongside her research interests in crystallography, catalysis, biominerals and supramolecular chemistry, she has developed and delivered events promoting inclusion in science to primary and secondary students, encouraging young people to enjoy and experience science. Her recent national school engagement project, Project M, engaged students at over 100 schools in a real science experiment, and has won awards from the Royal Society of Chemistry and the Costa Foundation. Murray is a member of the UK Research and Innovation External Advisory Group, the International Union of Crystallography Gender Equity and Diversity Committee, and she is a regular panellist for the peer review assessment process for the UK and Ireland Athena Swan Gender Equality Charter mark.

Christer Nordlund is a professor of history of science and ideas at Umeå University and member of the Royal Swedish Academy of Sciences. He was a Visiting Scholar at Uppsala University, the University of Cambridge, the Max Planck Institute for History of Science, and the Swedish Collegium for Advanced Study (SCAS). Nordlund's publications include the monograph *Hormones of Life* (2011) and the anthology *Understanding Field Science Institutions* (co-edited with Helena Ekerholm, Karl Grandin and Patience A. Schell, 2017), and articles in *Annals of Science*, *The British Journal for the History of Science*, *Environmental History*, *History and Technology* and *Studies in the History and Philosophy of Biological and Biomedical Sciences*.

Mary Mark Ockerbloom holds a BA in psychology as well as an MSc in computational science, and is the Wikipedian in Residence at the Science History Institute in Philadelphia, PA. Her position is mission-driven: to improve the science and history of science content on Wikipedia. She corrects errors and writes new Wikipedia articles, releases images from the Science History Institute's collections on Wikimedia Commons, and works with individuals in the Philadelphia area who want to learn more about Wikipedia. Mary became involved in Wikipedia as a result of running *A Celebration of Women Writers*, an online database that links to information about women writers and free online editions of works by women from all over the internet. As the editor of the *Celebration*, she has published over 400 freely readable editions of public domain books by women. Some of the most popular are Margaret Cavendish's *The Blazing World*, Rokeya Sakhawat Hossain's utopia *Sultana's Dream*, Nellie Bly's exposé *Ten Days in a Mad-House*, and Margery Williams Bianco's children's book *The Velveteen Rabbit*. Mary's personal favourites are the accounts of women travellers.

Donald L. Opitz, PhD, is an associate professor in the School of Continuing and Professional Studies at DePaul University, Chicago, with affiliations in DePaul's Department of History and LGBTQ Studies Program. He is the editor-in-chief of the journal *Endeavour*. Publications include the co-edited books *Domesticity in the Making of Modern Science* (2016, with Staffan Bergwik and Brigitte Van Tiggelen) and *For Better or for Worse? Collaborative Couples in the Sciences* (2012, with Annette Lykknes and Brigitte Van Tiggelen). His research areas include women, gender and science, and science in Victorian culture.

Ian Rae is an organic chemist (PhD Australian National University, 1965) who lives in Melbourne, Australia. Ian expanded his latent interest in history of chemistry and chemical technology when he left the laboratory to become dean of a science faculty (1990–1995) and then a university vice-president (1995–1997). Since then he has been an adviser on chemicals in the environment to Australian governments and to the United Nations Environment Programme while developing yet another career as a historian of science. He is an honorary professorial fellow at the University of Melbourne, and an editor of the Australian Academy of Science's journal *Historical Records of Australian Science*.

Marelene and Geoff Rayner-Canham have co-authored a wide range of research papers in the history of science, have been invited contributors to compilations on women in chemistry, and invited speakers at a variety of conferences. They are currently working on a new book: *Pioneering British Women Chemists: Their Lives and Contributions*. **Marelene** is a full-time researcher in the history of women in chemistry since her retirement from the Physics Department of the Grenfell Campus of Memorial University, Corner Brook, Newfoundland in 2003. She has co-authored or co-edited with Geoff Rayner-Canham the following books: *Harriet Brooks — Pioneer Nuclear Scientist* (1992); *A Devotion to their Science: Pioneer Women of Radioactivity* (1997); *Women in Chemistry: Their Changing Roles from Alchemical Times to* *the Mid-Twentieth Century* (1998); *Chemistry was their Life: Pioneering British Women Chemists 1880–1949* (2008); and *A Chemical Passion: The Forgotten Saga of Chemistry at British Independent Girls' Schools, 1820–1940* (2017). **Geoff** is a professor of chemistry at the Grenfell Campus of Memorial University. In addition to his work on the history of women in science with Marelene Rayner-Canham, he has published in the field of chemistry education. He is the co-author, with Tina Overton, of *Descriptive Inorganic Chemistry*, which is currently in its 6th edition and has been translated into seven other languages.

Maria Rentetzi, professor of history and sociology of science and technology at the National Technical University of Athens (NTUA), Greece, has published widely on the history of nuclear sciences with an emphasis on radioactivity, radiation protection and nuclear diplomacy. She was recently awarded an ERC Consolidator Grant to study the history of radiation protection and the role the International Atomic Energy Agency has played as a diplomatic and political international institution in shaping radiation policies and nuclear diplomacies. She currently runs InSciDe's (Inventing a shared Science Diplomacy for Europe) Security and Science Diplomacy work package, an EU Horizon 2020 funded project on science diplomacy. Rentetzi is the author of *Trafficking Materials and Gendered Experimental Practices* (2007) and the co-author (with Susanne Bauer and Martina Schlünder) of *Boxes: A Fieldguide* (2019). In addition to directing the Laboratory of History and Philosophy of Science and Technology at the NTUA, Rentetzi is a member of the board of the ERC Association, the president of the Commission Women and Gender in Science, Technology and Medicine, and a member of the board of the Division of History of Science and Technology, International Union of History and Philosophy of Science and Technology (IUHPS/ DHST).

Richard E. Rice received a PhD in physical chemistry from Michigan State University (1982). After being awarded an NSF Professional Development Award in STS in 1989–1991, he taught history of science and science-studies courses, as well as chemistry and integrated physical science, at several universities in the US. He was chair of the Division of the History of Chemistry of the American Chemical Society in 2001–2002. His main research interest is nineteenth-century chemistry. He is retired and lives in Montana.

Ann E. Robinson is a historian of science who dwells in the borderlands between chemistry and physics. She received her PhD in History from the University of Massachusetts Amherst, where her dissertation examined the roles of pedagogy and research in the development of a standard form of the periodic table. Her other major project explores the controversies surrounding the discovery and naming of the superheavy elements, and the role played by the International Union of Pure and Applied Chemistry in resolving the controversies. She was a fellow at the Consortium for History of Science, Technology and Medicine and at the Science History Institute, both in Philadelphia. Currently she chairs the Physical Sciences Forum of the History of Science Society and works in reference services at Harvard University's Widener Library.

Xavier Roqué is a senior lecturer in history of science at the Universitat Autònoma de Barcelona (UAB). He directed the University's Centre for the History of Science between 2000 and 2010, and coordinated the master programme in history of science between 2006–2010, and 2013–2018. His research deals with the history of the modern physical sciences, gender and science, and the relations between science and industry. Together with Soraya Boudia, Roqué edited *Science, Medicine, and Industry: The Curie and Joliot-Curie Laboratories*, a special issue of *History and Technology* (1997). Edited books include *La física en la dictadura: Físicos, cultura y poder en España*, 1939–1975 (with Néstor Herran, 2012), and *De la Guerra Fría al calentamiento global: Estados Unidos, España y el nuevo orden científico mundial* (with Lino Camprubí and Francisco Sáez de Adana, 2018).

Matthew Shindell is a historian of science and a curator of space history at the Smithsonian's National Air and Space Museum in Washington, DC. He curates the Museum's collections related to planetary science and exploration. Shindell holds a PhD in history of science and science studies from the University of California, San Diego (2011), an MS in life sciences/biology and society from Arizona State University (2004), and an MFA in creative writing from the University of Iowa Writers' Workshop (2001). He has held fellowships at the Chemical Heritage Foundation (Science History Institute), Philadelphia, PA, the University of Southern California, and Harvard University. His biography of the American chemist Harold C. Urey, *The Life and Science of Harold C. Urey* is forthcoming (2019). He is also the co-author of the book *Discerning Experts: The Practices of Scientific Assessment for Environmental Policy* (with Michael Oppenheimer, Naomi Oreskes, Dale Jamieson, Keynyn Brysse, Jessica O'Reilly, and Milena Wazeck, 2019). His current project is a study of the US National Research Council.

Ignacio Suay-Matallana is an assistant professor of history of science at the University Miguel Hernández (Spain), and a researcher at the "López Piñero" Interuniversity Institute. He was a long-term postdoctoral fellow at the Chemical Heritage Foundation/Science History Institute of Philadelphia (2014–2015), and a postdoctoral fellow at the Interuniversity Center for the History of Science and Technology of Lisbon (2015–2017). He also serves as the secretary to the EuChemS Working Party on History of Chemistry. Suay-Matallana's research has been published in journals like *Annals of Science*, *Journal of Chemical Education* and *Dynamis*. In 2015 he was awarded the New Scholars Award by the History of Alchemy and Chemistry (SHAC), and in 2016 the Early Career Scientist Grant by the European Society for the History of Science (ESHS). His main research interests are related to history of science and chemistry (1850–1950), especially customs laboratories, sites of chemistry, material culture, textbooks, scientific experts and regulations.

Brigitte Van Tiggelen is the director for European operations and Senior Fellow of the Center of Historical Research at the Science History Institute, Philadelphia, PA, and a member of the Centre de Recherche en Histoire des Sciences, Université catholique de Louvain, Belgium. A graduate in both physics and history, she focussed her PhD on chemistry in eighteenth-century Belgium. She has been chairing the Working Party on the History of Chemistry (EuChemS) since 2013 and the Commission for the History of Chemistry and Molecular Sciences

(IUHPST) since 2017. To promote the history of science to the general public and secondary school teachers, she founded Mémosciences asbl. Van Tiggelen's publications include co-edited books *From Bench to Brand and Back: The Co-Shaping of Materials and Chemists in the Twentieth Century* (with Pierre Teissier and Cyrus Mody, 2017); *Domesticity in the Making of Modern Science* (with Donald L. Opitz and Staffan Bergwik, 2016); *For Better or for Worse? Collaborative Couples in the Sciences* (with Annette Lykknes and Donald L. Opitz, 2012); *Madame d'Arconville (1720–1805): Une femme de lettres et de sciences au siècle des Lumières* (with Patrice Bret, 2011) and *The Public Image of Chemistry* (with Joachim Schummer and Bernadette Bensaude-Vincent, 2007).

Jessica Wade is a scientist with an enthusiasm for equality. She works in the Centre for Plastic Electronics at Imperial College London on organic light-emitting diodes that emit circularly polarised light. She has been involved in several projects aimed at improving gender diversity in science, and strongly supports the Institute of Physics' work in encouraging young people to study physics A-Level. She has won the IOP Daphne Jackson Medal and the Jocelyn Bell Burnell Medal, and is on the council of the Women's Engineering Society (WES). In 2017 Jess was selected for a US State Department International Visitor Leadership Program for women scientists and co-led the UK Team at the 2017 International Conference for Women in Physics. She is a keen Wikipedian, and is helping to upload the biographies of women, LGBTQ+ and POC scientists — creating one every day since the beginning of 2018.

Jennifer Wilson graduated in chemistry and worked as an industrial chemist before teaching chemistry in both further and higher institutions, specialising in organic chemistry. In 2010 she pursued her interest in the history of science by completing an MSc at Imperial College, followed by a PhD at University College London with a thesis on *Crystallographer and Campaigner: The Life and Work of Dame Kathleen Lonsdale (1903–1971)*. Wilson's publications include "Celebrating Michael Faraday's Discovery of Benzene" (2012) and "Dame Kathleen Lonsdale (1903–1971): Her Early Career in X-Ray Crystallography" (2015).

Element Index
(old and new elements)

Name Index

General Index

Abegg's rule, 192
Abitur, 187
absorption curves, 326
academies, 5 (*see also* (in Name Index) Academie (Royale) des Sciences, Austrian Academy of Sciences, Hungarian Academy of Sciences, National Academy of Sciences, Prussian Academy of Sciences, Royal Academy of Science, Royal Academies of Pharmacy and Sciences, Royal Irish Academy, Spanish Academy of Science, Swedish Academy of Sciences, Royal Astronomical Society, Royal Society in London, Royal Society of Chemistry)
accelerators, 391
accident, 458
acidification, 90
actinium precursor or mother substance, 327, 328, 329
activation energy, 228
activist, 46, 153, 197, 406, part 7

advanced instrumental methods, 42
affinity, 16
    affinity table, 16
African American woman scientist, 45, 441, 444 (*see also* non-white women)
agricultural work, 442
aircraft, 408
air, studies of, 17
alchemical quests, 11
alchemy, 271
allotrope, 239, 432
allotropic form, 237
alloys, 44, 45, 178, 405–413, 444, 448
aluminium production plant, 409
amateurs, 7
American Atoms for Peace Program, 355
American industry, 468
ammines, 177
analytical chemistry, 19, 26, 42, 128–130, 132, 171
    analytical-chemical methods, 33
    analytical-chemical skills, 18

519

Dobson spectro(photo)meter, 44, 432, 438
dolerites, 427
domestic, 464
domestic and social arrangements, 260, 266
domestic pollution, 153
domestic science, 458, 461

earth sciences, 430
earth, three forms
   terra fluida, 12
   terra lapidea, 12
   terra pinguis, 12
Edward C. Pickering Fellowship, 204
elasticity, 78
electrical properties of salts in solution, 24
electricity, 20, 21, 78, 104
electroaffinity, 24, 185, 188, 189, 191–193
electrochemistry, 160
electrolysis, 20, 107, 408
electrolytic conductivity, 137
electrometer, 25, 260, 261, 353
  (*see also* electroscope)
   electrometric balance, 261
   electrometric devices, 260
   piezo-electric quartz electrometer, 27
electron, 287, 365
electron diffraction, 375, 377
electronegative element, 158
electronegativity, 185, 191
electronic configuration, 112
electropositivity, 191
electroscope, 308, 317, 318, 325, 328
   Elster-Geitel electroscope, 130

   gold-leaf electroscope, 319
   Laborde electroscope, 315
element as an abstract entity, 3
element
   element, highly reactive, 446
   element, "negative-empirical" definition of, 14
   element, notion of a chemical, 85
   element, old concept of, 93
Elkem's patent office, 411
emanation, 269, 270, 314–316, 318, 353, 354
   actinium emanation, 34, 318, 319, 354
   radium emanation, 27, 271, 304
   thorium emanation, 34, 313–321
enamel, 473
enamel dust, 473
energy, 81, 354, 365, 442
enthalpy, 442, 445, 447
enzymes, 480
experimental philosophy, 74

Falu mine, 139
family cooperation, 259
family support, 8, 266
feminism, 196
fertilizers, 406
Fiskaa verk, 408, 409
fission, 382 (*see also* nuclear fission)
fission of uranium, 231
fission products of uranium, 386, 387
flame calorimeter, 448
   flame calorimetric systems, 448
   flame calorimetry, 441
   fluorine flame calorimetry, 450
fossil shells, 418
four-element theory, 12

IUPAC Periodic Table of the Elements and Isotopes

Element Background Color Key

Standard atomic weights are the best estimates by IUPAC of atomic weights that are found in normal materials, which are terrestrial materials that are reasonably possible sources for elements and their compounds in commerce, industry, or science. They are determined using all stable isotopes and selected radioactive isotopes (having relatively long half-lives and characteristic terrestrial isotopic compositions). Isotopes are considered stable (non-radioactive) if evidence for radioactive decay has not been detected experimentally.

Element has two or more isotopes that are used to determine its standard atomic weight. The isotopic abundances and atomic weights vary in normal materials. These variations are well known, and the standard atomic weight is given as lower and upper bounds within square brackets, [ ]. Conventional atomic weight, such as for trade and commerce, is shown in white.

Element has two or more isotopes that are used to determine its standard atomic weight. The isotopic abundances and atomic weights vary in normal materials, but upper and lower bounds of the standard atomic weight have not been assigned by IUPAC or the variations may be too small to affect the standard atomic weight value significantly. Thus, the standard atomic weight is given as a single value with an IUPAC assigned uncertainty that includes both measurement uncertainty and uncertainty due to isotopic abundance variations.

Element has only one isotope that is used to determine its standard atomic weight. Thus, the standard atomic weight is invariant and is given as a single value with an IUPAC evaluated uncertainty.

Element has no standard atomic weight because all of its isotopes are radioactive and, in normal materials, no isotope occurs with a characteristic isotopic abundance from which a standard atomic weight can be determined.

element name — **cadmium**
element symbol — **Cd**
atomic number (number of protons) — 48
conventional atomic weight for elements with pink background — 112.414g
standard atomic weight — (112.414 ± 0.004)

isotope mass number (number of protons + neutrons)
114: black number indicates the isotope is stable
116: red number indicates the isotope is radioactive
Isotopic abundance (mole fraction of isotope)
uncertainty (in last digit)

www.tableofisotopes.com

The Periodic Table of the Elements and Isotopes; Copyright Sara Głąckowski 2016–2018

Values are the latest IUPAC values as of October 2018

INTERNATIONAL UNION OF PURE AND APPLIED CHEMISTRY

www.ingramcontent.com/pod-product-compliance
Lightning Source LLC
Chambersburg PA
CBHW052116230326
41598CB00079B/3703